W Morgan's

The Elements of Structure

An Introduction to the Principles of Building and Structural Engineering

Revised by Ian G. Buckle

Department of Civil Engineering, University of Auckland, Auckland, New Zealand

Second edition

Longman Scientific & Technical
Longman Group UK Limited
Longman House, Burnt Mill, Harlow,
Essex CM20 2JE, England
and Associated Companies throughout the world.

© W Morgan, 1964 and Ian G Buckle, 1978

All rights reserved; no part of this publication may be reproduced, stored in a retrieval system, or transmitted in any form or by any means, electronic, mechanical, photocopying, recording, or otherwise without either the prior written permission of the Publishers or a licence permitting restricted copying in the United Kingdom issued by the Copyright Licensing Agency Ltd, 33–34 Alfred Place, London, WC1E 7DP.

First published in Great Britain by Pitman Publishing Limited 1964
Reprinted 1970, 1971, 1973, 1975
Second edition 1978
Reprinted 1979, 1981, 1982, 1984
Reprinted by Longman Scientific & Technical 1986, 1988, 1990

ISBN 0-582-99485-3

Produced by Longman Singapore Publishers (Pte) Ltd
Printed in Singapore

Contents

Preface

While preparing this second edition, I have been constantly aware of the basic purpose that the late William Morgan had in mind when he wrote the original work more than 12 years ago. He intended that the book be useful to first year architectural and engineering students and at the same time be readable by builders and laymen who are curious to know more about the structures they work and live in and see around them.

I have tried not to disturb Morgan's successful format and have resisted the temptation to include more mathematics. Therefore, analysis has again been minimized and the intuitive approach to structural principles has been retained. It is hoped that in this manner the book will continue to satisfy the curiosity of many people who simply wish to know how structures work.

Even the serious student of structural engineering will benefit from this approach. There is ample time in later years of a university degree course for him to learn analytical techniques, but now is the time to develop a physical 'feeling' for the way structures sustain load. The mathematics to quantify this feeling can follow later. In this way, common sense should always prevail and blind faith in mathematical solutions should thereby be minimized. A better educated structural engineer is the net result.

Fundamental to this new edition has been the change to SI units and some care has been taken to introduce these units gradually and carefully. Many of the examples appeared in the first edition and in the conversion to SI units numerical values have been rounded off but dimensions of actual structures have been converted as closely as possible to their metric equivalents. At the same time expanded definitions of fundamental concepts have been included along with new sections on high-rise buildings and earthquake effects on structures. Chapters 11, 14, 17 and 19 have been updated with data on recently completed structures and the photographic section has necessarily been revised and distributed throughout the text for easier reference.

Many people have helped make this new edition possible. Most of the work was done while I was on leave from the University of Auckland and visiting the University of California in Berkeley. I am therefore grateful for the use of the library facilities and the excellent secretarial assistance of Mary Laue during my time in Berkeley. No author can succeed without the encouragement and patience of his wife and family and in this regard I am no exception. For their faithful support I owe a great debt. I am also indebted to G. R. Penrose, whose excellent diagrams for the first edition have been used again in this edition. Last but not least, the publisher and his staff should be cited for their endless patience and exceptional skill while coordinating the various phases of the publishing process. This book can be seen to be the product of many people, extending back almost 15 years. I am privileged to have had a part to play in its evolution.

Ian G. Buckle
Auckland 1977

Introduction to SI Edition

In an attempt to keep an elementary text as simple as possible, it is tempting to avoid the difficult concepts and concentrate on those that are easier to understand and easier to explain. This must inevitably lead to confusion. For example, a clear distinction between mass and weight and between stress and strain is necessary for any reader to make progress with structural analysis. But common usage gives these terms equal meaning and infers they can be interchanged for one another. The difference is not easily explained and if treated superficially, endless confusion results. Nor is the problem simplified by changing to SI units.

Mass is the quantity of substance contained within a body whereas weight is that force exerted on the body by the earth's gravitational field. The fundamental SI units used to measure these two different quantities are the kilogram for mass and the newton for weight (force).

The newton (symbol: N) is defined as that force required to accelerate a mass of one kilogram (symbol: kg) by one metre per second per second (symbol: m/s²). Alternatively we can say that

$$1 \text{ N} = 1 \text{ kg} \times 1 \text{ m/s}^2 = 1 \text{ kg m/s}^2$$

Therefore the weight of a body of mass 1 kg is 9.81 N since

$$\begin{aligned} \text{weight} &= \text{mass} \times \text{acceleration due to gravity} \\ \therefore \text{weight of 1 kg} &= 1 \text{ kg} \times 9.81 \text{ m/s}^2 \\ &= 9.81 \text{ kg m/s}^2 \\ &= 9.81 \text{ N} \end{aligned}$$

However, a further complication arises because the metric system used in the market-place has adopted the kilogramforce (symbol: kgf) to measure weight, rather than the newton. The kilogramforce is simply defined as the weight of a body of mass one kilogram and therefore

$$1 \text{ kgf} = 9.81 \text{ N}$$

Certainly the kilogramforce (often condensed to 'kilo' for convenience) is easier to use since it is larger than the newton by almost a factor of 10 and because it automatically means that a mass of x kg weighs x kgf. We could of course say that it weighs $9.81x$ N but the convenience of the kilo (kgf) is immediately obvious. Once accepted for this purpose it is tempting to use the kilogramforce to measure all forces whether they be associated with weight or not. Herein lies the dilemma because the kilogramforce is not a 'preferred' unit for force measurement in the SI system particularly for scientific or engineering purposes. Therefore in this text we shall use the newton for all forces with one exception. The kilogramforce has been used in those examples where the weights of people are involved, on the assumption that the reader will more easily recognize the scale of forces in the problem if a set of units he is already familiar with has been used.

A word about stress and strain might also be useful at this point, although these quantities are both discussed in Chapter One. Stress is a force intensity and is measured by the magnitude of the force divided by the area over which it acts. So that if a force of 1N acts over an area of 1 square metre (1m^2) it gives rise to a stress of 1N/m^2. Strain describes the behaviour of a specimen of material when it is subjected to stress. It is defined as the change in length of the specimen due to the stress divided by the original length of the specimen. Since strain is a ratio of lengths, it is independent of unit systems provided that both the change in length and the original length are measured in the same system of units. The relationship between stress and strain is also discussed in Chapter One where it is seen to be a property of the elasticity of the given material.

The Newton is a relatively small unit of force and multiples of the Newton have been defined to help overcome the problem. The two multiples most useful in structural engineering are the kilonewton (kN) and the meganewton (MN) which are 10^3N and 10^6N respectively. Similarly, the square metre is not a convenient unit of area, particularly when used to measure stress. Again multiples are defined to give more reasonably sized units to work with. The square metre is too large a unit of area and the square millimetre is often used in its place. Since $1\text{m} = 10^3\text{mm}$, then $1\text{m}^2 = 10^6\text{mm}^2$ and therefore a stress of say, $1\ \text{MN/m}^2$

$$= 1 \times 10^6\ \text{N/m}^2 = 1 \times 10^6\text{N}/10^6\text{mm}^2 = 1\text{N/mm}^2.$$

Also we can see that a stress of $800\ \text{kN/m}^2$

$$= 800 \times 10^3\text{N/m}^2 = 800 \times 10^3/10^6\text{mm}^2 = 0.8\text{N/mm}^2.$$

The unit of stress: N/m^2 has been given a special name: the pascal (symbol: Pa). Consequently, a stress of 1MN/m^2 could be written as 1MPa (megapascal) and a stress of $800\ \text{kN/m}^2$ could be described as 800 kPa (kilopascal). The letters 'kPa' are easier to say than 'kN/m^2' and we can expect these derived units to become more popular as time goes by. However, in this book the pascal has not been used so as to avoid the danger of introducing too many new ideas in an elementary text.

There is little doubt that SI units are easier to learn than those in the Imperial system, and are less prone to error during manipulation. But confidence with these units will only be achieved by repeated practice and in engineering this means solving numerous problems in analysis and design. This book is not intended to exercise a student in this manner, and SI units are used here simply to give the reader a feeling for the size and scale of the elements of those structures that we are going to talk about. If competence in engineering calculation is required, there are many books available for this purpose and some of these are listed in the section on Further Reading at the back of this book.

1 Some Fundamental Definitions

Short definitions can sometimes be misleading, but keeping this fact in mind, it may be said that structural engineering is concerned with the strength, stiffness and stability of structures such as buildings, bridges, dams, and walls for retaining earth. The structural engineer, with the aid of applied mathematics, is concerned with the design, construction and maintenance of such structures.

Although not specializing to the same extent as structural engineers, architects and builders must have a sound knowledge of structural principles. On all but the smallest projects, engineers, architects and builders are in close collaboration before and during the building of the structures.

Structural engineering is a specialized branch of civil engineering; the name 'civil engineer' was first adopted in Britain in the 18th century to distinguish the engineer who was engaged solely in work of a civilian character from the military engineer, who was concerned with all works and machines necessary for attack and defence.

As early as 1771, it was resolved at a meeting in London that 'the Civil Engineers of this Kingdom do form themselves into a Society'. This Society was the forerunner of both the Institution of Civil Engineers founded in 1818 and the American Society of Civil Engineers, which was founded in 1852.

Civil engineering can be divided into many branches such as railways, harbours, docks, rivers and canals, waterworks, roads, municipal, mechanical, mining, structural, etc. With the great scientific advances of the 19th century and the growth of specialization, other professional societies were formed to concern themselves with particular branches of engineering. For example, the Institution of Mechanical Engineers was founded in 1847, and the Institution of Electrical Engineers in 1871.

At the beginning of this century, when concrete and reinforced concrete were being developed as important constructional materials, a number of engineers decided that there was scope for a professional society composed of members who were concerned with the specialized type of construction in the new materials, and in 1908 the Concrete Institute was formed. Before many years had passed it was decided to enlarge the scope of the Institute to deal with structural engineering generally. Its title was therefore changed in 1922 to the 'Institution of Structural Engineers.'

Statics

In his calculations the structural engineer is concerned generally with the applications of the principles of *statics* and is constantly occupied with *forces* and their effects. The word 'statics' is derived from the Greek science which dealt in particular with the balancing of levers and in general with problems relating to bodies in a state of equilibrium under the action of a number of forces. Statics is one of the main branches of *mechanics* and deals with forces on bodies which are 'at rest.' *Dynamics,* the other main branch, is concerned with moving bodies. When a structure is said to be in equilibrium, this means that it and all its parts are at rest. This distinguishes a structure from a machine, which consists of moving parts. Thus buildings and bridges are structures (static), whilst locomotives and aeroplanes are machines (dynamic), although in the latter some structural design is necessary.

Force

Everyone has some idea of what is meant by *force,* but the average man might find it difficult to give a satisfactory definition. Until the time of Sir Isaac Newton (1642–1727), force was associated solely with the idea of the pressure exerted by one body on another. Newton extended this meaning by defining force as any cause which changes or tends to change the state of rest of a body or its state of uniform motion in a straight line.

Units of force and weight

Using Newton's definition for force, units for the measurement of force can be derived. For example, the force required to accelerate a mass of one kilogram (kg) by one metre per second per second (m/s^2) is defined as the newton (N) in honour of this man who first proposed the idea that force was proportional to mass and acceleration.

The weight of a body is a special type of force because it is the gravitational force or attraction exerted on the body by the earth which if not resisted will cause the body to move vertically downwards accelerating as it does so at the rate of $9{\cdot}81\ m/s^2$. This acceleration due to gravity is approximately constant in magnitude over the entire earth's surface, and is sometimes called the gravity or gravitational constant (G). A careful distinction between mass ad and weight needs to be made: mass is the quantity of substance contained within a body and weight (as just noted) is the force exerted on that mass by the earth's gravitational field. Therefore, the weight of a body of mass one kilogramme is

$1\ kg \times 9{\cdot}81\ m/s^2 = 9{\cdot}81\ N$ (by definition of the newton)

However, the common everyday unit for weight is not the newton, but rather the kilogram-force (kgf) or kilo for short. The relationship between these two units is the gravitational constant since the kilo is defined as the weight of a mass of one kilogram and therefore

$$1 \text{ kgf or kilo} = 1 \text{ kg} \times 9{\cdot}81 \text{ m/s}^2 = 9{\cdot}81 \text{ N}$$

The kilogramforce is seen to be a larger unit of force than the newton (almost ten times) and is universally accepted as a measure of weight for everyday shopping and commercial transactions. Neverthless, the newton is the preferred unit for all scientific and engineering calculations and in this book the newton is used to measure all forces and weights with one exception: kilogramsforce are used for the weights of people.

Examples of forces

Figure 1.1 shows the wind blowing horizontally onto a building with a total uniform load of 100 kN. The direction of the wind forces is shown by the arrows. Note that words such as 'pressure', 'weight', 'load', 'pull', and 'push' are frequently used as alternatives for the word 'force'. However, not all are correct substitutes and 'pressure' in particular should be reserved to describe force intensities (i.e. force per unit area) as discussed on page 7.

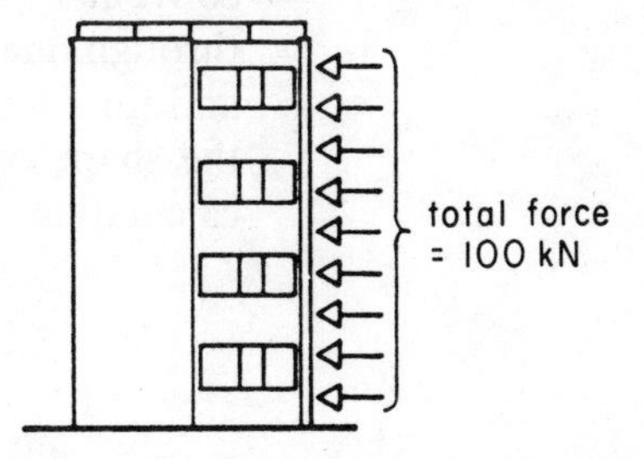

Fig. 1.1

A stationary object exerts force by virtue of its weight. In Fig. 1.2 (*a*) the two men, by merely standing on the beam, are exerting active forces in a downward direction, and this can be represented conventionally as at (*b*). The passive resisting forces or ***reactions*** exerted by the supports are in an upward direction.

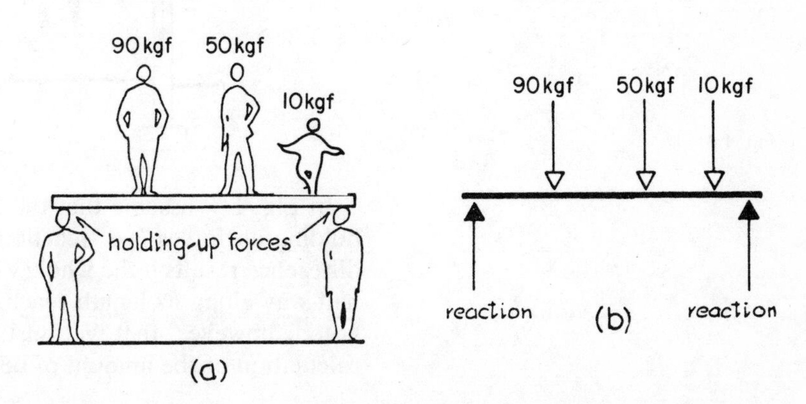

Fig. 1.2

Water exerts force in all directions, and its force on any surface is always at right angles to that surface. The magnitude of the force increases with the depth of the water (Fig. 1.3).

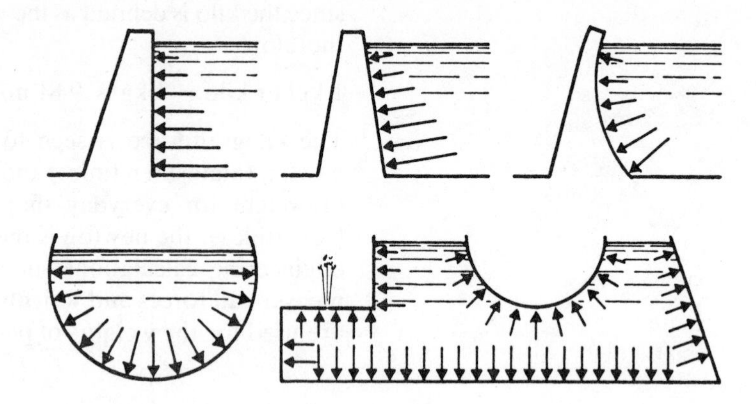

Fig. 1.3

Centre of gravity

Since gravity attracts vertically downwards the weight of any body such as a building or any part of a building acts vertically downwards and can be represented by arrows pointing in that direction (Fig. 1.4). Every particle of which a body is composed is attracted downwards in the same direction, but for certain calculations it is convenient to assume that the whole weight of the body is acting through one point. This point is called the *centre of gravity* (c.g.), and for a homogeneous body the position of the point depends on the shape of the body. Textbooks on mechanics give methods of calculating the positions of the centres of gravity of various shapes.

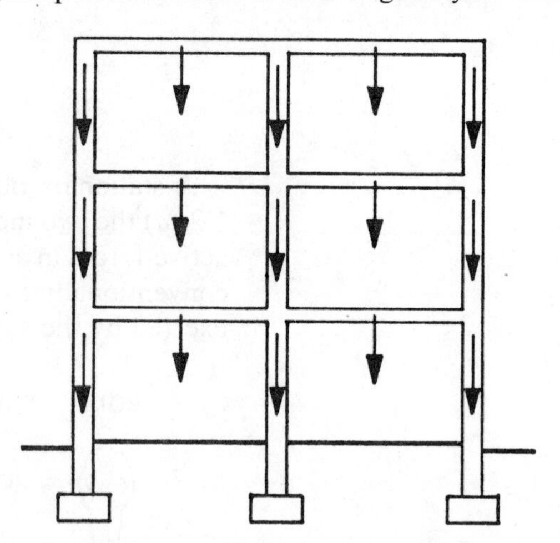

Fig. 1.4

In Fig. 1.5, assume that the beam weighs 10 kN. As far as the holding-up forces (i.e. reactions) of the walls are concerned, no difference results if the whole weight of the beam is assumed to act half-way along its length (each wall holds up 5 kN). It should be noted, however, that we must not assume this condition for the calculation of the amount of bending of the beam, since a load of

10 kN concentrated at mid-span will bend the beam to a greater extent than a load of 10 kN spread uniformly along its whole length.

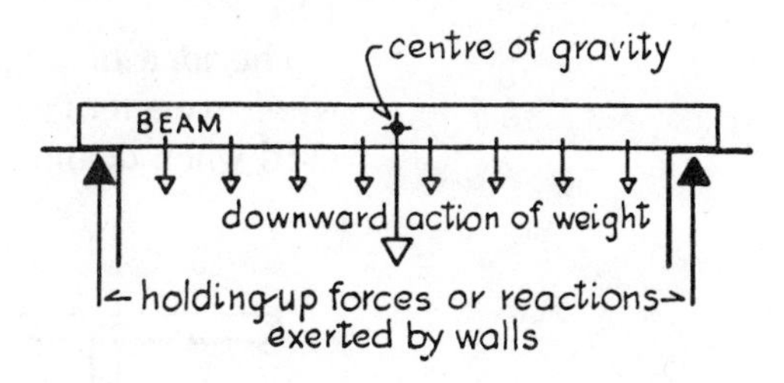

Fig. 1.5

If a triangular block of brickwork weighing 30 kN is supported by a uniform beam weighing 10 kN, the reactions on the supports will be the same if the forces (i.e. the loads) are assumed to act through the centres of gravity as in Fig. 1.6. Calculations will show that the holding-up force of the left support is 15 kN and that of the right support is 25 kN. It must again be emphasized that the

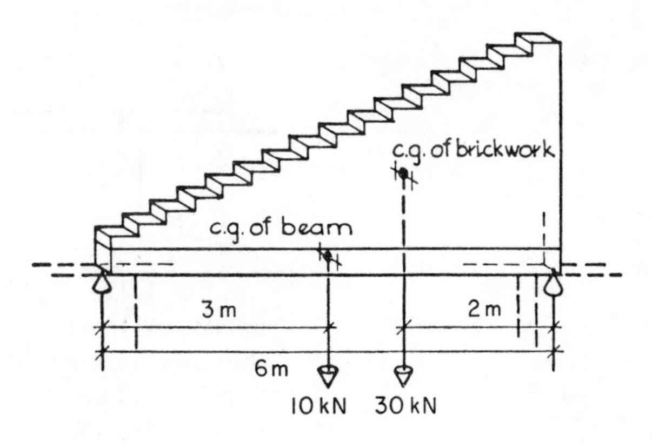

Fig. 1.6

amount of bending of the beam cannot be found by replacing the beam and brickwork by two point loads acting as shown.

The centre of gravity is an imaginary point and need not be within the body itself (Fig. 1.7).

It was stated earlier that structural engineering is concerned with the strength, stiffness and stability of structures; the concept of the centre of gravity is useful when considering stability. If a

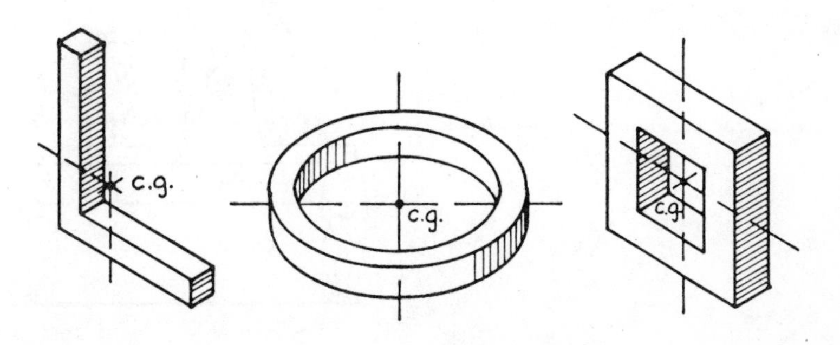

Fig. 1.7

vertical line through the centre of gravity falls outside the base upon which the body relies for stability, overturning will result unless precautions, such as tying down the base, are taken (Fig. 1.8).

The idea of a centre of gravity as being a point at which the whole weight of a body may be assumed to be concentrated can be used when dealing with a number of forces.

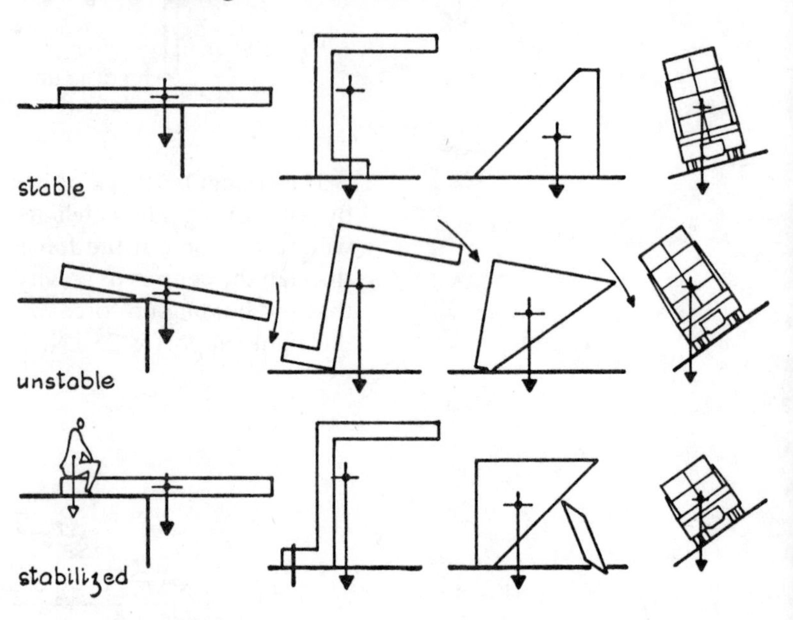

Fig. 1.8

In Fig. 1.9, it can be proved by mathematics that the load on the support at A (and therefore the upward reaction at A) is 90 kgf, whilst the reaction at support B is 70 kgf. The reactions at the supports will be unaltered if the three loads are assumed to be all concentrated at one point as shown in Fig. 1.10. This point is called the centre of gravity of the loads or forces, and is the point at which all the forces may be assumed to act so as to have the same effect on the *equilibrium* or stability of the body. The effect on the bending of the beam is *not* the same as that of the loads acting in their correct positions. The one force of 160 kgf is said to be the *resultant* of the three forces shown in Fig. 1.9.

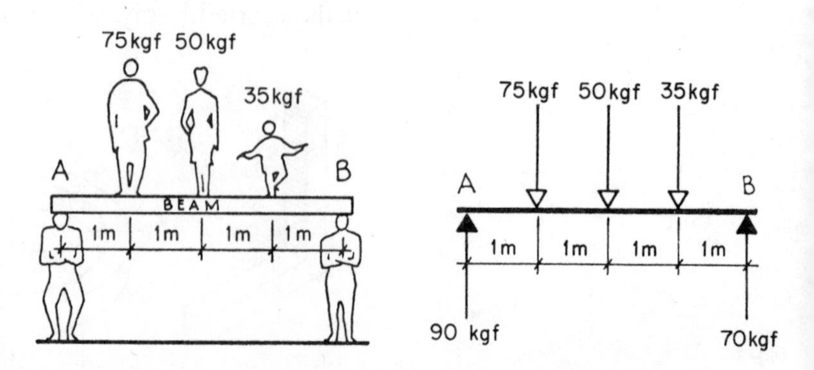

Fig. 1.9

A further example is illustrated in Fig. 1.11. Unless the feet at *A* are tied down [Fig. 1.11 (*a*)], the position shown is impossible. In (*b*) the resultant of the two weights falls inside the 'base' and equilibrium can be maintained.

The resultant of the weights 1, 2 and 3 in Fig. 1.12 must fall inside the base *AB* unless the crane is tied down to its support.

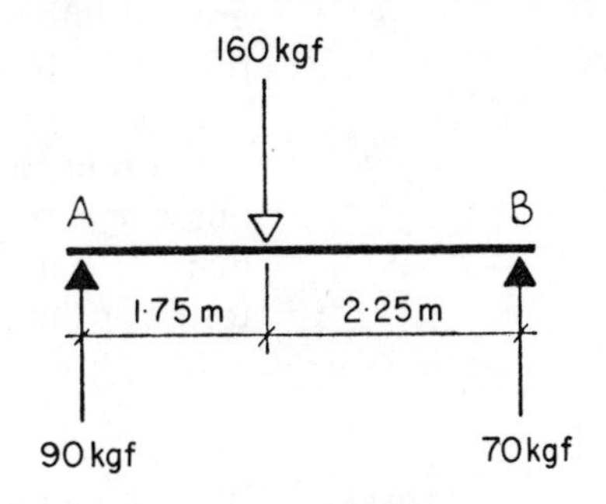

Fig. 1.10

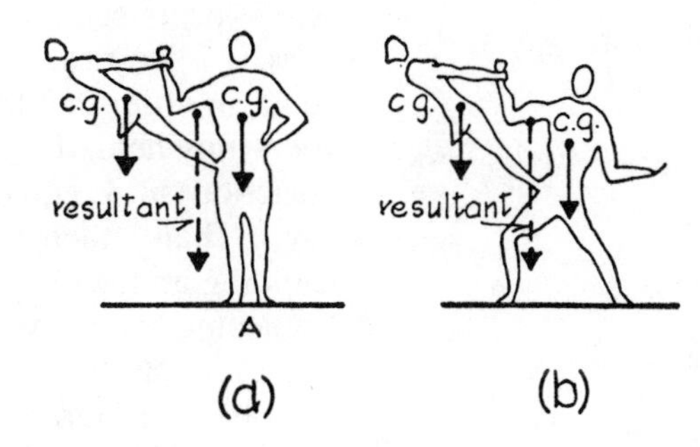

Fig. 1.11

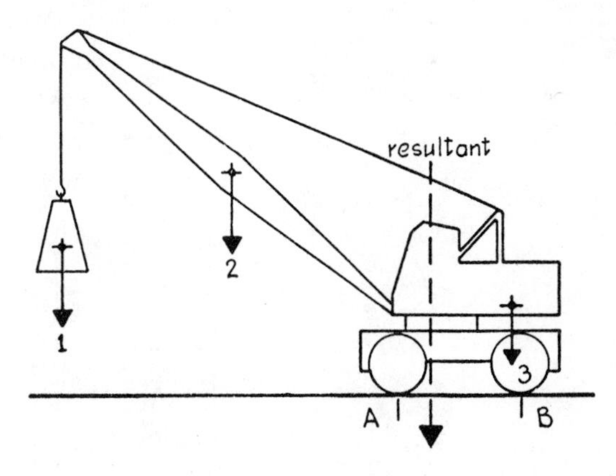

Fig. 1.12

Stress

When designing a structure, the engineer must estimate all the forces acting on the structure and its component parts. The effects of the forces are then considered in relation to the stability of the building as a whole, and the component parts are made strong enough to fulfil their particular functions. The forces to be considered depend on the type and purpose of the structure. In buildings, for example, the forces will be due to the weight of the structure, its occupants and contents as well as to the pressure exerted by wind, earthquakes and snow. In a wall for retaining earth, the forces will be due to the weight of the wall, the pressure exerted on the wall by the earth behind it, and the pressures due to any buildings or roadways near the top of the wall.

The internal fibres or particles of the various parts of a structure are put into a state of *stress* by the forces they are called upon to resist, the basic forms of stress being *tension*, *compression*, *shear* and *torsion* (i.e. twisting).

In order to grasp the idea of 'stress' it is helpful to compare the fibres of a material subjected to force to the bones and muscles of a man. If a man carries a heavy load, his body is 'distressed' as a result of the efforts he has to make to bear the load.

Tension

When a member is being stretched by the forces acting upon it the stress produced is *tensile*. In Fig. 1.13, the action of the weight (i.e. force) is vertically downwards and the string is stretched. Note that, in order that the string may stretch, the top of it (at A) has to be firmly held. There is in fact a holding-up force (i.e. an upward reaction) at A, equal to the downward pull. If the string is replaced by a chain, then each link is subjected to tension and will be holding up the one below it and pulling down on the one above, as indicated by arrows in Fig. 1.14. (The maximum stress is in the top link.) Similarly in a string or solid rod each particle of the member can be considered as a link in a chain and will be in a state of tension.

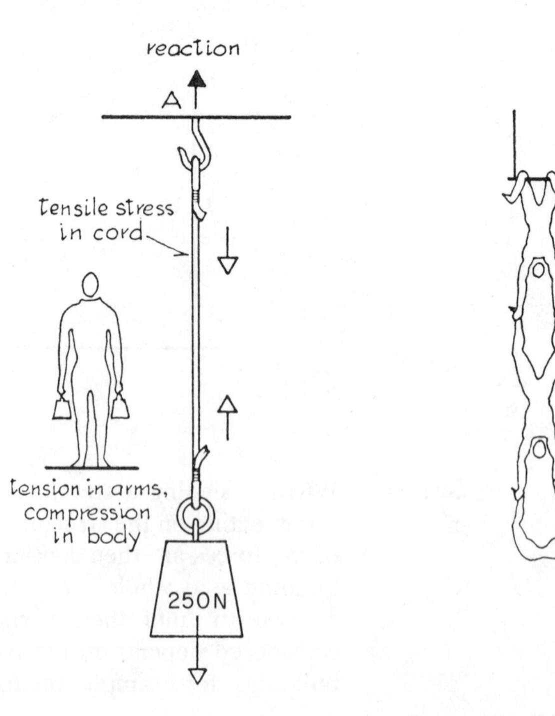

Fig. 1.13 Fig. 1.14

Compression When a member is being compressed or squeezed by the forces acting upon it, the stress produced is *compressive*. In Fig. 1.15 (*a*) there is compressive stress in head and body.

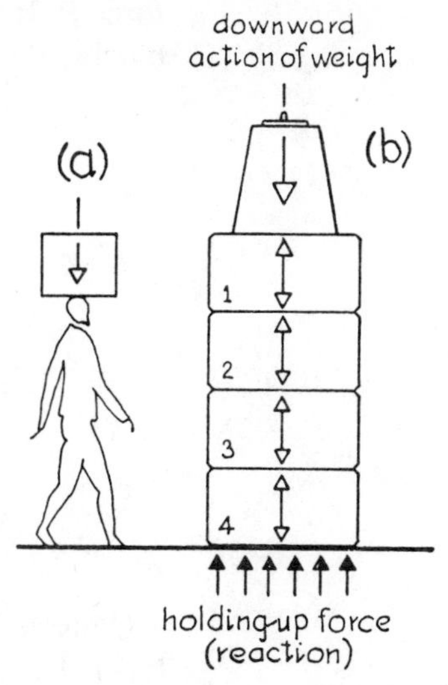

Fig. 1.15

Fig. 1.15 (*b*) represents a member built up of four blocks. Block 1 is holding up the load and pressing down on block 2 as indicated by arrows. Block 2 is holding up block 1 (and the load) and pressing down on block 3, and so on. The holding-up force at the ground must, for equilibrium, be equal to the total downward force, i.e. the superimposed weight or load plus the weight of the blocks. If the pier consists of one block of concrete instead of separate blocks, the principle still applies, which means that every particle of the concrete is being stressed in compression.

Shear When sliding occurs or tends to occur as a result of application of force [Fig. 1.16 (*a*)] the stress produced in the material is called *shear stress*. If the frictional resistance between blocks 1 and 2 were greater than that between blocks 2 and 3, sliding would occur as shown at (*b*). In fact sliding could occur at any level, including

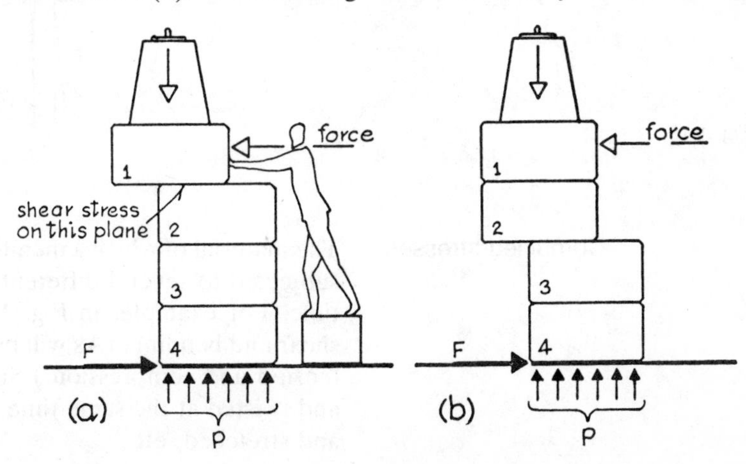

Fig. 1.16

the point of contact with the ground, and for equilibrium there must be a resisting force F (supplied in this example by friction between the blocks and the ground) in addition to the holding-up force P. If the structure consists of one member instead of separate blocks, the tendency to slide at any level is still present.

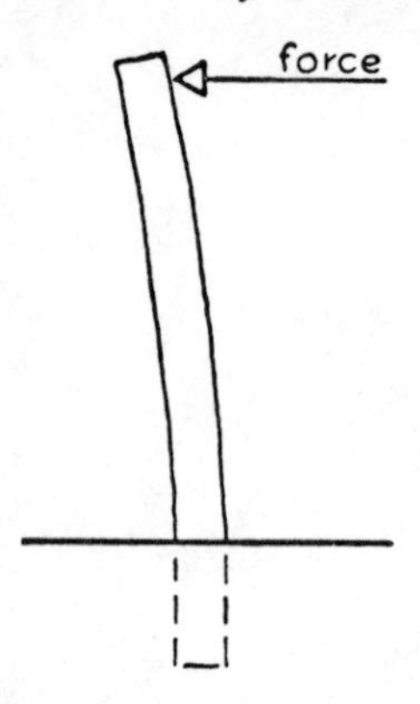

Fig. 1.17

Consider a force applied to a post embedded in the ground (Fig. 1.17). The force causes bending (*see* page 62), and also shear as in Fig. 1.16, but if the resistance of the material to shear is greater than its resistance to bending, the post will fail due to the bending action.

Torsion

When a member is being twisted by the forces acting upon it, the material is said to be stressed in *torsion* (Fig. 1.18).

Fig. 1.18

Combined stresses

The material of which a member of a structure is composed may be subjected to several different types of stress at one and the same time. For example, in Fig. 1.17 the material is stressed in both shear and bending. (As will be shown later, bending itself involves tension and compression.) Similarly a member may be stretched and twisted at the same time, or compressed and twisted, or bent and stretched, etc.

Cross-sectional area and calculation of stress

If a cut is made (in imagination) across a member, the *cross-section* is the plane or area exposed. Normally it is convenient to consider the cross-section at right angles to the longitudinal axis. In Fig. 1.19 (*a*) the *total* resisting force in the member is equal to the applied force or pull of 100 kN, but since this is resisted by an area of 1250 mm² (square millimetres) the stress is 80 N/mm² (newtons per square millimetre sometimes called megapascals or

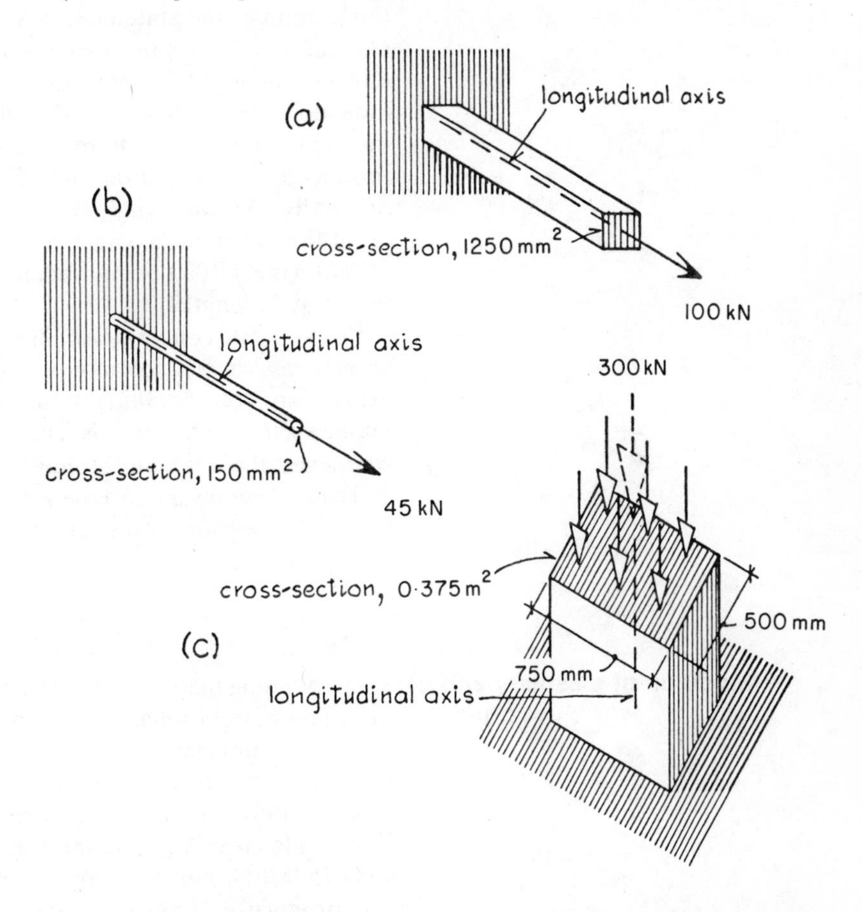

Fig. 1.19

MPa). We can observe that stress is a measure of force intensity since it is calculated from the force divided by the area over which it acts. At (*b*) the tensile stress in the bar is 300 N/mm² since the 45 kN load is being resisted by 150 mm². The stress in (*c*) is 0·8 N/mm² or 800 kN/m².

Tensile and compressive stresses are uniform over the cross-sectional areas of members when the load is axially or centrally applied as in Fig. 1.19; i.e. when the line of the force coincides with the centre-of-gravity axis of the member.

Strain

A material cannot be subjected to stress without being *strained*, i.e. without having its dimensions altered (although the dimensional variations in structural materials are usually very slight). A member stressed in tension becomes longer and thinner, a member stressed in compression becomes shorter and fatter, and so on. Normally, the structural engineer is more concerned with the small alterations in length than with minute changes in thickness. Just as it is convenient to express stress by dividing the load by the area of cross-section, so it is convenient to define *tensile* or *compressive strain* as the elongation or shortening per unit length. For example, the statement 'a stress of 200 N/mm² produced an elongation of 2·5 mm' does not give sufficient information. This elongation might be permissible in a long member but might indicate lack of *stiffness* (lack of ability to resist being stretched) if the member were very short. The statement 'a stress of 200 N/mm² produced an elongation of 2·5 mm in a member which was originally 2500 mm long' gives more exact information. This could be abbreviated to 'a stress of 200 N/mm² produced a strain of 1/1000' (the 1/1000 being obtained by dividing the elongation by the original length).

In everyday conversation, the word 'strain' is often confused with 'stress' and a common example of this incorrect use is the expression, 'the breaking strain of the steel cable was so many kilonewtons'. The words 'breaking stress' should be used whenever the ultimate strength of a material is implied. It is extremely important in engineering calculations to have a clear distinction between stress and strain.

Modulus of elasticity and stiffness

Some building materials, for example steel and timber, are *elastic*; i.e. upon being stretched or shortened by force they immediately revert to their original dimensions when the force is removed (provided that the force is not too great). Another property of an elastic material is that stress is proportional to strain. For example, if a tensile stress of 30 N/mm² in a steel bar produces a strain of 0·00015, 60 N/mm² will produce a strain of 0·00030, 120 N/mm² will produce a strain of 0·00060, and so on, provided that the material remains elastic.

Robert Hooke (1635–1703) discovered this very important law, now usually expressed as 'stress is proportional to strain', and first mentioned it in a lecture in 1678. Hooke was a great scientist, inventor and architect and a friend of Sir Christopher Wren, the designer of St Paul's Cathedral. Among other activities Hooke was very interested in springs, clocks and watches, and it was as a result of his experiments on springs that he discovered the famous law.

If stress (tensile or compressive) is proportional to strain, then for any given material, stress divided by strain will be a constant. This constant is taken as a measure (*modulus*) of the elasticity of the material; it is called *modulus of elasticity*, or *Young's modulus*, and is usually denoted by the symbol E. This extension of Hooke's law was due to Thomas Young (1773–1829), a physician and physicist. Briefly,

$$\frac{\text{Stress}}{\text{Strain}} = \text{Modulus of elasticity} = E$$

In the example given above,

$$E = \frac{30\ \text{N/mm}^2}{0{\cdot}00015} \quad \text{or} \quad \frac{60\ \text{N/mm}^2}{0{\cdot}00030} \quad \text{or} \quad \frac{120\ \text{N/mm}^2}{0{\cdot}00060}$$

The value of E obtained from each of the calculations is 200,000 N/mm^2.

By loading specimens in tension or compression in a testing machine, and recording the stresses and the strains they produce, the values of E for various building materials can be calculated. For example, E for mild steel is approximately 200 kN/mm^2, E for aluminium is approximately 70 kN/mm^2, and E for timber is approximately 10 kN/mm^2. These figures do not relate to the strengths of the material but to their *stiffnesses*, i.e. the relationships between the stresses and the strains they produce.

The figures show that it is easier to stretch aluminium than it is to stretch steel, and much easier to stretch timber. If three identical members of steel, aluminium and timber are tensioned with equal loads, the elongation produced in the aluminium member will be 2·9 times as great (200/70) as that produced in the steel member, and the extension of the timber will be 20 times as great (200/10).

To sum up: if a material has a high value of E, it is difficult to stretch or shorten it, i.e. a large stress is required to produce a small strain. If a material has a small value of E, it is easy to stretch or shorten it, i.e. only a small stress is required to produce a large strain.

An important use of the modulus of elasticity of a material is for calculating elongations of structural tension members, the amounts of shortening in compression members, and the amounts by which beams bend.

Permissible or working stresses

When making calculations to determine the sizes of structural members to resist tension, compression, shear, bending, etc., the engineer refers to standard specifications, codes of practice and building by-laws appropriate to the district in which the structure is to be built in order to obtain the permissible stresses. These *permissible* or *working stresses* are based on knowledge of the *ultimate* or *failing strengths* of the materials, obtained by testing them in tension, compression, shear, etc., and on research into the natures, properties and behaviour of the materials.

Factor of safety

The permissible stresses must, of course, be less than the stresses which would cause failure of the members of the structure; in other words there must be an ample safety margin. This is ensured by adopting a *factor of safety*, which is based sometimes on the failing stress of the material, sometimes on the stress which would cause excessive deformation (excessive elongation, bending, etc.), and sometimes on other considerations. For example, if the ultimate or failing stress of a particular type of concrete is 21 N/mm^2 in compression and the working stress is 7 N/mm^2, the factor of safety is

$$\frac{\text{Ultimate Stress}}{\text{Permissible Stress}} = \frac{21\ N/mm^2}{7\ N/mm^2} = 3$$

Load factor

When a member is loaded to the point of failure the manner in which the stress is distributed is different from that at ordinary working loads. Many calculations are now based on these final stress conditions, and the load factor, which serves the same purpose as the factor of safety in ensuring structural safety, is defined as the failing or collapsing load of the structure divided by the design or working load for the structure.

The tools of the structural engineer

A high proportion of the structural engineer's work consists of *structural analysis*—the application of the tool of mathematics to determine forces, stresses and strains in the structures he is called upon to design. The accurate and economic design of structures usually requires the use of a variety of formulae and considerable calculation. For some structures, the mathematics involved may be simple, but for others it can be of an advanced standard, involving laborious calculation. For many complex structures digital computers are used.

Another important tool is a sound knowledge of the nature, properties and behaviour of the structural materials. Draughtsmanship and surveying are also important.

Although mathematics is an important tool it must be remembered that structural engineering is not an exact science. Building materials and human beings do not behave in a perfect manner, and in developing mathematical theories for the design of structures, assumptions have to be made which may be very close to the truth for certain materials in certain circumstances but which may be wide of the mark in others.

One professor of engineering has said, "Don't expect the structure to make the same assumptions as you do." Professor A. N. Whitehead in his *Introduction to Mathematics* said, "There is no more common error than to assume that, because prolonged and accurate mathematical calculations have been made, the application of the result to some fact of nature is absolutely certain." On occasions, therefore, it may be necessary to interpret very carefully the results given by mathematical analysis and to apply to the problem sound common sense, intuition and a 'feel' for the structure. During the construction of a bridge of novel design on the Continent, many eminent engineers proved by mathematics that it could not stand up; the bridge is still there.

Intuition

Although mathematics has its limitations, intuition by itself is not sufficient. Dictionary definitions of intuition are: immediate apprehension by the mind without reasoning; immediate apprehension by sense; immediate insight.

For example, give to the average man, who is trying to prise a tree root from the ground, a long steel bar, and he will immediately use it correctly. This no doubt is the result of experience, and it is indeed questionable whether 'structural' or any other intuition is possible without previous experience or experiment.

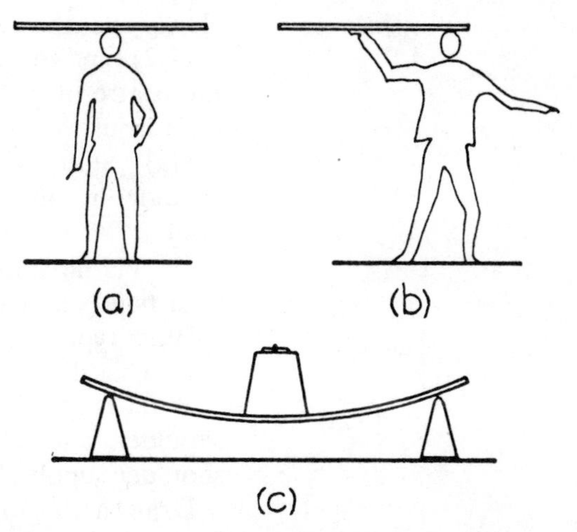

Fig. 1.20

Again, ask a man to support a plank, say 2 m long, on his head and he will at once get near the position of correct balance automatically [Fig. 1.20 (*a*)]. Ask him to support it with one end only of the plank on his head, and without conscious reasoning he should behave as shown at (*b*).

Intuition (or experience?) will predict that a thin piece of wood will bend under a weight as shown in Fig. 1.20 (*c*), but intuition is powerless to predict how much it will bend. An experiment is one way of finding out, and mathematics with proper assumptions and a knowledge of the size and properties of the timber will make the prediction without the beam being seen by the engineer.

An interesting definition of intuition is given by Agatha Christie in *A.B.C. Murders* (London, Collins). Hercule Poirot, the detective says—

> 'In a well-balanced reasoning mind there is no such thing as an intuition—an inspired guess! You *can* guess, of course—and a guess is either right or wrong. If it is right, you call it an intuition. If it is wrong, you usually do not speak of it again. But what is often called an intuition is really *an impression based on logical deduction or experience.*'

However, if a complex and important structure is under consideration, both mathematics and intuition may be insufficient to design the structure adequately (for example, the Sydney Opera House in Australia). In these situations a scale model will be made and tested in a laboratory (sometimes to destruction) to give the designer an indication of the stresses and deflections to be expected in the real structure. Careful modelling can give accurate predictions of prototype behaviour and of course confidence that the proposed structure is acceptable on both structural and economical grounds.

Exercises

(1.1) Find the approximate positions of the centres of gravity of simple objects, such as a book, a fork, a knife, by pushing them slowly over the edge of a table or by balancing them on a pencil.

(1.2) Find the approximate position of the vertical line in which the centre of gravity of a chair lies by tilting it over on two legs until it just overbalances.

(1.3) Stand sideways against a wall with one foot and one shoulder in contact with the wall. Try to lift the other foot from the floor.

(1.4) Hold your arm straight in front and ask someone to clasp your hand and pull hard.

What type of stress is in the arm?

In what direction is the resisting force at your hand?

What is the direction of the force that the arm applies to the shoulder, and in what direction is the resisting force which the shoulder supplies?

Draw a diagram showing by arrows the directions of these forces.

(1.5) Stand upright and get someone to press down heavily on your shoulders.

What type of stress is induced in your body?

In what direction is the resisting force offered by your shoulders?

In what direction are your feet acting on the ground, and in what direction is the resisting force supplied by the ground?

Draw a diagram showing, by arrows, the directions of forces and reactions.

(1.6) What type of stress is induced in a piece of string when being cut by scissors?

Can string resist tension, compression, shear, torsion?

(1.7) Hold your arm straight in front and get someone to twist it gently.

What happens to your arm?

How does it deform?

Break a stick of chalk by twisting it; study the fracture.

Get someone to pull your arm and to twist at the same time; observe the reactions of your muscles.

2 The Principal Structural Materials

The chief structural materials are brick, stone, concrete, timber and steel. In recent years aluminium alloys have been used for some types of structure, and even some plastics have been used structurally.

Bricks

Bricks have been used for many thousands of years. The ancient Egyptians made bricks of mud, mixed with chopped straw or reed to strengthen them and to prevent their crumbling after being dried in the sun. As recorded in *Exodus*, the Israelites were punished by being deprived of straw.

Most bricks today are made from clay which is pressed into moulds either by hand or machine. The bricks are then 'burnt,' that is they are heated in an oven called a kiln to a temperature of about 1000°C. Since there are many types of clay, there are many types of brick of varying strengths. The strongest bricks, the type frequently used to support the ends of bridges, are called *engineering bricks.* A well-known engineering brick is the Staffordshire blue, which is dark blue in colour. There are also red engineering bricks, which, like the blue bricks, are very hard, dense and strong.

All bricks are very much stronger in compression than in tension, and their tensile strength is so low that they are unsuitable for resisting forces which would produce tension. It is not possible to obtain an accurate estimate of the strength of brickwork in compression by testing to destruction individual bricks, since the nature of the mortar, the quality of the workmanship, all have an influence on strength. The Building Research Station in England has carried out extensive tests on brickwork, and developed codes of practice which give engineers a guide as to the safe loads to impose on various types of brickwork.

For example, a short brick pier (i.e. block) of engineering bricks can be safely stressed to about 3·5 N/mm^2, whilst short piers of ordinary building bricks can be allowed to support from about 0·5 to 2·0 N/mm^2 according to the type of brick used. For example, the Fletton brick, which is popular because it is cheap, may be allowed to be stressed in brickwork up to about 1·75 N/mm². Hard London stock brickwork (yellow) may be stressed up to about 1·25 N/mm².

Stones The use of stone in construction is probably older than that of brick. Natural stone is not often used nowadays for supporting loads (i.e. forces). In districts where stone is plentiful and cheap, small houses may still be built with walls of stone. Piers and abutments of bridges may be of stone, and many railway bridges built in the 19th century were supported in this way. When stone is used in large modern buildings, it is usually in the form of a thin skin applied to brickwork, the whole of the masonry construction being supported by a frame or skeleton of steel or reinforced concrete. Like bricks, stones are brittle and have a high strength when used in compression but are weak in tension. The safe compressive load for a stone pier or wall depends on the type and hardness of the stone. For example, granite may be allowed to support about 5 N/mm², whilst a good limestone may have a safe carrying capacity of about 1·75 N/mm². (The actual failing loads of these materials are much greater, an adequate factor of safety being adopted in order to find the 'safe' loads.)

Timber Timber, like stone, being of natural origin, has been used from earliest times for supporting loads or forces. The earliest bridge across a small stream was either a stone slab or a log of timber.

Since there are many types of timber, the load-carrying capacities vary. Unlike bricks and stones, which are granular materials and therefore brittle, timber is fibrous and elastic, and is strong in tension as well as in compression. The strength varies for any one type of timber according to the direction of the grain in relationship to the direction of the load. Softwoods such as whitewood, deal, spruce and pine are the woods most commonly used for joists supporting floors of houses. For these woods, the permissible stress in compression parallel to the grain, in short members, is about 6 N/mm², and the permissible stress in tension is about 8 N/mm². Pitch pine is stronger, but being more expensive is not used so frequently, and this remark applies even more to hardwoods such as oak. In recent years arch structures and portal frames have been built of laminated timber.

Concrete Concrete is made from a mixture of cement, fine aggregate, coarse aggregate and water, which sets to form a hard stone-like material. The cement is usually Portland cement (of which more will be said later). The fine and coarse aggregates are usually sand and gravel dug from gravel pits, and not contaminated by salts. For ordinary constructional work the sand is usually graded so that most of the particles will pass through a sieve having holes 5 mm square and be retained on a No. 100 sieve (hole size: 150 microns). Most of the particles of gravel will pass a 20 mm sieve and be retained on a 5 mm sieve (larger aggregates are often used in massive work such as dams). It is important that the particles should be of many different sizes, so that, on mixing, the smaller particles will fill the voids between the larger ones, thus giving a dense concrete with an economical amount of cement. A concrete much used in buildings is known as the *nominal* 1:2:4 *mix*. This means a mixture by volume of one part of cement, two parts of sand and four parts of gravel. The two parts of sand are more than sufficient to fill the voids in four parts of gravel, and the one part of cement (a fine powder passing a No. 170 sieve, hole size: 90 microns) is more than sufficient to fill the remaining voids and to coat all the particles. Because of its fine nature, cement is difficult to measure accurately by volume, and a nominal 1:2:4 mix is specified as 0.5 kN of cement, 70 dm^3 of sand and 140 dm^3 of gravel (based on cement weighing 15 kN/m^3).

The strength of concrete depends on many factors, a most important one being the water/cement ratio. Other factors being equal, however, stronger concretes can be obtained by using a higher proportion of cement, so for some purposes nominal mixes of 1:1½:3 and 1:1:2 are used. Concrete making can be compared to cooking. Given identical ingredients, one cook can prepare a first-class dinner, and another will produce an average or even an inferior meal.

Sometimes, concretes are made by using crushed granites or sandstones, whilst light-weight concretes are made from particles of pumice stone or clinker or other light material. These light-weight concretes are not normally used for load-carrying purposes but are used for heat and sound insulation.

Concrete as we know it today dates back only to the latter half of the 19th century and was made possible by the invention of Portland cement. The Romans used 'concrete' of a sort, but it usually consisted of small stones and of lime mortar in alternate layers, filled in between outer facings of brick and stone. Lime mortar made from white chalk or limestone is very weak, but the Romans discovered how to make stronger cementing materials by mixing pozzolanic materials wth the lime. The most common material of this sort was volcanic ash, or pozzolana, which takes its name from Pozzuoli in Italy.

Strong hydraulic cements (cements capable of setting under water) can be made by mixing pozzolanic materials with lime, and the use of such materials accounts to a large extent for the durability of Roman concrete walls and other constructions.

After the Romans left Britain, the art of good mortar and concrete making appears to have declined, and one of the first references to a study of cementing materials dates to about 1756, when John Smeaton (1724–92), a Yorkshireman, was commissioned to build a new Eddystone lighthouse. (The two previous lighthouses, both of wood, were destroyed, one by a storm in 1703 and the other by fire in 1755.) Now, lime mortar, obtained by burning white chalk or pure limestones, will not set unless it can receive carbon dioxide from the air and unless the water used for mixing can evaporate. Even if conditions are favourable for setting, the resulting mortar is very weak and 'crumbly.'

Smeaton was very conscious of this, and states—

> 'I seriously began to consider the great importance that it was likely to be of to our work to have a cement the most perfect that was possible to resist the extreme violence of the sea. . . . It seemed that nothing in the way of cement would answer our end but what would adhere to a moist surface and become hard without ever becoming completely dry.'

Smeaton collected samples of lime from many parts of the country for testing and chemical analysis. For example, he states—

> 'Having heard of a lime produced from a stone found at Aberthaw upon the coast of Glamorganshire that had the same qualities of setting in water as Tarras [a pozzolanic material], I was very anxious to procure some of the stone which I did and burnt into lime.'

As a result of his experiments in collaboration with a friend who was a chemist, Smeaton discovered that the property some limes possessed of setting under water was due to the presence of clay in the limestones, but came to the wrong conclusion as to the manner in which the clay acted—

> 'Whereas for some reason or other, when a limestone is intimately mixed with a proportion of clay which by burning is converted into brick, it is made to act more strongly as a cement.'

It must be remembered that the science of chemistry, like other sciences, was not very far advanced in the 18th century: put quite simply, the setting of a mortar made from a limestone containing clay is due to the action of the mixing water on various types of limeclay compounds formed during the burning of the limestone.

Smeaton eventually built his lighthouse of stone jointed with a hydraulic lime mortar, and the lighthouse stood until 1882, when it was replaced because of weakening of the foundations. Smeaton's lighthouse was re-erected as a monument on Plymouth Hoe, where it still stands.

It is reasonable to suppose that, someone having realized that a stronger cement could be obtained from a limestone containing clay than from a more pure (white) limestone, should have reasoned thus: "If a natural clayey limestone produces a good cement, why can't clay be mixed with white chalk or limestone to obtain the same result?"

There is no evidence for this supposition, but the fact remains that Joseph Aspdin, a Leeds bricklayer, patented such a cement in

1824 and called it Portland cement because of its supposed resemblance in colour when hard to Portland stone (the type of limestone used in St Paul's Cathedral, Buckingham Palace and many other public buildings). The original Portland cement was very little different in strength from hydraulic cements made from clayey limestones, but continuous improvements in manufacture have resulted in a cement which is far superior to hydraulic lime cements and to any cements used by the Romans. Portland cement factories are usually situated where there are ample supplies of chalk (or limestone) and clay, such as occur in the Thames estuary.

The chemistry of the setting action of Portland cement is very complex, but it may be stated simply that setting is due to the combination of the mixing water with the various chemical compounds of lime and clay produced by the high temperature to which the raw materials are subjected during manufacture. Portland cement will set and harden equally well when placed under water as when it is allowed to harden in air. Various types of Portland cement are available for special purposes, such as white cement, low-heat cement and sulphate-resisting cement.

Portland cement and steel are the two most important modern building materials.

Another important cement for making concrete is high-aluminous cement, which was originally developed in France and was first marketed in Britain in 1923. It is darker than Portland cement, being almost black, and is stronger, quicker in hardening, and more resistant to certain corrosive influences such as sulphates and sea water. It is manufactured from a mixture of chalk and bauxite (an ore of aluminium). High-aluminous cement is not used normally in the general construction of buildings, since Portland-cement concrete can usually be made of sufficient strength for the purposes required and is much cheaper. High-aluminous cement concrete is used, however, for special constructions and also for foundations and tunnels in soils containing sulphates which would attack Portland cement.

As stated earlier, the strength of concrete depends on many factors, but a 1:2:4 Portland-cement concrete when used in a short pier would be allowed to support a compressive load of about 7 N/mm^2.

More important than the use of concrete by itself is the combination of concrete with thin steel bars to form reinforced concrete and prestressed concrete; these composite materials are dealt with in later chapters.

Steel

The basic materials for the manufacture of all types of iron and steel are iron ores, which are rocks containing iron in chemical combination with other materials. To free the iron from the ore, heat is required. The ferrous metals can be broadly divided into three groups: cast iron, wrought iron and steel. Although these materials consist almost entirely of iron, the presence of small amounts of other elements has important influences on the properties and characters of the metals. A very important ingredient in irons and steels is carbon. When the percentage of carbon is small, the metal is soft (mild) and ductile; when the percentage is high, the metal is hard and brittle.

The manner in which iron was first extracted from its ores cannot be known, because iron was used by prehistoric man. Probably the first form in which it was obtained was a type of wrought iron pulled from the primitive furnace as a pasty mass and afterwards hammered into shape. Cast iron was probably discovered when as a result of heating at a high temperature it was found that the metal became molten and could be cast into shapes made in sand.

Although steel has been made for hundreds of years, until the latter half of the 19th century it was expensive and only produced in small quantities for swords and springs.

In modern methods of production, the first stage in the manufacture of any type of iron or steel is the heating of the ore in a blast furnace to produce pig iron. The pig iron, which runs from the furnace in a molten stream, is cast into moulds or taken direct to the steel furnace.

Pig iron contains a large percentage of carbon (which makes it brittle) as well as other impurities such as silicon, manganese, phosphorus and sulphur.

Cast iron is produced by the remelting of pig iron, and according to the refining process and added ingredients, various types of cast iron can be obtained, some less brittle than others.

Wrought iron also is made by refining pig iron, but the process is laborious and not so amenable to the use of mass-production processes as is steel. The temperature at which wrought iron is produced is not high enough to keep the metal molten, and the wrought iron is extracted from the furnace as white-hot spongy lumps which are hammered or rolled into plates, bars, etc. (Unlike cast iron, wrought iron is very malleable.)

Wrought iron has been used for many hundreds of years for ornamental ironwork, as for example, gates and grilles, as well as for links of chains.

To produce steel, pig iron, or pig iron together with steel scrap, is purified by heating in a furnace to drive off carbon, phosphorus, etc. To the liquid metal thus purified are added the exact amounts of the various ingredients (including carbon) necessary to give the steel the desired properties. The steel is tapped as a molten liquid into ingot moulds; the ingots are eventually reheated to make the metal soft, and rolled into plates, bars and other shapes. Alternatively, castings can be made by running steel into moulds of the required shape.

Structural steel (mild steel), as used for beams and columns in buildings and for bars in reinforced concrete, contains only a small amount of carbon (about 0·2 per cent). Tools for cutting must be hard and therefore contain a higher percentage of carbon. The adjective 'mild' denotes that the steel is soft, i.e. ductile.

For supporting loads in structures the first ferrous metal used was cast iron, which was not produced in any appreciable quantity in Britain until the 18th century, when Abraham Darby (1677–1717) introduced a process using coke produced from coal as the furnace fuel. The castings for the first iron bridge in England were made in 1779 in the foundries of a descendant of his, also named Abraham Darby.

In addition to greater malleability, wrought iron has a higher tensile strength than cast iron, and owing to improvements in manufacture and application, it gradually superseded cast iron towards the middle of the 19th century.

Both cast iron and wrought iron were eventually superseded by steel, thanks to the improved manufacturing processes starting with Sir Henry Bessemer's converter in 1856 and followed by the Siemens–Martin open-hearth process (1862–68). Much of the steel first produced by the Bessemer converter was brittle and crumbled to pieces during rolling. This defect was corrected as the result of a suggestion by Robert Mushet that manganese be added to the steel before casting. Troubles were not over, however, because the steel produced from certain ores was still brittle, and it was only after much research that the trouble was traced to the presence of phosphorus. It was not until 1878–79 that Sidney Thomas and Percy Gilchrist introduced a method for eliminating phosphorus by lining furnaces with materials capable of combining with the unwanted material and removing it from the iron. In a lecture by J. Gibson reported in the *Structural Engineer* of October, 1937, it is stated that:

> "Thomas, a junior law clerk interested in metallurgy, had attended a lecture on Cast Iron where the lecturer said, 'The man who eliminates phosphorus by means of the Bessemer converter will make his fortune.' In conjunction with his cousin Percy Gilchrist, a works chemist in South Wales, his experiments culminated in a successful practical demonstration of that attainment in April, 1879."

The present age of steel can justifiably be said to date from that year, and many types of steel of varying strengths, hardness, etc., can now be produced having reliable consistency of properties.

Steel was not readily accepted at first, and it is interesting to note that in 1859 the use of steel for the Charing Cross bridge, London, was prohibited by the Board of Trade; the restriction on the use of mild steel for bridges was not lifted until 1877.

Types of structural steel

The steel normally used for the frameworks of buildings and for bars in reinforced concrete (not prestressed concrete) is 'ordinary mild steel.' This must have, according to a British Standard specification, an ultimate or failing stress of 450–500 N/mm^2. Almost as important as the failing stress of mild steel is its *yield stress*, which may be about 250–300 N/mm^2.

The explanation of yield stress is as follows. If a bar of mild steel is tested in tension by applying to it a gradually increasing load, the bar will stretch uniformly until a certain load is reached. For example, a bar 5000 mm long will stretch 1 mm when a stress of 40 N/mm^2 is applied, 2 mm when the stress is 80 N/mm^2 and 4 mm when the stress is 160 N/mm^2 and so on. (Steel is *elastic*, an elastic material being one which stretches equal amounts for equal increments of load or force, and reverts to its original length when the load is removed.) As the stress increases, however, a point is reached (*elastic limit*) at which the steel loses its elasticity and suddenly begins to stretch a great amount compared with the previous small elongations. This point is called the *yield point* of the steel. If the bar is left for a while without further stress being applied, it 'recovers' to a certain extent and more load can be applied. The extensions now become increasingly greater, the steel being in a plastic state, and the bar finally snaps at a stress of about 450–500 N/mm^2. The final length of the bar will be about 1000 to 1250 mm, representing a total elongation of 20 to 25 per cent. The yield stress is very important, and in designing buildings the factor of safety must be such that there is no danger of the steel yielding, as this would cause excessive bending of beams, etc. On the other hand, the excessive elongations which occur at the yield stress (which is smaller than the failing stress) give warning of imminent collapse of a structural member, whereas a failure in a brittle material such as cast iron occurs suddenly and without warning.

For ordinary mild steel in tension the permissible or working stress is 150 N/mm^2 (compare with a yield stress of about 250 to 300 N/mm^2). The factor of safety based on the yield stress is therefore about 2.

Medium-tensile and high-tensile steels

By incorporating certain ingredients such as copper during the manufacture the strength of steel can be increased. Medium-tensile structural steel has a failing stress of 500 to 575 N/mm^2, and high-tensile steel has a failing stress of 575 to 675 N/mm^2.

The permissible stresses for these steels are higher than those of ordinary mild steel, thus permitting the use of smaller structural members.

Steel for prestressed concrete

The steel for prestressed concrete usually consists of cold-drawn steel wire of smaller diameters than those used for bars in ordinary reinforced concrete. The smaller the diameter to which the wire has been drawn out, the higher is the failing stress, which ranges from about 1500 to 1600 N/mm^2 for wire of 7 mm diameter to 2200 to 2300 N/mm^2 for wire of 2 mm diameter.

Aluminium alloys

The most common source of aluminium is the ore called bauxite, which was first found in 1821 near the village of Les Baux in France. Aluminium is too soft for structural purposes, but its strength can be increased by combining other metals with it to form alloys, the chief alloying elements being copper, magnesium, silicon, nickel and zinc. Nowadays it is possible to obtain aluminium alloys which approximate to the tensile strength of ordinary mild steel, and since the weight of these alloys is about 27.5 kN/m^3, as compared with steel at 77.0 kN/m^3, lighter structures can be designed in aluminium alloys than in mild steel. Two disadvantages, however, are the higher cost of manufacture and the fact that the elastic modulus (*see* Chapter 1) is much lower than that of mild steel. Since aluminium alloys deform (i.e. stretch, shorten, etc.) more readily under load than mild steel, they are best used in structures where these larger deformations do not necessarily mean using an increased amount of material, or in rigid structures where deformations are small. For certain types of large-span roofs there is a great potential for the use of aluminium alloys (*see* Chapters 14–16).

Summary

Brick, stone and concrete are granular, brittle materials and therefore have little resistance to tension and shear but great resistance to compression. They are used for piers, walls and foundations when the loads induce only compressive stresses or, at the extreme, very small tensile stresses.

Timber is strong in tension as well as in compression and is elastic, but its tensile, shear and compressive strengths are very much less than those of structural steel.

The permissible stresses for all these materials are given in standard specifications and building by-laws, and are based on adequate factors of safety in relation to the stresses which would cause failure or yield.

Exercises

(2.1) Break a stick of chalk in tension. Try to break a stick of chalk in direct compression. Is chalk brittle or ductile?

(2.2) Obtain a dressmaker's pin, a needle, a hacksaw blade (all of steel) and if possible a piece of cast iron, and try to bend them until failure occurs. Which metals are 'soft' and which are 'hard' and why?

(2.3) Fill a box to capacity with gravel (or stones).

Find how much water you can pour into the box before it overflows. Estimate the amount of voids in the gravel.

(2.4) Fill a box (say 150 mm × 150 mm × 150 mm, i.e. 3·375 dm^3) with sand. Take double this quantity of gravel (6·75 dm^3). Mix the sand and gravel together and estimate the volume of the mixture. Prove that one plus two does not necessarily make three.

(2.5) Study the direction of the grain in timber. What is the best direction of the grain for resisting tension?

3 Effects of Forces: Moments: Force Diagrams

The purpose of this book is not to teach methods of structural calculation; there are many books which do so. Nevertheless, in order to understand how structures behave it will be necessary on occasion to introduce some theoretical principles and to illustrate them with simple calculations.

Two basic principles apply to all types of structural design:

(1) The structure must be stable, i.e. all forces must be in equilibrium.

(2) The stresses caused in the various members of the structure as a result of the external forces, and the consequent reactions from other members or from the ground, must not exceed the values previously decided as being safe.

Equilibrium

Vertical equilibrium

In Fig. 3.1 (*a*) the weight of the block is 10 kN acting vertically downwards. In order to maintain stability, the upward resisting pressure (the reaction) from the soil must also be 10 kN—not a fraction more nor a fraction less. If the weight is increased to 20 kN, as at (*b*), the resistance of the earth will increase to balance the load exactly. At (*c*) it is assumed that the surface soil has reached the limit of its resistance and is unable to support a load of

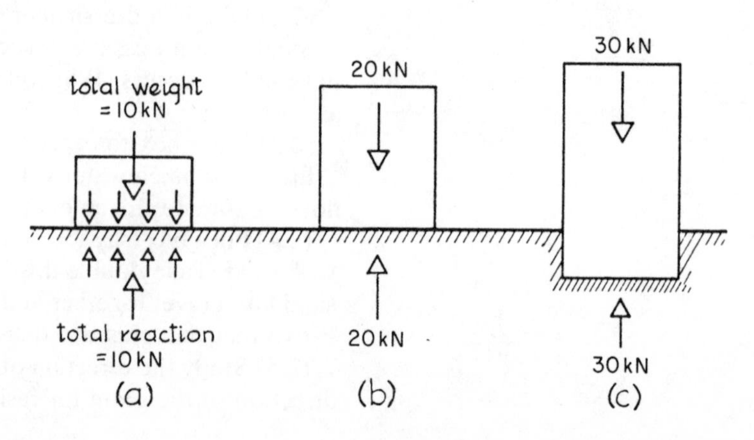

Fig. 3.1

30 kN. The block will therefore sink and compress the earth until it is capable of supporting the load, so reinstating equilibrium. The block is once again stable, but it has failed to perform as originally intended.

If a heavy man sits on a weak chair which collapses, he will come to equilibrium on the floor, which is capable of supporting his weight. A bridge which collapses also finishes up on the ground in a state of equilibrium, but, of course, it is no longer a bridge.

Horizontal equilibrium

Fig. 3.2 (*a*) represents a condition where vertical equilibrium is satisfied but horizontal equilibrium is not. At (*b*), friction between the block and the rough surface prevents horizontal movement or sliding by inducing a horizontal reaction equal to the horizontal forces.

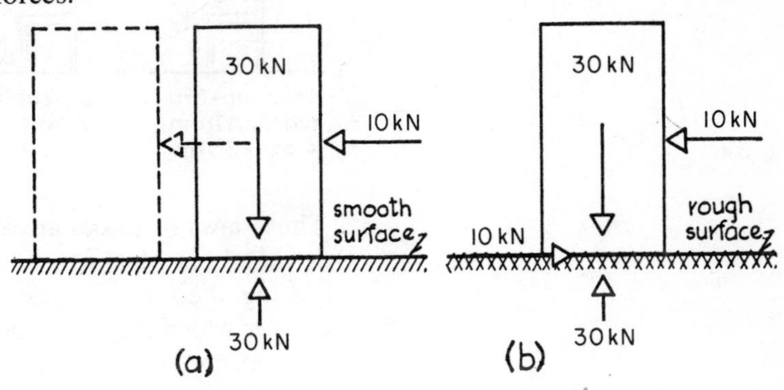

Fig. 3.2

Turning equilibrium

Although vertical and horizontal conditions of equilibrium are satisfied in Fig. 3.2 (*b*), there is a possibility of overturning (as in Fig. 3.3) with point *O* as the pivot. The body will begin to overturn when the anti-clockwise *turning effect* of the horizontal force *A* just balances the resisting clockwise effect of force *B* (due to the weight of the block).

It should be noted that for the conditions shown in Fig. 3.3 all the weight of the block (30 kN) is resisted at the point *O*. This means a heavy concentration of load on the soil at this point.

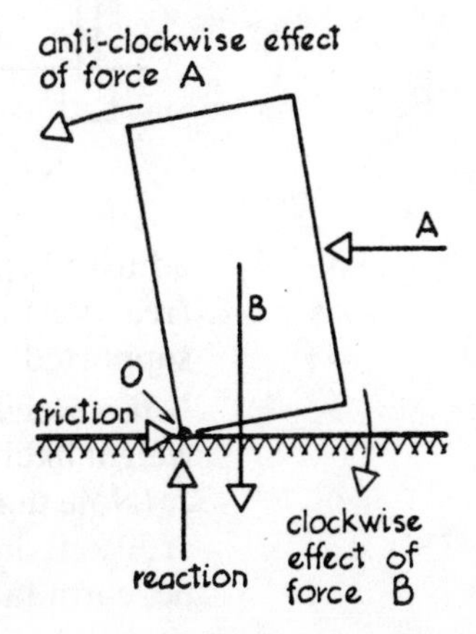

Fig. 3.3

Figs 3.1, 3.2 and 3.3 demonstrate that for forces in the same plane, equilibrium results:

(1) When upward forces equal downward forces.

(2) When horizontal forces to the left equal horizontal forces to the right.

(3) When turning effects in a clockwise direction are balanced by equal turning effects in an anti-clockwise direction.

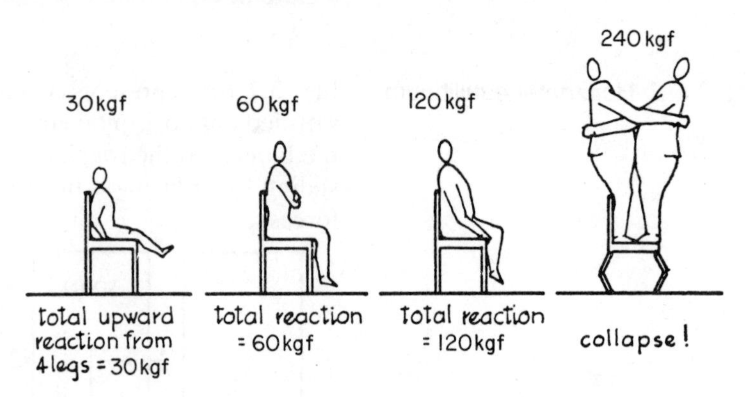

Fig. 3.4

These laws of statics are always true, but design should ensure that members of a structure and their supports are capable of adjusting their resistances so that they can take the maximum loads for which they have been designed.

For example, assuming the whole weight of a man to be supported entirely by a chair (usually some weight is supported by the feet resting on the ground), Fig. 3.4 shows how the chair legs

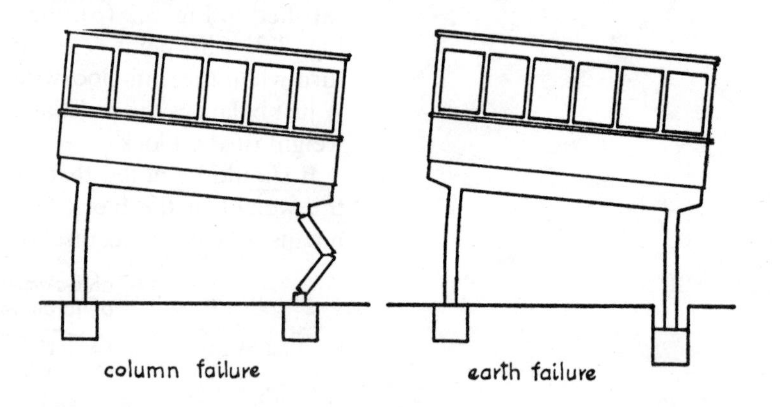

Fig. 3.5

adjust themselves as load is increased. The upward resistance (reaction) of the chair legs must always be equal to the weight supported.

If a weight of 240 kgf would cause collapse and the maximum weight likely to be supported is 120 kgf, there is a factor of safety of 2. (Note that failure of the 'structure' could occur at 120 kgf or even at 30 kgf, due to the chair legs sinking into soft earth, but this would be 'earth failure,' not 'chair failure.')

Figure 3.5 indicates two possible failures in a building.

Figure 3.6 shows how the frictional resistance increases with increasing height of water behind a wall. If the maximum hydrostatic force behind the wall is 30 kN, and if the earth under the wall has a maximum frictional resistance of 60 kN, there is a factor of safety of 2 against sliding of the wall.

Note that the frictional force must always be equal to the force causing sliding. If it were less the wall would slide forward: if it were greater the wall would be pushed backwards against the water! In designing this wall calculations would also be necessary to ensure that the earth beneath could support the wall (vertical equilibrium), and that there was no possibility of the wall overturning (turning equilibrium).

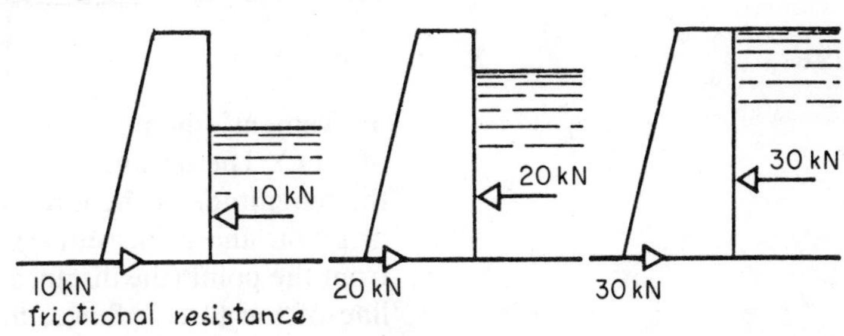

Fig. 3.6

The principle of the lever—moments of forces

Reference has already been made (*see* Fig. 3.3) to the turning effect or leverage of a force. Aristotle (384–322 B.C.) and Archimedes (287–212 B.C.) dealt with the theory of the lever, and during the 11th or 12th century Jordanus de Nemore also concerned himself with problems on levers.

Use is often made of the fact that a small force, if applied in a particular manner, can have a big effect; familiar examples of levers are given in Fig. 3.7.

Figure 3.8 shows a stability failure due to the small force at *B* exerting a bigger clockwise turning effect than the anti-clockwise turning effect exerted by the larger force at *A*.

The seesaws shown in Fig. 3.9 are in a state of balance or equilibrium (if somewhat unstable) as the result of clockwise turning effects about the pivot being balanced by anti-clockwise turning effects.

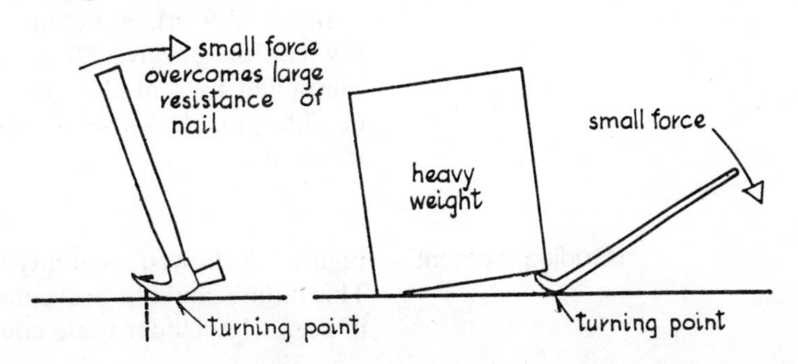

Fig. 3.7

It can be verified by a simple experiment that a small force (30 kgf) at a large distance (2 m) can have a turning effect or leverage equal to that of a larger force (60 kgf) at a smaller distance (1 m).

In structural work, the turning effect of a force is called the *moment of a force.* This must not be confused with 'moment' as applied to time. The word is derived from a Latin word meaning

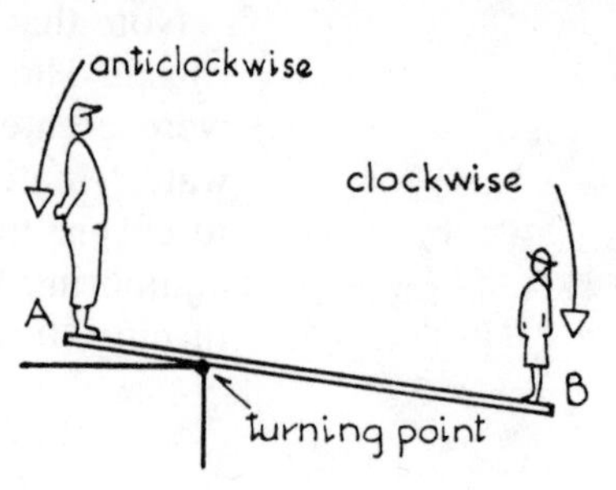

Fig. 3.8

'movement' (moments of time refer, of course, to the 'movement' of time). The turning effect or moment of a force depends on both the magnitude of the force and its distance from the turning point or pivot, and is measured by multiplying the force by its distance from the point (the distance being at right angles to the directional line of the force). Since, to obtain the measure of a moment, a

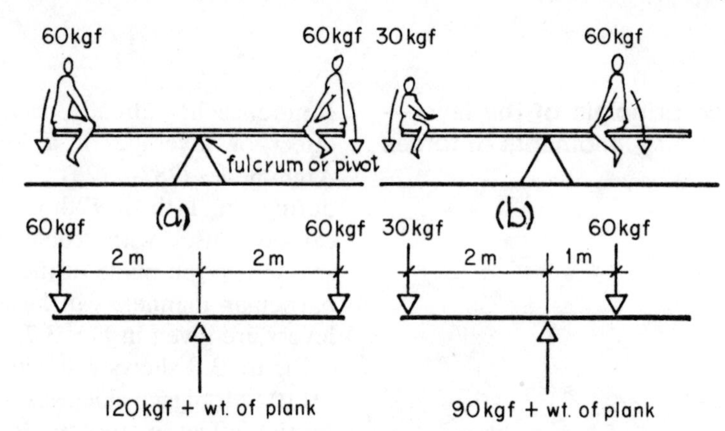

Fig. 3.9

force (in kgf or kN) is multiplied by a distance (in metres), the answer cannot be expressed in units of force or distance alone, but by a combination of units, for example kgf m or kN m (kilo-gram metres or kilonewton metres).

In Fig. 3.9 (*b*), 30 kgf multiplied by its distance (2 m) from the pivot or fulcrum gives 60 kgf m (anti-clockwise moment) and 60 kgf multiplied by 1 m also gives 60 kgf m (clockwise moment). For equilibrium, clockwise moments must always equal anti-clockwise moments.

Bending moment

Figure 3.8 showed a stability failure due to the moment of a force. This failure could be prevented by making the force at *A* larger, as in Fig. 3.10. Under these conditions, the clockwise moment is 20 kN multiplied by 2 m, i.e. 40 kN m, and the anti-clockwise resisting moment is 160 kN (acting through its centre of gravity) multiplied by 0.5 m, i.e. 80 kN m. This is twice the moment required for equilibrium; therefore there is a factor of safety of 2 against overturning and no danger of a stability failure. There is, however, a danger of structural failure; if the beam is weak, it will bend excessively under the moment exerted by the 20 kN load and might break at point X (Fig. 3.11), where the leverage is greatest. In brief, the beam has to be capable of resisting a *bending moment* of 40 kN m.

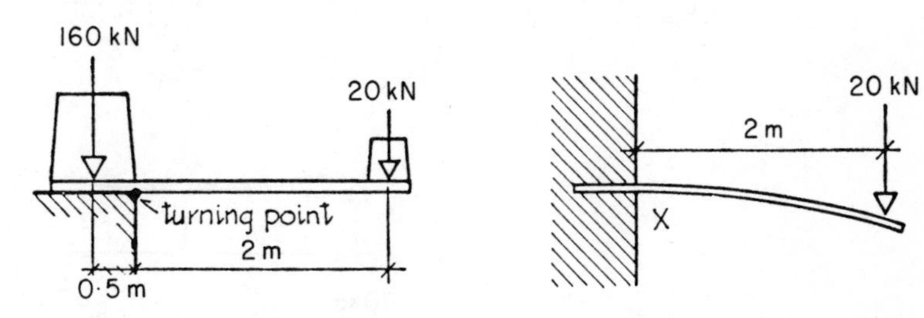

Fig. 3.10 Fig. 3.11

The plank in Fig. 3.12 (*a*) has to be capable of resisting safely a maximum bending moment of 60 kgf multiplied by 2 m, i.e. 120 kgf m, and if it is not strong enough it will break at point C.

In Fig. 3.12 (*b*) the maximum bending moment is the leverage action of the 30 kg weight at 2 m from the pivot (60 kgf m), *or* the leverage action of the 60 kg weight at 1 m from the pivot (60 kgf m). The bending moment is *not* the sum of the two moments (120 kgf m), since they act in opposite directions, clockwise and counter clockwise, and furthermore if one load is removed, bending cannot be caused at all (Fig. 3.13).

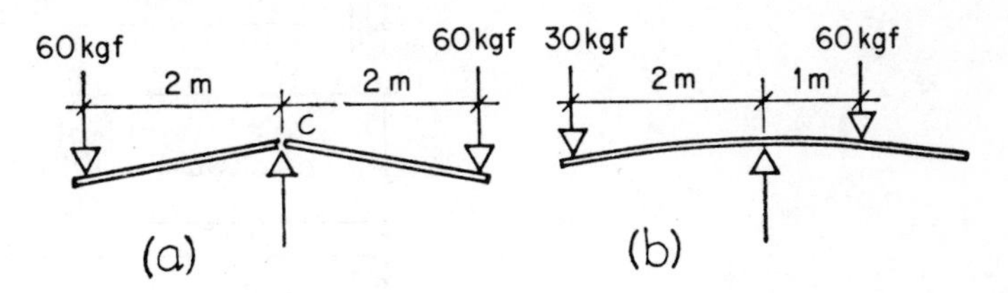

Fig. 3.12

Compare Fig. 3.14 carefully with Fig. 3.12 (the weights of the planks being ignored). If Fig. 3.14 is looked at upside down it will be seen that the conditions of loading, etc., are identical with those of Fig. 3.12. Beam (*a*) would therefore be designed to resist a bending moment of 120 kgf m, and beam (*b*) a bending moment of 60 kgf m.

The reactions at the supports, the way beams bend, and the values of the maximum bending moments, depend on many

Fig. 3.13

factors such as amount and arrangements of loads, the number of the supports, and the manner of fixing at the supports. Methods for calculating reactions and bending moments can be studied in

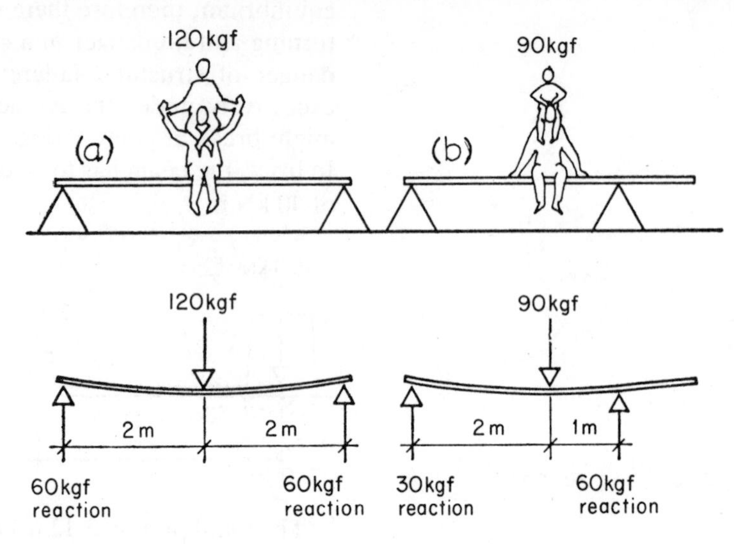

Fig. 3.14

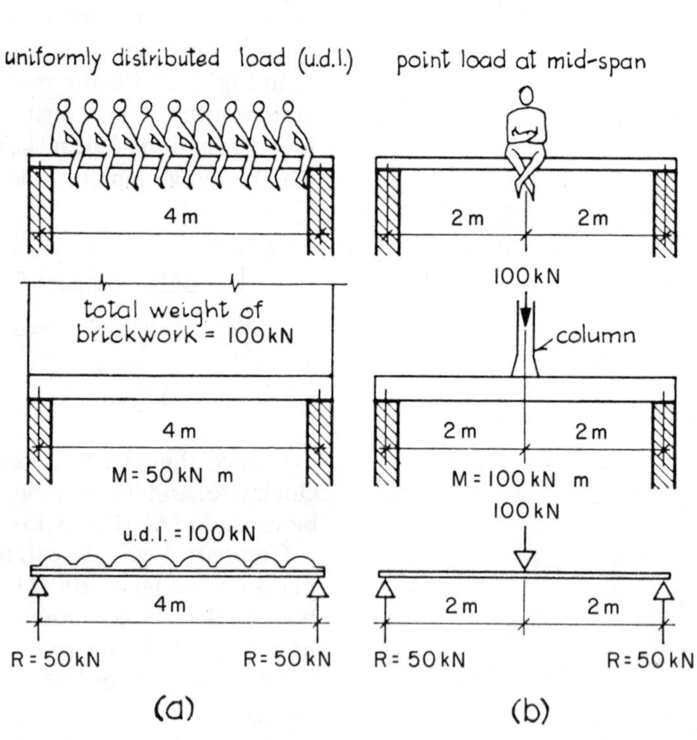

Fig. 3.15

the appropriate textbooks. A few general examples are given in Figs 3.15–3.18.

In Fig. 3.15, the maximum bending moment in the beam at (*b*) is twice that in the beam at (*a*), and a bigger beam would be required although the loads and reactions are equal. (The weights of the beams are ignored in this simple description but not in structural calculations. The weight of a uniform beam is taken as a uniformly distributed load.)

The reason for the difference in the values of the maximum bending moments can be explained by reference to Fig. 3.16, where the amount of bending is much exaggerated. If a cantilever beam (*a*) is pushed upwards by a force of 50 kN, the maximum bending moment (at the support), ignoring the weight of the

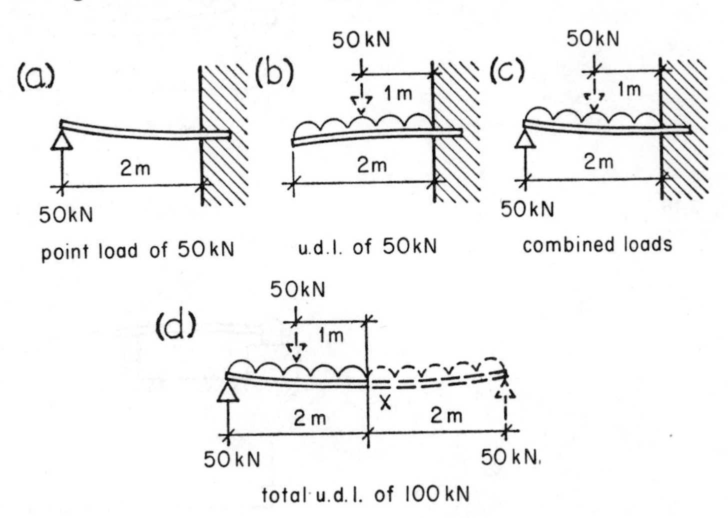

Fig. 3.16

beam, is 50 kN x 2 m = 100 kN m. The maximum bending moment due to a uniformly distributed load (u.d.l.) of 50 kN (*b*) is 50 kN × 1 m = 50 kN m. (The u.d.l. can be assumed to act at the centre of gravity of the load, i.e. at 1 m from the support.) If the upward point load of 50 kN and the downward u.d.l. of 50 kN act simultaneously as at (*c*), the maximum bending moment is the

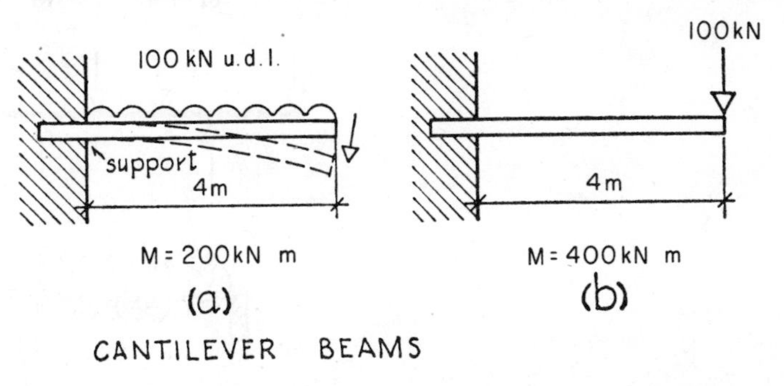

Fig. 3.17

difference between the two maximum values, i.e. 100–50 or 50 kN m. The conditions of bending for (*d*) are identical with those for (*c*); therefore the maximum bending moment (at mid-span) is 100 kN m (clockwise with respect to point *X*) minus 50 kN m (anti-clockwise), giving 50 kN m.

In the cantilever beam of Fig. 3.17 (*a*) the maximum bending moment is at the support and is four times the bending moment in Fig. 3.15 (*a*). The maximum bending moment in Fig. 3.17 (*b*) is eight times that in Fig. 3.15 (*a*).

For the fixed-end beam of Fig. 3.18 (*a*) the maximum bending moments occur at the supports, and it can be proved that these moments are each two-thirds of the maximum bending moment of the beam in Fig. 3.15 (*a*).

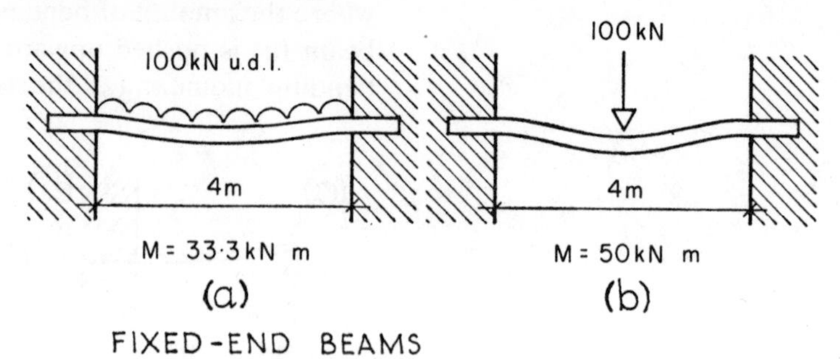

Fig. 3.18

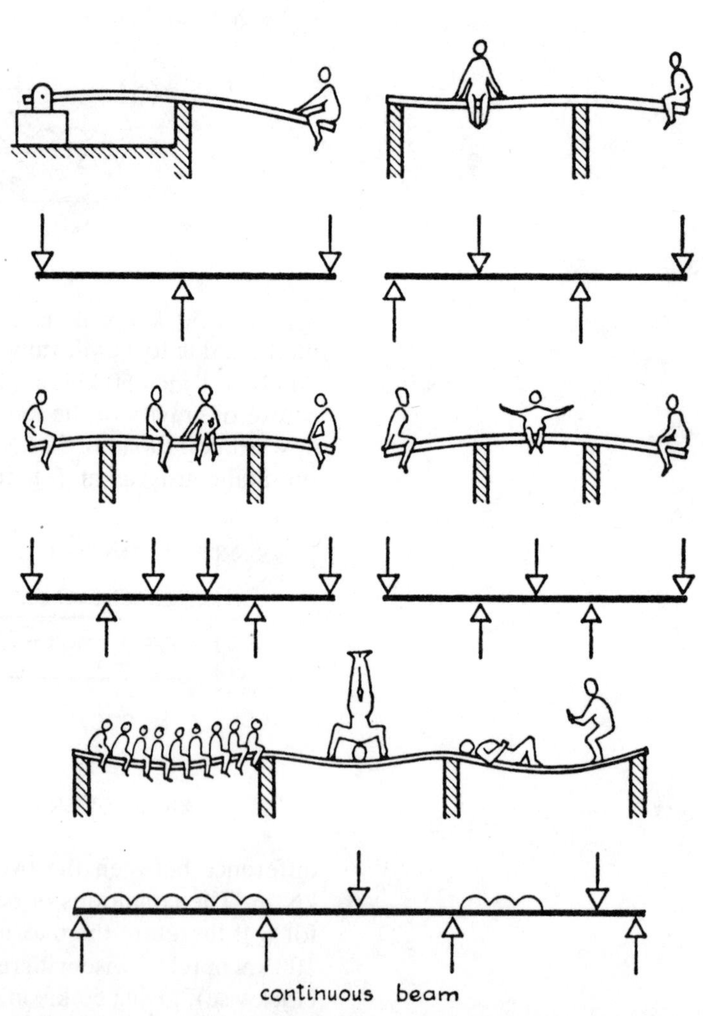

Fig. 3.19

Some other ways in which beams may bend are shown in Fig. 3.19. When a beam rests freely on two supports or is fixed at one end only, the maximum bending moment can be obtained very easily. However, when a beam is fixed at both ends or is continuous over several supports, the calculations are more complex.

Forces can have leverage effects. The leverage effect of a force depends on the amount of the force and its distance from a given point, and is called the *moment of the force.* Turning effects or moments acting on beams are called *bending moments.* It is essential to know the bending moments for any given conditions of loading, etc., so that beams can be made large enough to resist them. (The bending of beams has been greatly exaggerated in the diagrams. In practice, the amount of bending is usually undetectable by the unaided eye.)

Overturning moment

Another example of the use of moments is in problems relating to the stability of walls or piers. A very simple illustration is given in Figs 3.20 and 3.21, which also demonstrates how the intelligent use

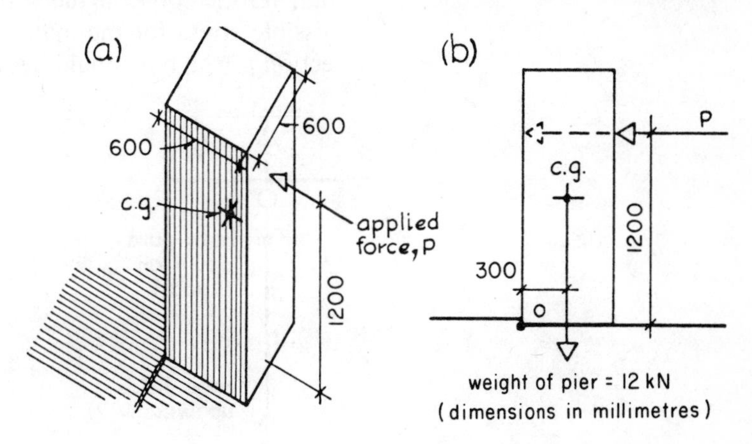

Fig. 3.20

of available material can result in greater efficiency in the resistance of forces.

In Fig. 3.20 (*a*) the pier weighs 12 kN and is subject to a horizontal pushing force. Figure 3.20 (*b*) shows the elevation of the pier, which is all that is necessary for calculation purposes. So far as the stability of the pier is concerned, the whole weight of 12

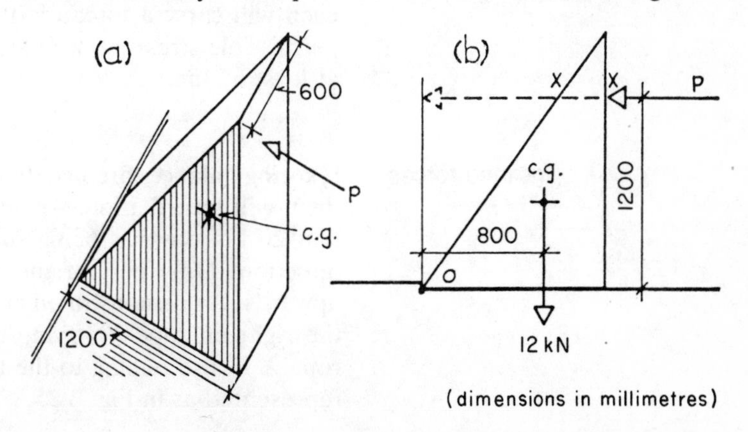

Fig. 3.21

kN can be assumed to act through the centre of gravity. The pier will begin to overturn, pivoting about O, when the anti-clockwise moment, $P \times 1200$ mm, just equals the clockwise moment, 12 kN $\times$ 300 mm, supplied by the weight of the pier. The pier therefore begins to be unstable when P is 3 kN.

The triangular pier of Fig. 3.21 contains exactly the same amount of material as that in the rectangular pier. Its base is 1200 mm, and it can be proved that its centre of gravity is 800 mm from O. The resisting moment of the pier is therefore increased to 12 kN $\times$ 800 mm, which means that the applied force, P, can be increased to 8 kN before instability occurs. Of course, at the level XX there is less material to resist shear than in Fig. 3.20, and the strength in shear would have to be checked.

Members in tension

Consider the simple structural problem shown in Fig. 3.22. The total tensile force in the rope or bar is 20 kN, and if the permissible stress for the material is 150 N/mm², the area of cross-section of the bar should be at least 20,000/150, i.e. 133 mm².

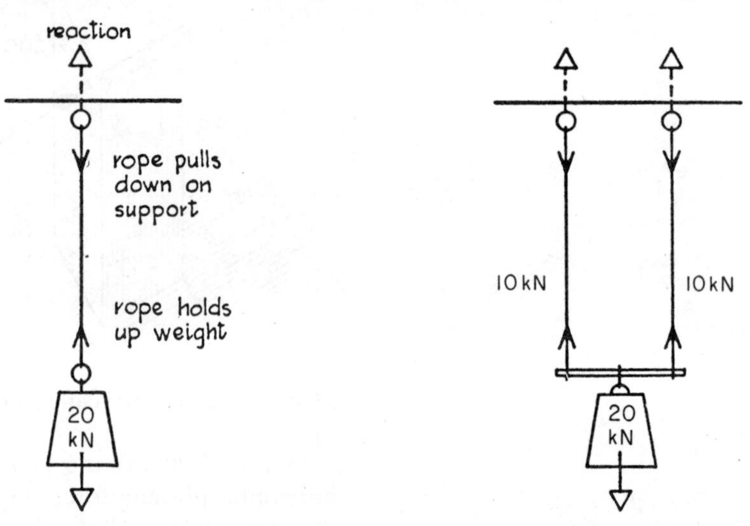

Fig. 3.22 Fig. 3.23

In Fig. 3.23, assuming the two ropes are identical in all respects, each will carry a force of 10 kN and in order not to exceed the permissible stress, each rope should have a cross-sectional area of at least 67 mm².

Inclined forces

If spring balances are inserted in the ropes of Fig. 3.24 (a) each of them will record a force of about 14·14 kN. It appears, therefore, that 20 kN down is being supported by 28·28 kN in an upward direction. Only part of this force, however, is acting *vertically* upwards. As demonstrated at (b) and (c), rope Y is also pulling to the right on rope X (trying to regain the vertical position), and rope X is also pulling to the left on rope Y; the condition can be represented as in Fig. 3.25.

Note that the two relevant conditions of equilibrium (vertical and horizontal) are satisfied at the meeting point of ropes X and Y and the rope suspending the load or weight of 20 kN. The horizontal pull of each rope on the other is given in Fig. 3.25 as 10 kN; this will be explained later. [*See* Parallelogram of Forces Law

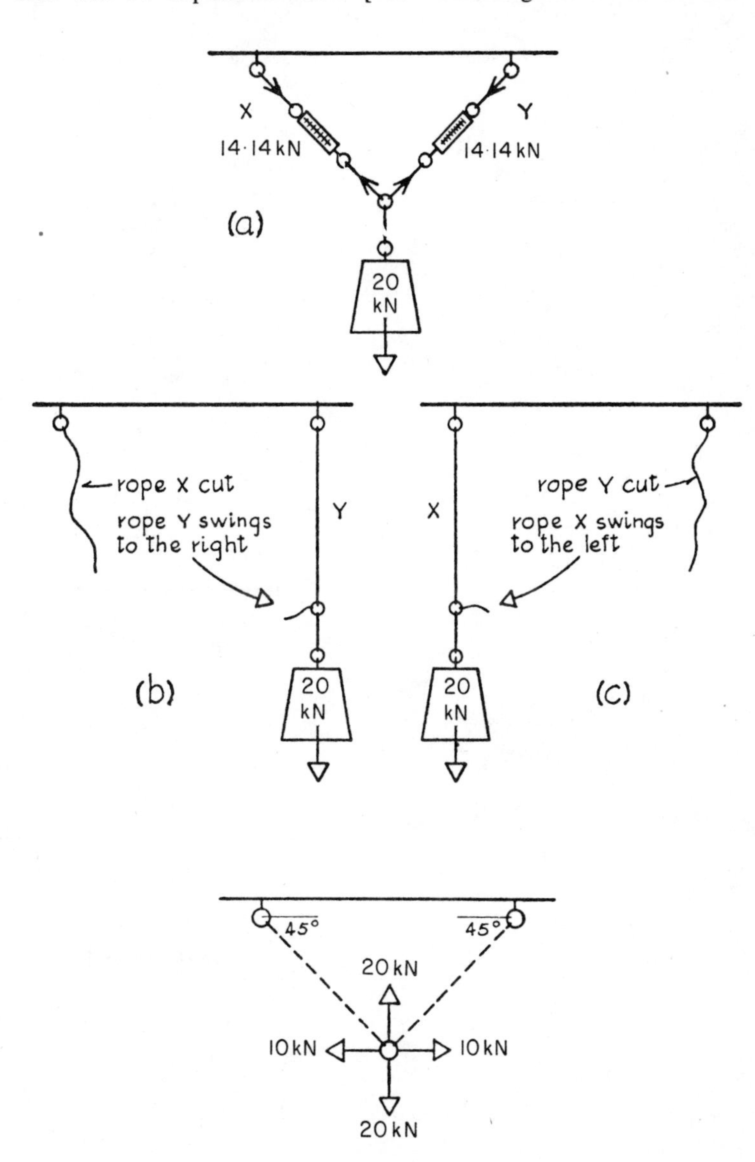

Fig. 3.24

Fig. 3.25

(below), and Resolution of Forces (page 41).] Figure 3.26 gives some further examples. The given forces can be verified by the student after reading pages 39 and 41.

The above diagrams illustrate that the smaller the angles the ropes make with the horizontal, the greater will be the forces in the ropes. They also show that the force in each rope can be much greater than the vertical weight which is being supported. For example, in Fig. 3.26 (c) the force in each rope is 38·64 kN.

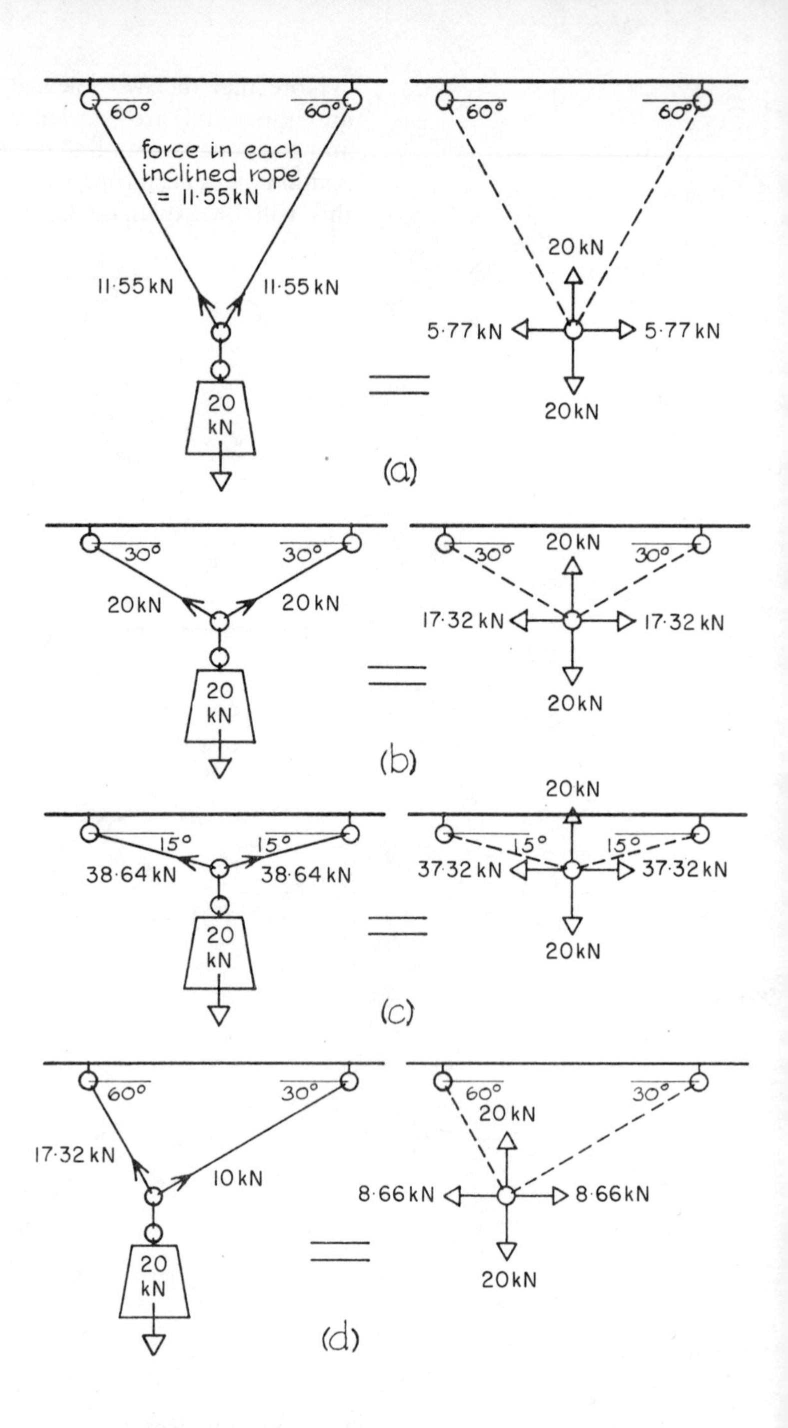

Fig. 3.26

Parallelogram of forces law

The discovery of the law which is demonstrated in Fig. 3.27, and can be proved by experiment, is attributed to Simon Stevins (1548–1620), who was an engineer in the Dutch Army. He is reputed also to have been the first person to represent the magnitude of a force by a line on paper drawn parallel to the direction of the force, the length of the line representing the amount of the force. The general statement and deduction of the law is also

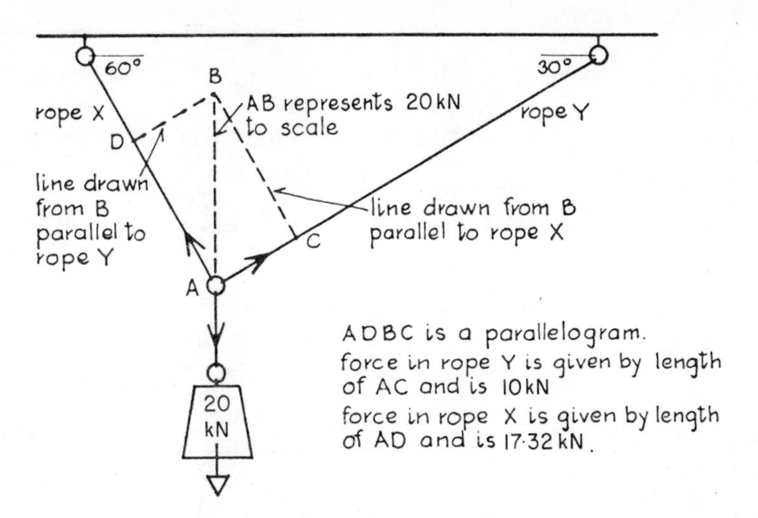

Fig. 3.27

attributed to Sir Isaac Newton, who stated it in connexion with the velocities of moving bodies.

Triangle of forces law

This law, which gives answers identical with those obtained by the parallelogram of forces, is sometimes a more convenient method of dealing with three forces meeting at a point (Fig. 3.28).

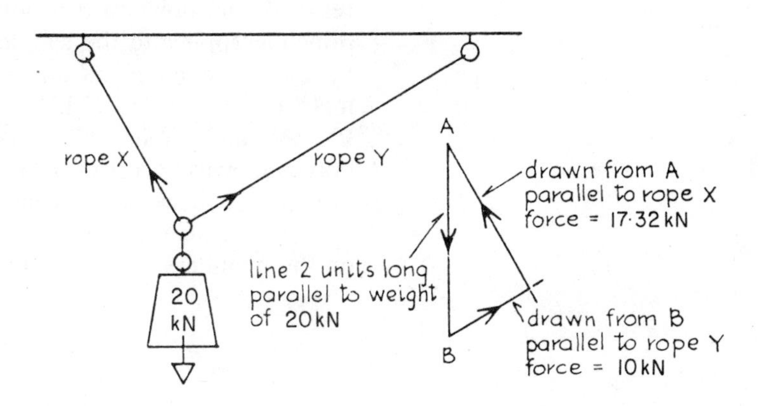

Fig. 3.28

Polygon of forces law

Imagine four men pulling on a ring as indicated in Fig. 3.29. What is the resultant pull on the ring? In other words, if one man only were to pull, what should be the force exerted and in what direction, to have the same effect on the ring as the combined pulls of the four men?

Line *ab* is drawn parallel to rope 1, and its length is made to represent 540 N. The other forces are similarly drawn as shown, finishing at point *e*. The line joining the first point *a* to the last point *e* gives the resultant force as 1710 N. If the ring is not to be

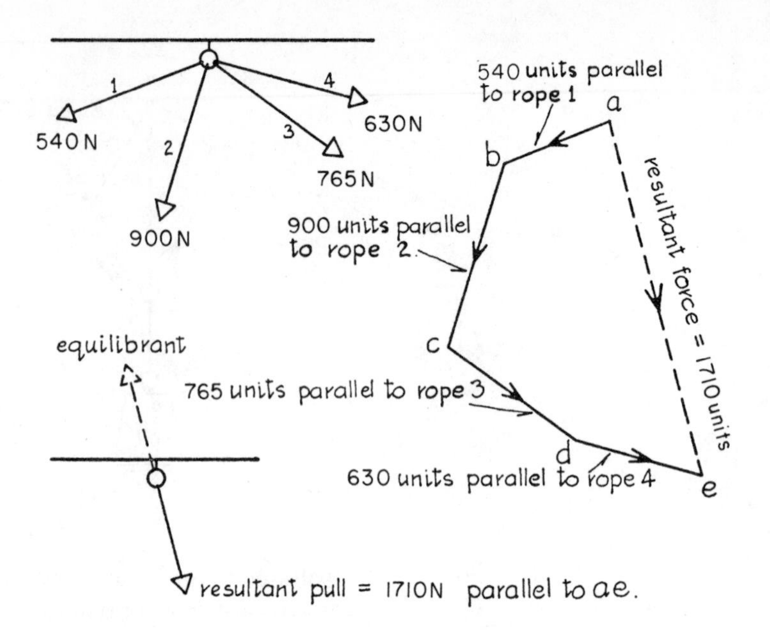

Fig. 3.29

pulled out from its support, there must be a reaction (or equilibrant) equal and opposite to the resultant.

Members in compression

In Fig. 3.28, the two ropes *X* and *Y* are stressed in tension as a result of the downward action of the load of 20 kN. The same principle applies to the simple structure shown in Fig. 3.30, but here the two members are put into compression. The force in member *X* will be 17·32 kN, and the force in member *Y* will be 10 kN. Assuming mild steel, the sizes of the members cannot be obtained by using a stress of 150 N/mm^2 as for the tension member of Fig. 3.22, because the slenderness of the members would cause them to buckle at a much lower stress (*see* Chapter 9). The arrows on the members represent the directions of the internal resisting

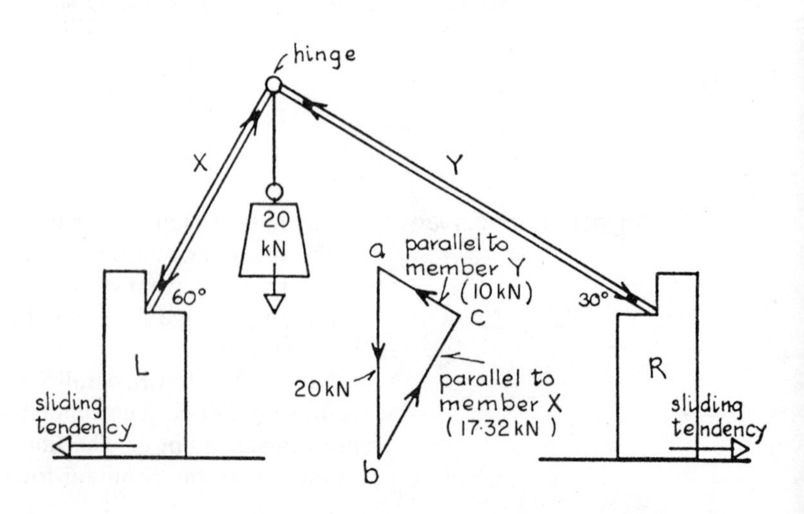

Fig. 3.30

forces; i.e. a member which is being compressed has to exert outward resisting forces to prevent its being crushed.

The two supports L and R prevent the two members X and Y from flattening out, but there is a danger of the blocks either sliding, or being overturned as in Fig. 3.31. For example, the inclined force exerted by member X has a vertical effect and a horizontal effect.

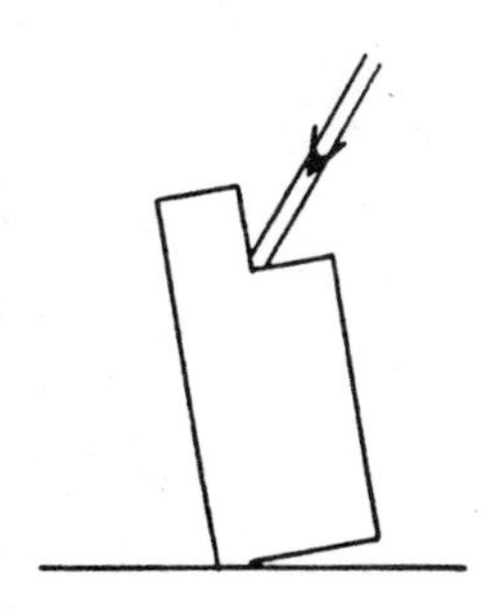

Fig. 3.31

Resolution of forces

For convenience it is quite usual in structural calculations to *resolve* an inclined force into vertical and horizontal components, i.e. to replace it by two imaginary forces which together have the same effect as the inclined force. This resolving can be done by the parallelogram of forces law.

If, in Fig. 3.32, line AB is drawn to represent 17·32 kN, which is the force in member X, then the length of line AC will represent

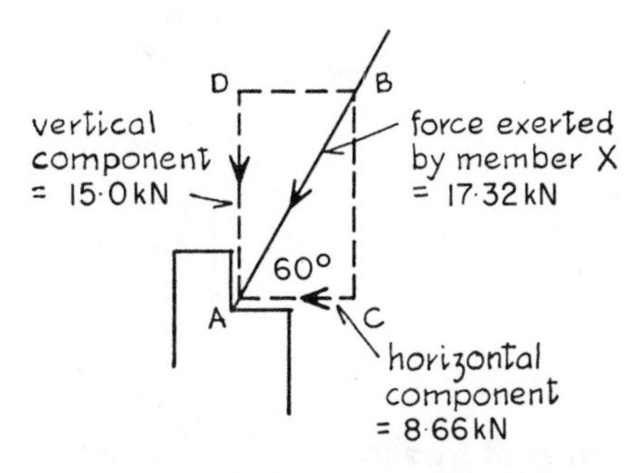

Fig. 3.32

the horizontal effect of the force (8·66 kN), and the length of line AD will give the vertical effect (15·0 kN). The vertical force of 15·0 kN is the force tending to cause crushing of the material of the support, and the horizontal force of 8·66 kN tends to cause overturning or sliding. The smaller the angle of member X with the horizontal the greater is the horizontal thrust AC. This will be referred to later with regard to the thrust exerted by an arch on its supports.

The introduction of a *tie* connecting the lower ends of members X and Y (Fig. 3.33) prevents them from spreading apart and thus takes over from the supports the work of resisting the horizontal thrust. The force in the tie will be 8·66 kN, and the load on the supports is now entirely vertical as indicated by the dotted lines.

There is now no tendency for overturning or sliding to occur at the supports. Compare this structure with the common household steps, where a rope is used to prevent spreading.

The diagrams in Fig. 3.34 demonstrate the way in which horizontal thrust from the untied structure increases as the horizontal angle decreases.

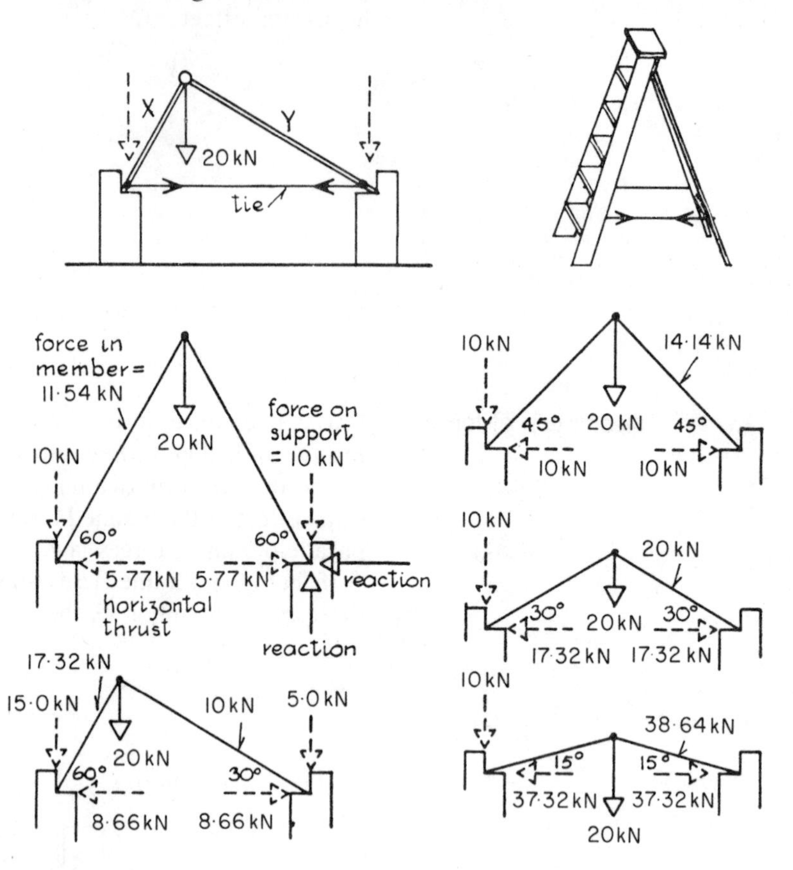

Fig. 3.34

Forces in roof trusses

The principles of the parallelogram, triangle and polygon of forces are much used in determining the forces in members of roof trusses and other framed structures, one example being given in Fig. 3.35. The following description may be ignored, if desired, during a first reading of the book.

Each joint in the structure is the meeting point of a number of forces. the vertical lines represent the forces due to the weight of the roof covering. The circles show how the truss can be split up into a number of problems involving the triangle or polygon of forces. Consider joint 1: there are three forces meeting at this point, the holding-up force or reaction supplied by the support, and the forces in the two members *a* and *b*. The reaction at the support will be known (12·5 kN), and a triangle of forces can be drawn which will give the forces in *a* and *b*. At joint 2 there are four forces: two of these are known, i.e. the vertical force of 5 kN due to the roof covering, and the force in member *b* (found from joint 1). A polygon of forces can be drawn for joint 2, and forces in members *c* and *d* are obtained. The next joint (members *e* and *f*) can be treated in a similar manner knowing the vertical load and the force in *c*, and so on throughout the truss. With a knowledge of the forces all the members have to resist, the members can be made of sufficient size. (It must be remembered that, in addition

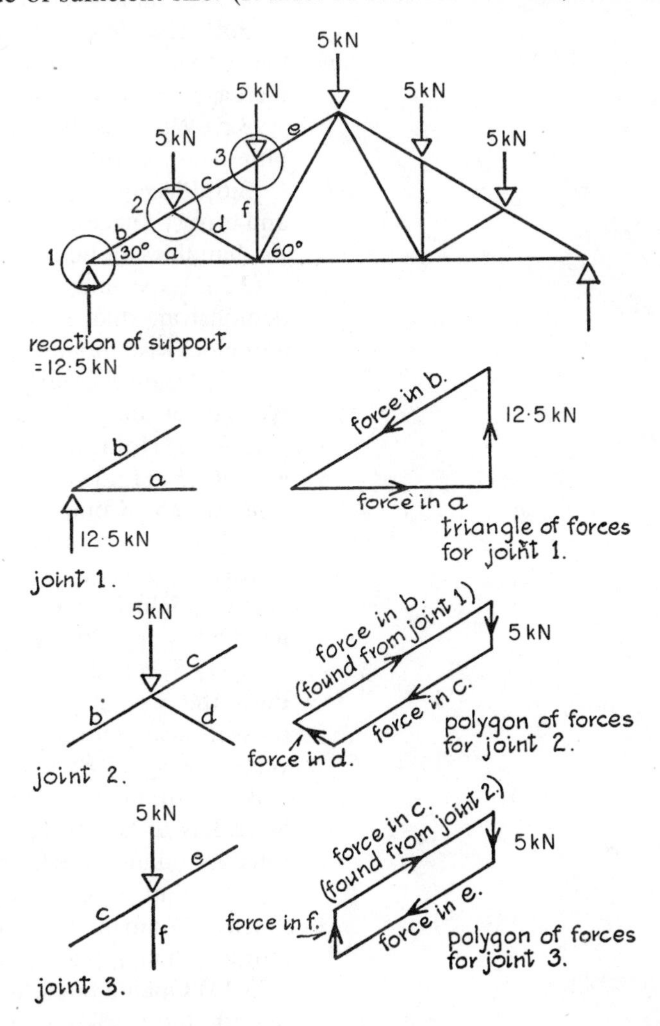

Fig. 3.35

to supporting the weight of the roofing material, the members have to resist pressures due to wind, and these wind forces would have to be considered before deciding on the actual sizes of members.)

Exercises

(3.1) Hold your hand out flat and get someone to rest a book on it. Observe the 'reaction' of your hand and its direction. Place another book on top of the first and so on until you have to give up. Observe how your resistance automatically increases.

(3.2) Place a book on a smooth table and push horizontally with a finger. Make the weight increasingly greater by piling up more books on top of the first one, and prove that the frictional resistance increases with the weight.

(3.3) Place an object such as a tin of beans or a tea-caddy on a smooth table and try to cause, by application of a horizontal force, (*a*) sliding, (*b*) overturning. Which type of failure is more likely to occur, (*c*) on a smooth surface, (*d*) on a rough surface?

(3.4) Assume a leap-frog position. Get one person to sit on your back, then two persons, and so on. Observe how your resistance increases until collapse occurs.

(3.5) Where is the best position for a door handle—particularly on a heavy door? Why?

(3.6) Place a 250 N (or 25 kg) weight in a wheelbarrow. By trial and error (if necessary) determine the position of the weight when the handles can be lifted with the least effort.

(3.7) Experiment with a length of wood and weights, and demonstrate that, for equilibrium, clockwise moments must equal anti-clockwise moments.

(3.8) Using thin strips of wood which bend easily, demonstrate types of bending of beams illustrated in this chapter.

(3.9) (*a*) Hold your arm vertically by your side and grasp a heavy weight. (*b*) Hold your arm out horizontally and try to support the same weight. Compare the stresses in your arm.

(3.10) Clasp your hands together close to your body so that the width of your shoulders and your two arms make a triangle. Hold a heavy weight in your hands; relax your shoulders and take one hand away from the weight. Observe what happens.

(3.11) Suspend a weight on a string from an overhead support. Push the weight out of the vertical by applying a horizontal pressure with your finger. Is your finger supporting any part of the vertical weight? By pushing further, increase the angle the string makes with the vertical and observe whether or not the horizontal force has to be increased. Could you determine this horizontal force by taking measurements, etc.?

(3.12) Get a heavy rectangular box and prevent it sliding by chocks. Compare the horizontal forces required to cause overturning when applied at different points.

(3.13) Open a book, and with the spine uppermost place it on a smooth table so that it forms a triangle (a roof truss). What happens? How can you prevent it happening?

(3.14) With a book as in the previous example, prove that as the 'roof' gets shallower, the thrust increases. (This can be done by using small objects to prevent sliding and by increasing the number of objects.)

4 The Masonry Arch, Vault and Dome

As mentioned in Chapter 1, a great deal of the structural designer's work involves structural analysis, which requires a good knowledge of mathematics. Theoretical analysis of the behaviour of structures was practically unknown before the end of the 18th century, and it was probably not before the middle of the 19th century that structural analysis began to achieve any importance comparable to that which it has today.

For example, in the first half of the 19th century a government inquiry was held as the result of certain failures of cast-iron beams. The engineer, Fairbairn, in his report includes this statement: "There is a wide difference between loading a beam in the middle and loading it along its whole length. In the latter case it would carry just twice the weight."

This, of course, is a very simple case of calculation of bending moments, and the fact mentioned by Fairbairn is familiar to a present-day 1st-year engineering or architectural student. It is obvious, however, from Fairbairn's report that at least some of the designers of structures in his day were unaware of the simplest facts relating to the bending of beams. What of the structures which were designed and erected before theoretical principles and mathematics began to be used extensively by engineers and architects? Monumental structures, many hundreds of years old, still stand as evidence of the skill and ingenuity of their designers and builders. As far as is known, however, they had no mathematical methods of calculating the sizes and strengths of the members of their structures. They relied upon trial and error, intuition, and craft experience and knowledge passed on from one generation to another. The 'intuitive feeling' of structure that many of these engineer-architects possessed enabled them to construct cathedrals in the materials available to them, in a manner which commands our admiration. It should be remembered, however, that the old structures which are still standing are the successful ones. Because of the trial-and-error methods there were many structural failures and many (if not most) of the buildings were very uneconomical in the use of materials.

Apart from timber, the only materials available to the builders of, for example, the great churches of the Middle Ages were stone and brick. These materials are only suitable for resisting compression, and structures had to be designed so that no tensile stresses (or only very small ones) were produced. This explains the widespread use of the *arch*, the *vault* and the *dome*. Using the traditional materials of stone and brick, these methods of construction are not of great importance today, but similar shapes using the more recent materials—steel, aluminium, reinforced concrete and prestressed concrete—are now used. (The principles of design of these modern structures are in the main different from those of traditional masonry structures.)

The arch

One of the fundamental problems in structural engineering is the bridging of openings. A simple solution is a horizontal beam supported by walls or columns (Fig. 4.1 and frontispiece). This

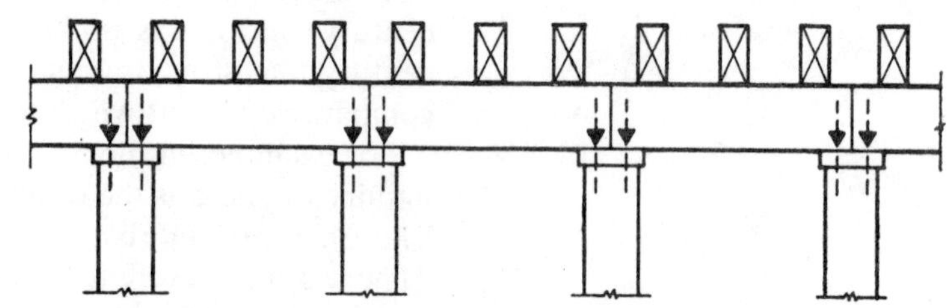

Fig. 4.1

method has been used since time immemorial by employing beams of stone and timber. Stone, however, being granular and brittle, breaks very easily in tension, and as tension is caused by bending (*see* Chapter 5) only very small spans can be obtained by using stone lintels. Longer spans can be bridged by timber, and, of course, timber is still used in this way today. It is, however, unsuitable (in the solid form) for very long spans because of the difficulty of obtaining timber of the necessary size. (Modern construction using laminated timber has enabled larger openings to be bridged than was possible only a few years ago.) The ancient builders no doubt gave a great deal of thought as to how they could utilize the

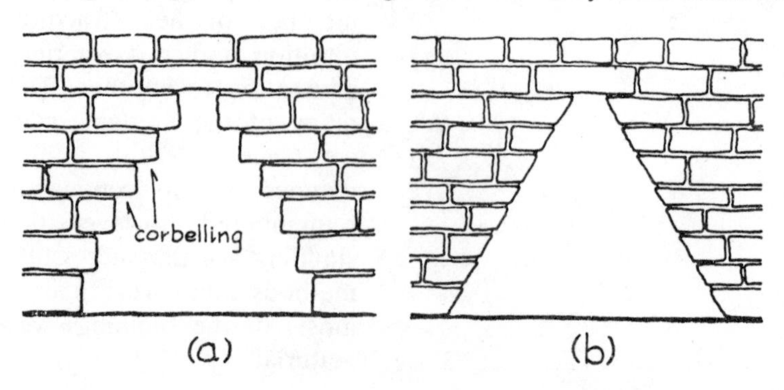

Fig. 4.2

great strength of stone and brick in compression without having to be limited to its very small strength in tension. The arch is a satisfactory solution to this problem, but before the true arch came into existence it was probably preceded by preliminary intuitive attempts as depicted in Fig. 4.2.

Temple at Paestum

By *corbelling* out the masonry blocks as shown in Fig. 4.2, a greater span could be bridged than by a single block used as a beam. This method of construction was used for roofing internal chambers of the Egyptian pyramids.

The next step in the evolution of the arch was probably the use of two inclined stone beams, as shown in Fig. 4.3, a method which also was used in the pyramids. The two stone members act in a similar manner to the two members of Fig. 3.30. With adequate depth of the stone 'beams' no tensile stresses are produced, and this structure can therefore support heavier loads than can be supported by a horizontal beam. A new problem arises, however. With a horizontal beam the reactions on the columns or other supports are purely vertical. With the inclined beams an inclined thrust is produced which must be resisted if failure is not to occur, as shown in Fig. 4.3 (*b*).

The next step in the evolution of the arch was probably the introduction of a third stone, as shown in Fig. 4.4 (*a*). As long as there is sufficient resistance supplied to the thrust by the abutments, the stones can be said to support one another by their

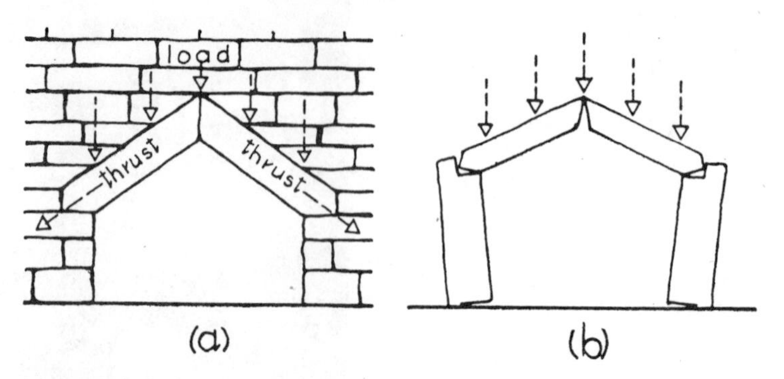

Fig. 4.3

mutual compression and cannot fall into the space below them. This can be said to be an intuitive approach to the problem. By using still more stones or bricks, longer spans can be bridged, as in Fig. 4.4 (*b*).

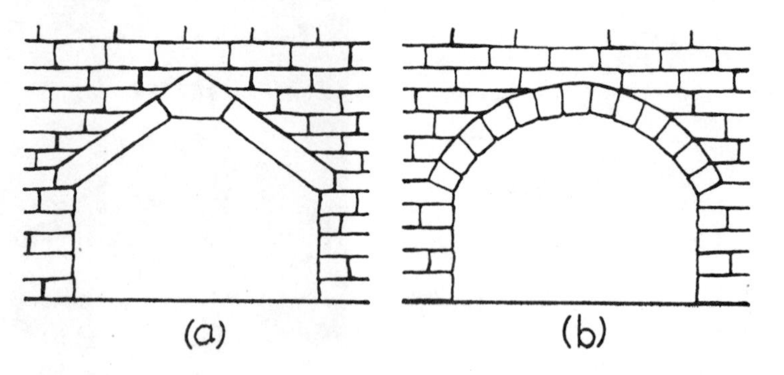

Fig. 4.4

The discovery of the arch was truly a great achievement in that the arch utilized the ability of stones and bricks to resist compression and avoided subjecting these materials to tension for which they were unsuitable.

Arches were used in Egypt and Iraq centuries before the birth of Christ and were used extensively by the Romans for the construction of bridges, and for the aqueducts built from about 300 B.C.

Pont du Gard, Nimes (*French Government Tourist Office*)

The thrusts at the supports (*abutments*) are greater with shallow arches (*see* Chapter 3) than with arches of steeper curvature, and methods of resisting arch thrusts gave rise to great ingenuity in the design of arches, vaults and domes. An Arab proverb states that 'the arch never sleeps,' and this could be taken as meaning that the arch is constantly exerting thrust and trying to 'flatten out.'

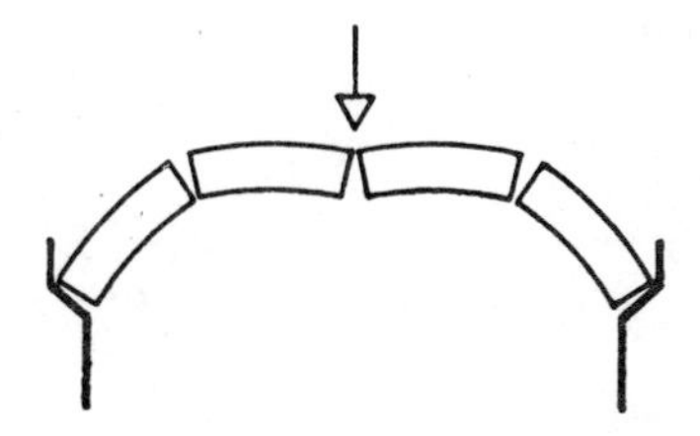

Fig. 4.5

The avoidance of tensile stresses in the stones or bricks of the arch is accomplished by making the arch ring of sufficient thickness. If the arch is too thin it could fail by the joints opening as indicated in Fig. 4.5.

The determination of the pressures in an arch and their directions is not an easy problem, and many engineers and mathematicians have investigated the behaviour of arches. As far as is known, mathematicians in the 17th century made the first attempts

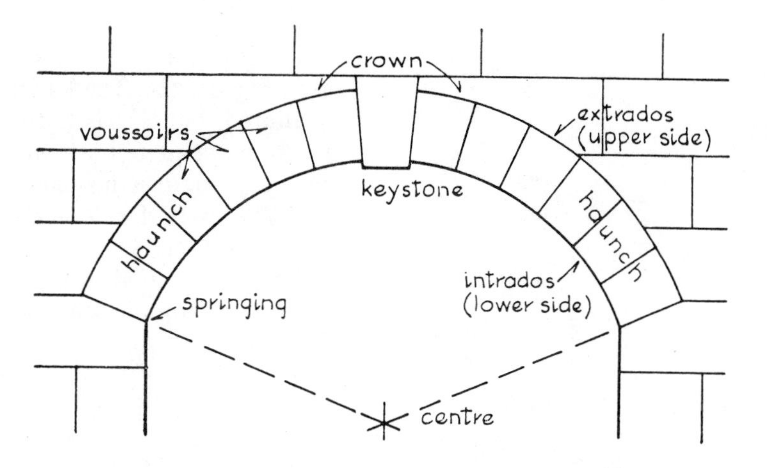

Fig. 4.6

at a theoretical solution. De la Hire (1640–1718), Robert Hooke and Rondelet, among many others, tackled the problem, but even in 1951 it was stated, 'A general survey of the methods in use for the design or analysis of a voussoir arch indicated that these were largely of an empirical nature.' This statement is in the introduction to Research Paper No. 11, 'A Study of the Voussoir Arch' (National Building Studies), published by H.M. Stationery Office. An arch made up of separate small blocks is called a *voussoir arch*; other terms used to describe the various parts of an arch are given in Fig. 4.6.

Arch abutments

If the abutment is not sufficiently big and heavy it will overturn. In Fig. 4.7, *oa* represents to scale the thrust *T* of the arch, *ob* represents the weight *W* of the abutment, and *oc* the resultant thrust *R*. If the resultant thrust falls outside the base, as shown at (*a*), the abutment will overturn.

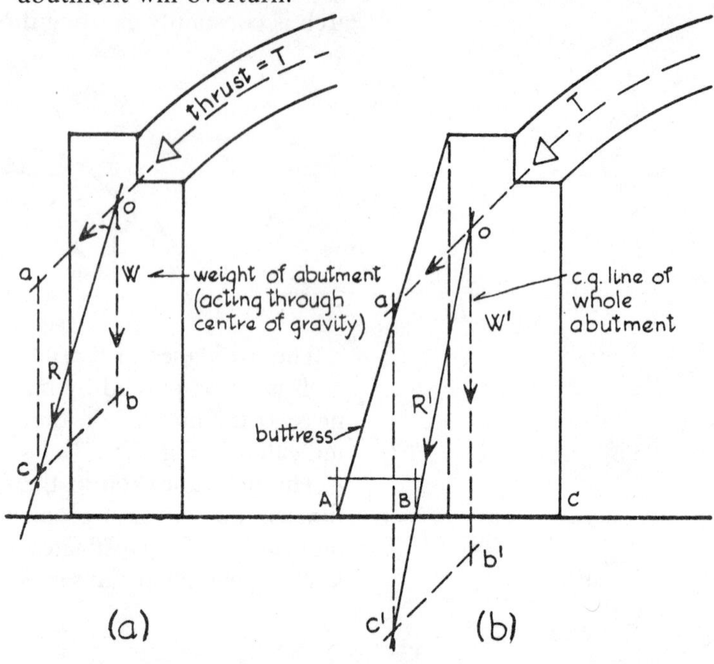

Fig. 4.7

The wall can be made thicker and heavier by adding a buttress as in Fig. 4.7 (*b*). The length *oa* representing the thrust *T* remains unaltered, but *W* has been increased to *W'*, represented by *ob'*. With an adequate buttress the resultant thrust *R'* falls inside the base and there is no danger of overturning. In fact, to prevent tensile stresses developing in the masonry near the base, the distance *AB* should be at least equal to one-third of the distance *AC*.

Another method of giving added weight to an abutment is by building on top of it, as shown in Fig. 4.8.

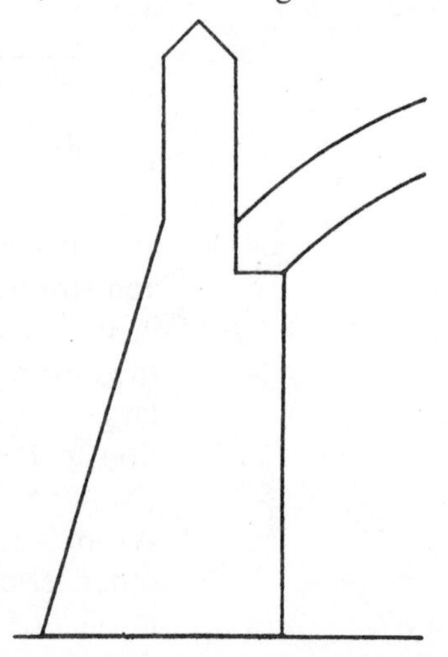

Fig. 4.8

With a number of arches, as in Fig. 4.9, the supports *C* and *D* can be made smaller than support *A*. If 1 represents the thrust from arch *AC*, and 2 the thrust from arch *CD*, the resultant reaction *R* on the supporting pier is vertical, since the push of arch *AC* to the right is exactly counteracted by the push of arch *CD* to the left. This, of course, is assuming that the loading over the arch is

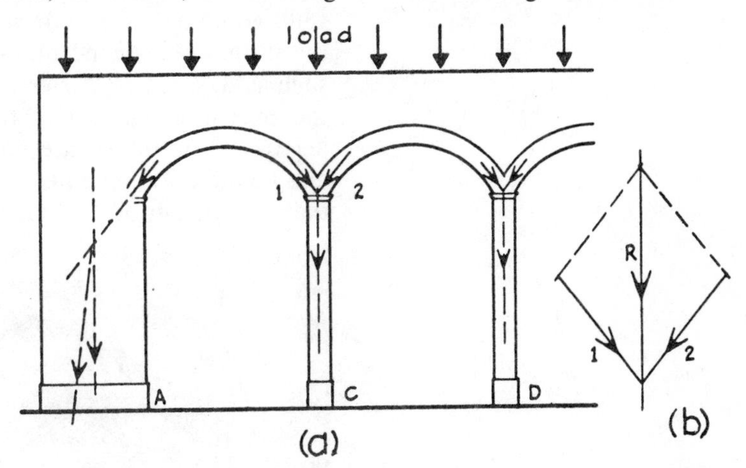

Fig. 4.9

uniformly spread. If the loading is unsymmetrical, thrust 1 will be different from thrust 2 and the resultant *R* will not be vertical. The thrusts, however, will still counterbalance each other to some extent, thus reducing the required size of the pier. The thrust at *A* is not, of course, neutralized by a thrust in the opposite direction and so *A* must be a more massive abutment.

The vault

To many people a vault implies a bank strong-room or a burial place. In architecture, however, a vault means a roof or floor constructed in an arch form. The simplest type of vault, shown in Fig. 4.10, is known as a *barrel vault.* This can be considered as a long arch thrusting against the supporting walls along their whole length. These walls have to be thick in order to resist the thrusts,

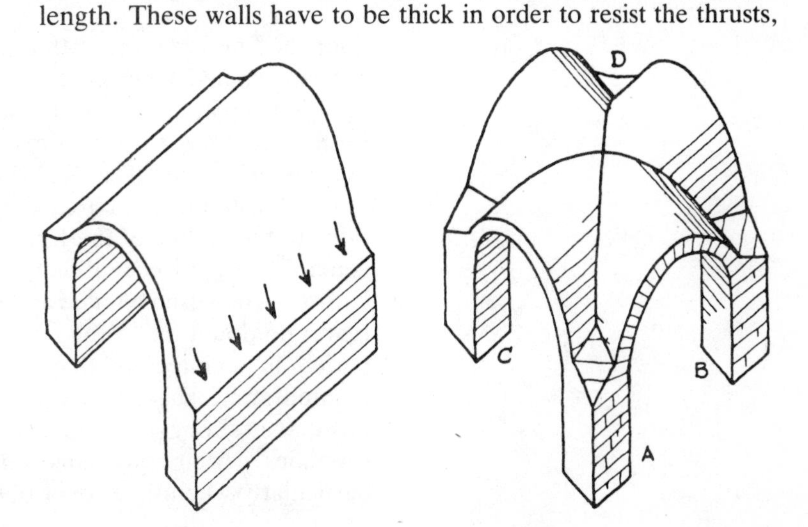

Fig. 4.10 Fig. 4.11

and any openings for windows, etc., must be small so as not to reduce the efficiency of the abutments. The barrel vault was known to the Egyptians and Assyrians, and both barrel and *intersecting vaults* were used by the Romans.

A great structural advance was made by the introduction of the intersecting vault or groined vault (Fig. 4.11). By this means, the walls could be dispensed with as supports for the arched roofs, all the thrusts being concentrated at the points where arches spring, such as *A, B, C, D*. A slender column is not adequate for resisting the overturning effect of these thrusts, so heavy piers plus (if necessary) buttresses are required. The development of vaulting and methods adopted for resisting arch thrusts are responsible for that great period in architecture known as Gothic, which

Groined Vault, Westminster Abbey Crypt (*Walter Scott, Bradford*)

flourished between the 11th and 15th centuries. Full descriptions can be obtained in books on the history of architecture. The object of this chapter is merely to outline the evolution of structural forms and to trace attempts made to make the best possible use of the building materials available.

The pointed form of arch is typical of Gothic architecture, but, prior to the middle of the 12th century, the semicircular form was generally employed, although the Egyptians, Assyrians and Etruscans had used pointed arches many hundreds of years earlier. It was no doubt the development of vaulting during the medieval period in architecture that led to such widespread use of the pointed arch. This has several advantages over the semicircular form. It is stronger and exercises less thrust, and furthermore it simplifies certain problems which arise when vaults intersect, particularly when they cover oblong instead of square spaces.

Ribbed Vault, Durham Cathedral
(*Walter Scott, Bradford*)

The reasons for the ultimate adoption of pointed arches are as follows. Referring to Fig. 4.12 (an example of semicircular arches), where the two vaults meet, curved intersecting lines called *groins* are formed. These lines give an arch form as shown by *CDB*. Assuming a square bay and semicircular transverse and longitudinal arches, the span of each arch is equal to the length of side of a square (such as *AC* or *AB*). The span of the 'arch' *CB*,

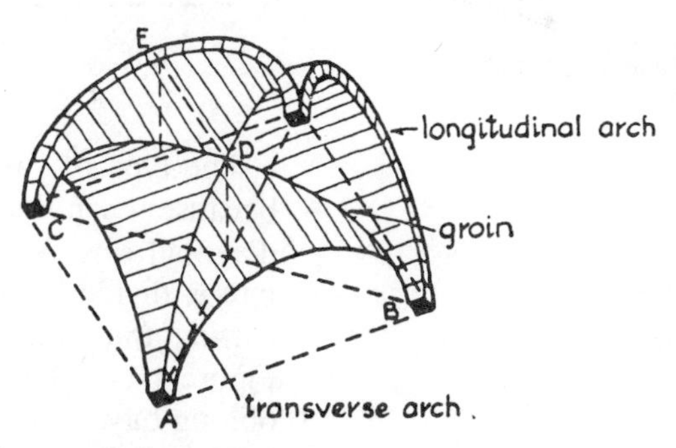

Fig. 4.12 Groined vault
(view looking up)

however, is equal to the diagonal of the square, and if the rise of the diagonal 'arch' formed by the groin is equal to the rise of the transverse and longitudinal arches (i.e. if points *D* and *E* are at the same height), then 'arch' *CDB* is a semi-ellipse. It is therefore flatter than arch *AB*. In the type of construction shown this does not matter very much because the vault acts as a single unit in conducting the thrusts to the supporting columns.

A new development in vaulting technique, however, made the groins very important structurally. The development consisted in constructing ribs along the groin lines so that these ribs formed load-bearing arches spanning diagonally (Fig. 4.13). It was now an advantage to make these diagonal ribs or arches semicircular so as to reduce the thrust in these arches. The transverse and longitudinal arches had to be constructed so as to give a satisfactory structural and aesthetic roof. One expedient was to stilt these arches, i.e. to make them semicircular by having their springing points higher than those of the diagonal ribs or arches. (Figure 4.13 is drawn as seen from below.)

In this type of ribbed vault all the loads are taken by the constructional framework or skeleton formed by the ribs or arches, which transfer the loads and thrusts to the supporting piers. The vault was completed by filling in between the ribs, and since this infilling had only its own weight to carry, it could be made only a few inches thick. The infilling of the vaults of Notre-Dame, Paris, is said to be less than 100 mm thick.

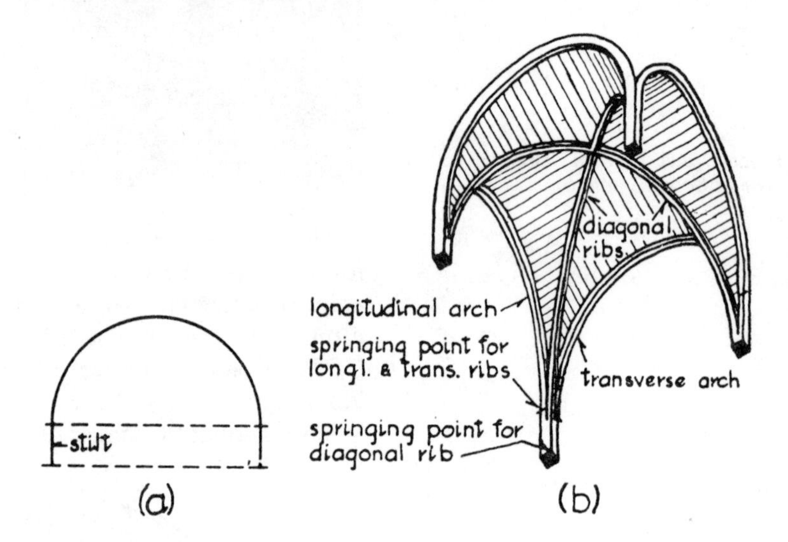

Fig. 4.13 Stilted ribbed vault (view looking up)

The type of vault shown in Fig. 4.13 is known as *quadripartite* because each bay is divided into four compartments by the diagonal ribs. *Sexpartite* vaults were also constructed where each bay is divided into six compartments by using a transverse arch cutting the diagonal ribs where they intersect [Fig. 4.14 (*a*)]. Figure 4.14 (*b*) is another example of a rib framework covering an oblong bay.

The difficulties encountered in constructing ribbed vaults led eventually to the use of the pointed arch, which is such a feature of Gothic architecture. As mentioned earlier, pointed arches are stronger and exert less thrust than semicircular ones. In addition, all the arched ribs, transverse, longitudinal and diagonal, can be of the same height (Fig. 4.15) without recourse to stilting, even in oblong bays. (Although all the arches are pointed they are not, of course, identical.) It appears that the pointed arch was not used constructionally in England before about A.D. 1140, although examples of its use in France date to between 1100 and 1120.

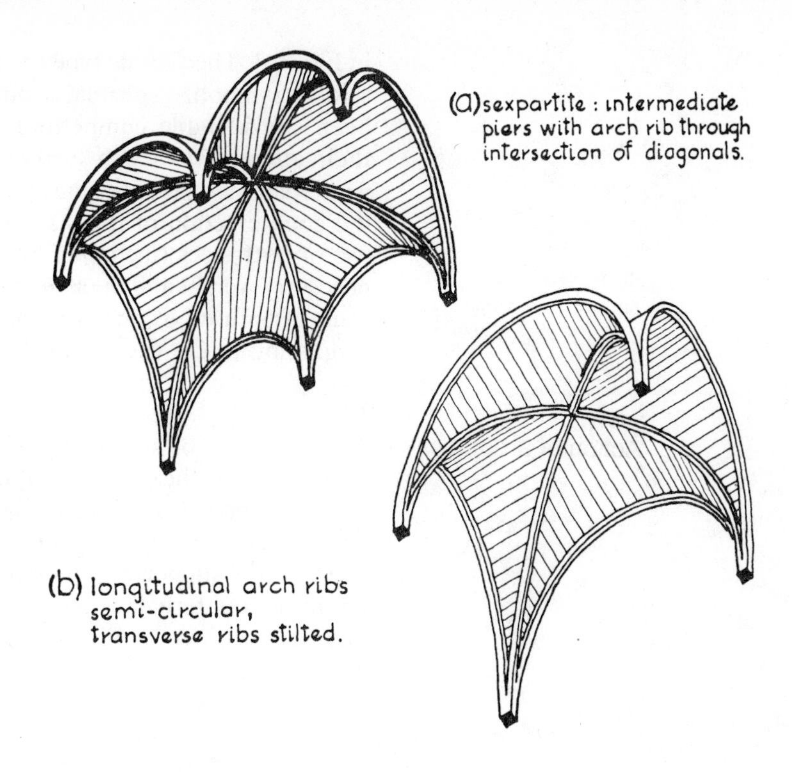

Fig. 4.14 Ribbed vaulting to oblong bays (view looking up)

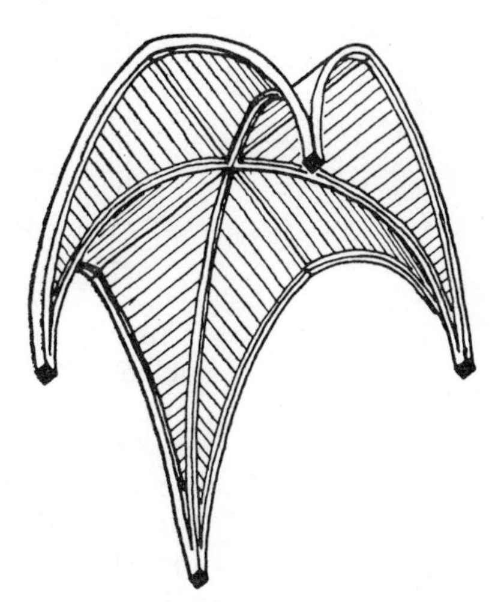

Fig. 4.15 Pointed arch: Ribbed vaulting to an oblong bay (view looking up)

For further examples illustrating the development of vaulting, including *fan vaulting*, reference may be made to books on the history of architecture.

The buttress

Another feature which is distinctive of Gothic architecture is the *buttress*. The thrust of vaulted roofs is all directed through the transverse, longitudinal and diagonal ribs or arches to the common meeting points from which the arches spring, and these thrusts exert overturning effects (moments), already demonstrated in Fig. 4.7. The Gothic type of architecture, with its lofty naves, is said to symbolize spiritual aspiration, and there appears to have been considerable competition among the medieval builders in their attempts to reach further towards the sky. The desire for wide spans and great heights gave rise to great problems relating to the counteraction of thrusts. These problems were solved by the use of buttresses (including *flying buttresses*), which provided the necessary weight and thickness to counteract the thrusts. Pinnacles were often used to increase the vertical weight and so reduce the inclination to the vertical of the resultant thrust.

Action of a buttress

All the weight of a building, whatever its shape, must eventually come down to the ground. All thrusts also which are not neutralized internally (as in Fig. 3.33) must be brought down to the ground. As already mentioned, a thrust exerts an overturning effect on the supporting column or pier, and weight is necessary to supply a counterbalancing moment. Shape also is important, as was demonstrated by Figs 3.20 and 3.21, and further illustrated by Fig. 4.16.

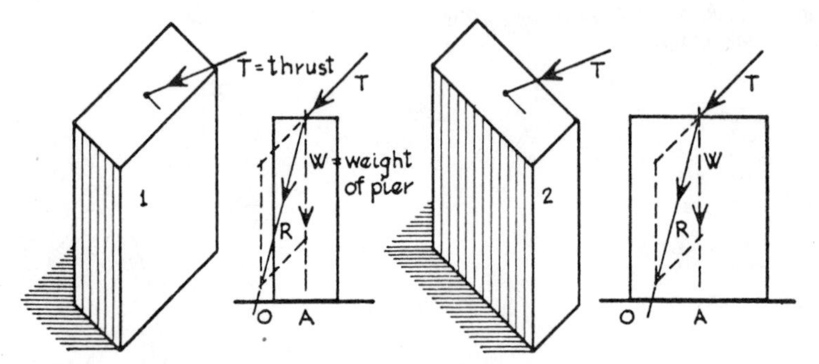

Fig. 4.16

In Fig. 4.16 the thrusts T on piers 1 and 2 are equal, and the weights W are equal, but pier 1 is unsafe because the resultant thrust R falls outside the base, whilst pier 2 is safe because R falls inside the base. The resisting moment ($W \times OA$) of pier 2 is greater than that of pier 1. This is a simple example of how intelligent disposition of a given amount of material can give a satisfactory structural solution.

Figure 4.17 illustrates why church buttresses are increased in width as they approach the ground. Consider the tendency to overturn at levels AA, BB and CC in Fig. 4.17 (*a*). At AA, the stabilizing force is the weight of block 1. Combining this weight of 70 kN with the thrust of 60 kN gives a resultant R_1, which cuts the base AA as shown and there is no danger of overturning at this level (*oabc* is a parallelogram of forces). At level BB, the stabilizing force is 140 kN, and combining this with the thrust (parallelogram *oade*) the resultant R_2 is obtained which cuts the base BB dangerously close to the edge. The resultant R_3 given by the parallelogram of forces *oafg* falls outside the base CC, which means that the buttress will overturn.

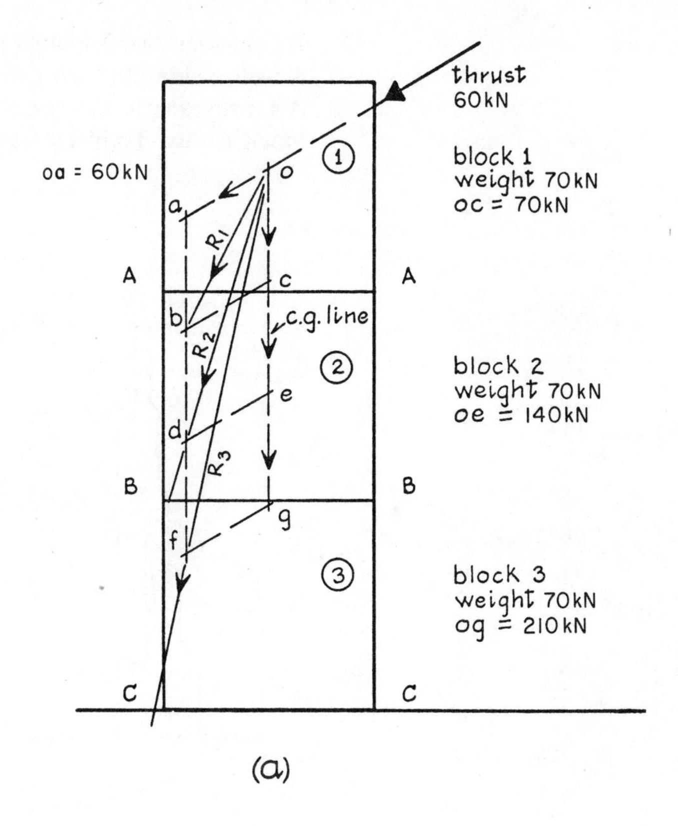
thrust
60kN
block 1
weight 70kN
oc = 70kN
oa = 60kN
o
a
R1
A
A
c
b
c.g. line
R2
block 2
weight 70kN
oe = 140kN
e
d
R3
B
B
g
f
block 3
weight 70kN
og = 210kN
C
C
(a)

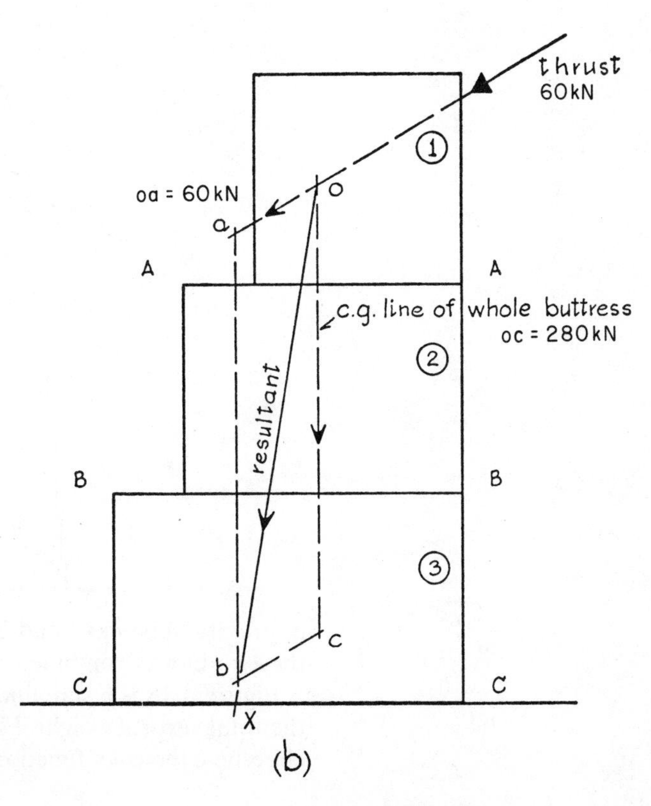
thrust
60kN
oa = 60kN
o
a
A
A
c.g. line of whole buttress
oc = 280kN
resultant
B
B
c
b
C
C
X
(b)

Fig. 4.17

Consider now Fig. 4.17 (*b*); the resultant falls at point *X*, which is well inside the base *CC*. The conditions of equilibrium at level *AA* are exactly the same as those in Fig. 4.17 (*a*), considering block 1 only. To investigate the conditions at level *BB*, the centre

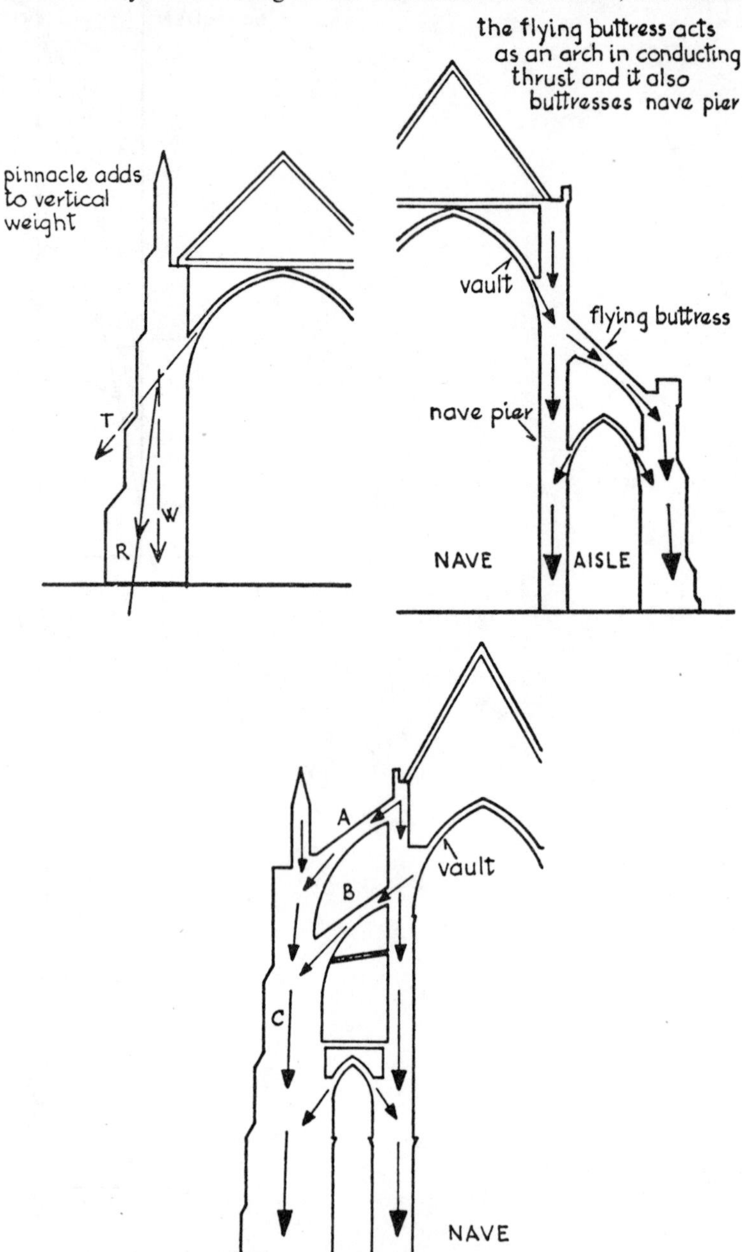

Fig. 4.18 Fig. 4.19

Fig. 4.20

of gravity of blocks 1 and 2 would be found and the total weight of the two blocks combined with the thrust of 60 kN.

Figure 4.18 is a simplified diagram of a church, *W* representing the total vertical weight. Flying buttresses were used in many large churches; the chief function of a flying buttress is to transmit thrust to the outside walls, thus leaving mainly direct vertical load on the internal piers. These need not then be so bulky and they allow side aisles to be closely related to the nave. The outside vertical buttresses can be as wide as required without blocking circulation or internal view (Fig. 4.19).

In Fig. 4.20 the top flying buttress *A* is not intended to resist the thrust from the vault but helps to support the wall above the vault and to transmit any thrust from the timber roof. The buttress *B* transmits the thrust from the vault and carries it over to the external buttress *C*. The arrows represent roughly the directions of the individual thrusts and stresses in their journey towards the foundations.

The dome

Domes, like arches, have been used from ancient times. An arch exerts thrusts on its abutments. A vault, which can be considered as an arch extended to form a tunnel, exerts thrust over the whole length of the walls, which consequently have to be of considerable

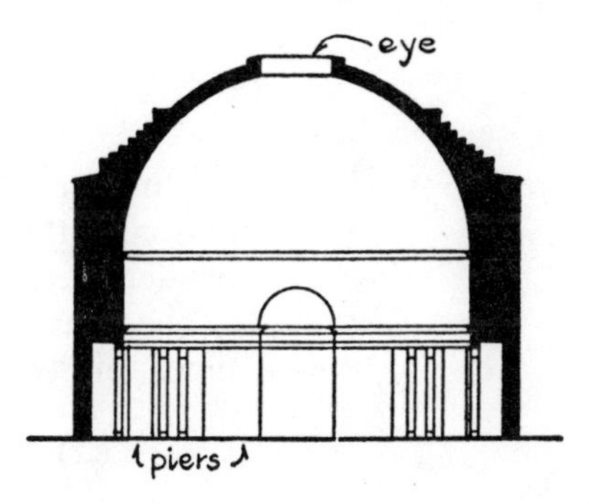

Fig. 4.21

thickness. A dome exerts thrust around the whole of its circumference since it acts like a series of radial arches which are restrained in part by a series of concentric rings. The tensile stress in these rings, which run at right angles to the arches, are greatest at about one-third of the distance from the springing to the crown and depend on the ability of the dome to transfer thrust to the supporting structure at its circumference. One method of counterbalancing the thrust is by means of weight in the supporting walls or piers. The Pantheon, Rome, is an example of this method; this remarkable dome, constructed in A.D. 120–124, has a diameter of 43·3 m and is supported by eight huge piers. Figure 4.21 gives an idea of the massiveness of the construction.

One of the most famous of domed churches is Santa Sophia, Istanbul, built A.D. 532–7. The span of the main central dome is about 30 m, but because the main dome is supported by semi-domes, there is a floor length of about 69 m unbroken by columns. This dome has a series of radial ribs (compare with vault ribs) to channel the thrusts to simple abutments instead of to a continuous massive wall (Fig. 4.22). Santa Sophia is also an example of a solution to the problem of covering a square space with a circular dome by using pendentives. A *pendentive* is a curved triangle (portion of a sphere) built in between the supporting arches of the dome.

Figure 4.23 gives three examples of the use of pendentives. Figure 4.23 (*a*) shows a combined dome and pendentives, and can be thought of as a dome with sections cut out rather than as a dome sitting on independent construction. At (*b*) four arches support the dome, and these leave four curved triangular spaces to

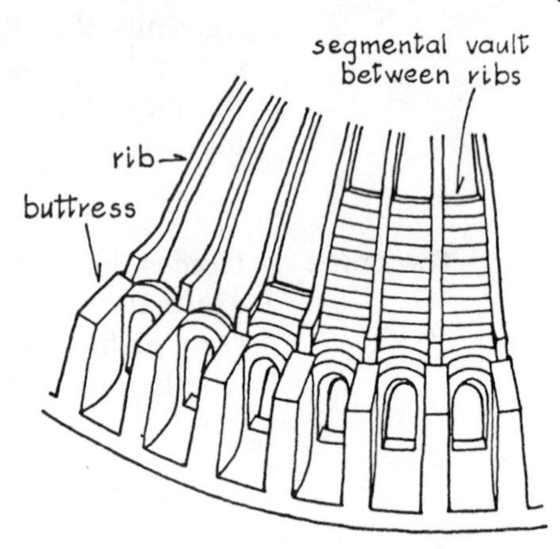

Fig. 4.22

be bridged. This bridging is accomplished by filling in with masonry, thus forming the pendentives. At (*c*) four arches with pendentives support a drum, which in turn supports the dome.

The dome was not popular in Western Europe in Medieval times, but two famous examples of its use during the Renaissance

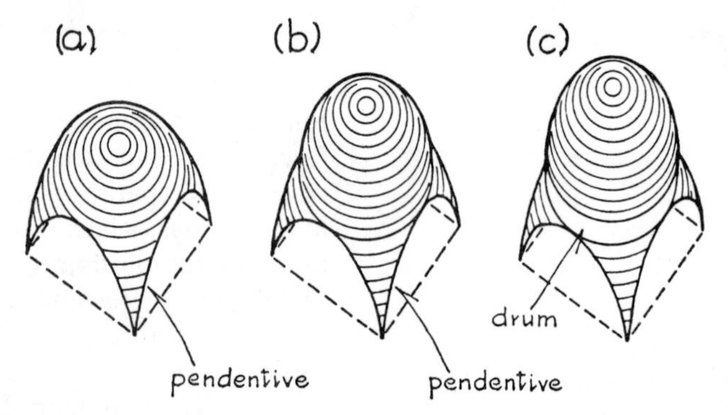

Fig. 4.23

period are St Peter's, Rome (finally completed in 1626), and St Paul's, London (built between 1675 and 1710). In both these churches the domes are surmounted by heavy lanterns. Now, the dome is not a suitable structure for carrying weight on its top, or crown. Besides increasing the thrust due to the weight of the dome, the heavy lantern will tend to cause the haunches to rise. In fact, during the construction of St Peter's, signs of failure began to appear and a great chain had to be made to surround the dome in order to counterbalance the thrust.

Wren solved the problem of supporting the huge weight of the lantern or cupola (about 8 MN of Portland stone) by carrying it on a cone of brickwork supported on piers, etc., near the base of the internal dome (Fig. 4.24). Therefore, neither the internal dome nor the external dome (which is of timber and could not possibly support any appreciable weight) contributes anything to the support of the heavy lantern. Wren also used an iron chain surrounding the internal dome to strengthen it and to prevent it bulging. During repairs, between 1925 and 1930, this chain was replaced by two chains of stainless steel. Models can be seen in the trophy room at St Paul's.

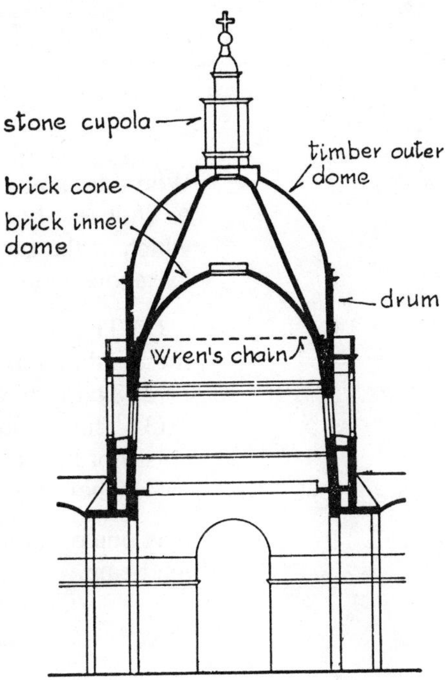

Fig. 4.24

5 The Beam: Timber, Iron and Steel

When designing a beam, i.e. when calculations are made in order to choose a beam which is strong enough to carry its loads, and which at the same time makes economical use of the material, at least four items must be considered—

(1) The beam must have a bending strength adequate to resist the bending moments.
(2) There must be no danger of failure due to shear forces.
(3) The amount the beam bends (the deflexion of the beam) must not be excessive.
(4) There must be no danger of lateral (sideways) buckling.

Any one of these items may be the deciding factor in the design of the beam.

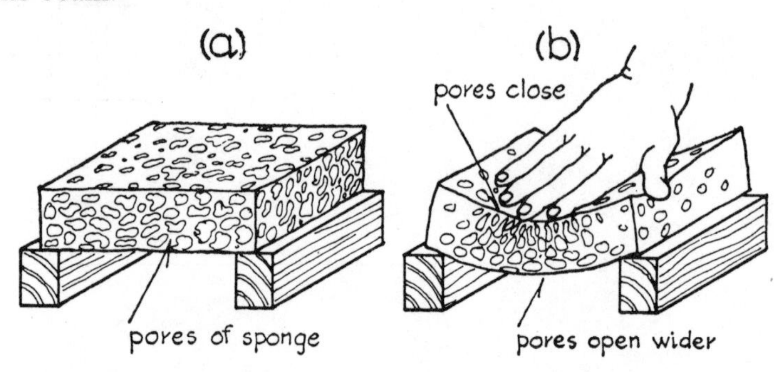

Fig. 5.1

Bending in beams

The bending strength will be considered first by studying the behaviour of a car sponge (say 150 mm × 100 mm × 50 mm). It will be observed than when force is applied, as shown in Fig. 5.1 (*b*), the pores in the top close indicating compression, and the pores at the bottom become wider indicating tension. Near the ends of the 'beam' the pores remain practically unaltered because the bending moments are very small compared with those at mid-span.

Now, consider the behaviour of a beam of elastic material (such as timber or steel). Imagine the beam to consist of layers of longitudinal fibres, as shown in Fig. 5.2, all layers being securely cemented together. (Beams in buildings bend by only a small amount, usually undetectable by eye, but for explanation purposes the bending in Fig. 5.2 is greatly exaggerated.) The longitudinal fibres near the top of the beam become shorter as a result of the bending and are therefore stressed in compression, whilst the fibres near the bottom become longer and are stressed in tension.

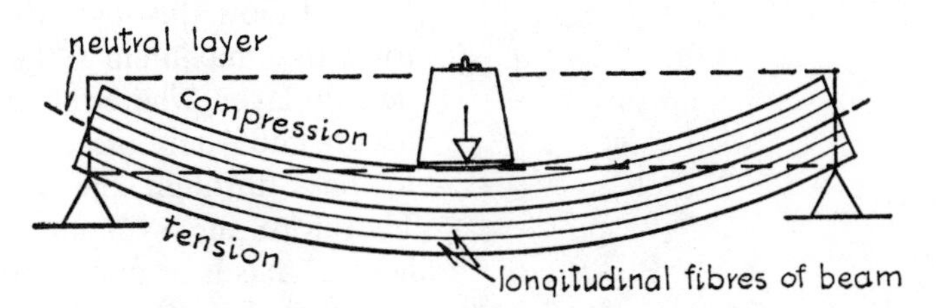

Fig. 5.2

Before dealing with these stresses, consider again an example of pure tension as in Fig. 5.3 (*a*). If we isolate in imagination any small portion (say *ABCD*) of the rod, the arrows indicate that it is pulling up on the portion of the rod below and pulling down on the portion above. (Compare with links of chains in Fig. 1.14.) The stress is uniform over the cross-section of the rod, so that, if the area is 1250 mm^2 and the pull is 100 kN, the tensile stress is 80 N/mm^2. Now, if we imagine a similar slice of a beam [Fig. 5.3 (*b*)] at the point of maximum bending moment where the beam is most

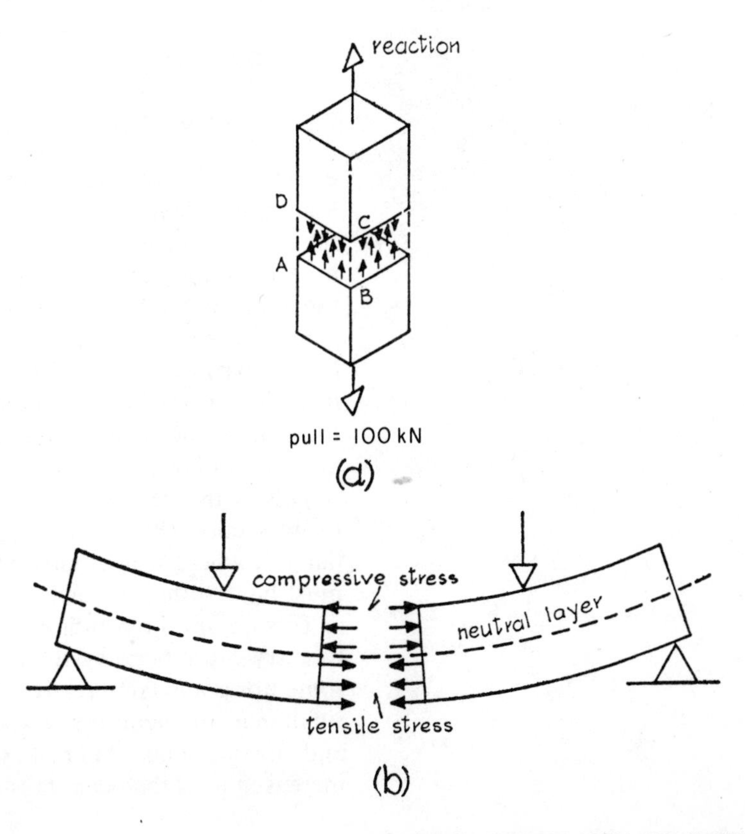

Fig. 5.3

likely to break, the stress will be compressive on part of the cross-section and tensile on the other part. (There will be one layer, called the *neutral layer*, which remains unstressed.)

Furthermore, the compressive stress will not be uniform over the portion of the beam above the neutral layer. It will be greatest in the longitudinal layer at the top of the beam, and the stress in the remaining fibres will get progressively less towards the neutral layer. Below this layer the stress, which is now tensile, increases towards a maximum at the extreme bottom longitudinal fibre. The neutral layer, where the change-over from compression to tension occurs and where there is no stress, can be proved by mathematics to pass through the centre of gravity of the cross-section. Therefore, in a symmetrical beam, such as a rectangular one, the neutral axis is at mid-depth.

Figure 5.4 illustrates the resisting forces within the beam material which are tending to compress the fibres near the top and stretch those near the bottom of the section. *C* is the total compression and is the one imaginary force which would have the same

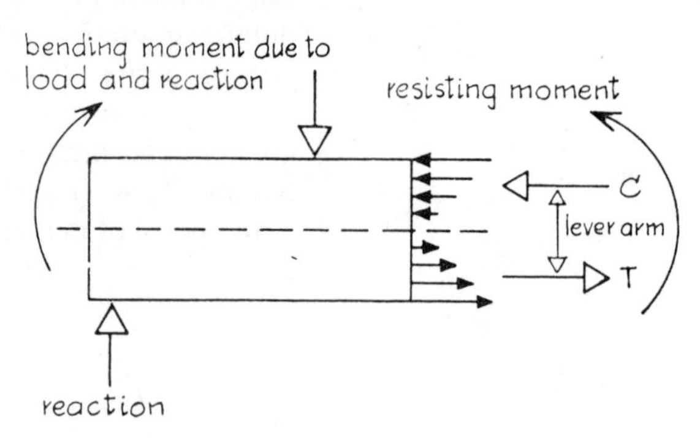

Fig. 5.4

effect as all the smaller compressive forces supplied by the various longitudinal layers or fibres above the neutral axis. Similarly, *T* is the total tensile force, and these two internal forces in the beam set up a resisting moment to oppose the bending moment caused by the loads on the beam and the resulting reactions. The greater the value of the lever arm (Fig. 5.4) the greater is the moment that can be set up by the resistance of the beam fibres. Also, since the fibres near the top and bottom of the beam are more highly stressed than those near the neutral axis, it is an advantage to have as much material as far as possible from the neutral axis.

This is the reason for the shape of the commonly used steel beam section (Fig. 5.5). Most of the steel is concentrated in the flanges, where it is of most effect in resisting bending. (The web must have sufficient steel to resist the shear forces.)

To sum up: the bending moment due to the loads the beam has to carry cause bending stresses in the material (tensile stresses in some fibres and compressive stresses in others). These stresses are small near the neutral axis and much larger at the extreme fibres, and if the loads (and therefore the bending moments) are increased until the beam fails, failure will occur either by crushing of the extreme fibres which are in compression, or by tearing of the extreme fibres which are in tension. The material of the beam is obviously important. A steel beam is much stronger than a timber beam of identical dimensions. The shape of the cross-section of the beam is also important, and the depth of the beam is more critical than the width for resisting bending.

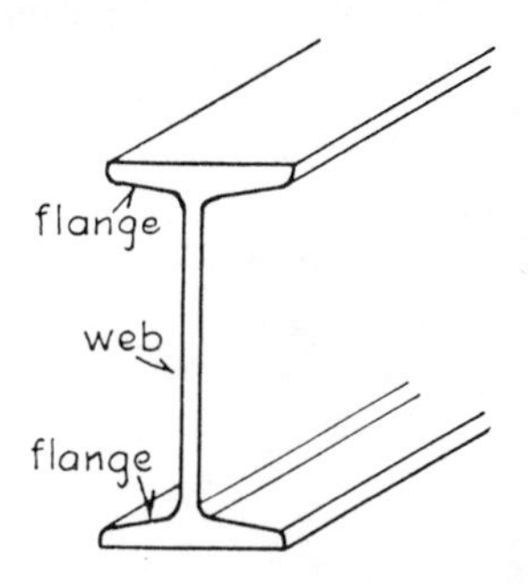

Fig. 5.5

The two timber beams of Fig. 5.6 have identical amounts of material and the spans are equal. Beam *B*, however, is capable of carrying twice the load that beam *A* can carry. It can be proved that the bending strength increases proportionally to the square of the depth but only in direct proportion to the breadth. Beam *A* is said to be bending with respect to axis *yy*, and beam *B* with respect to axis *xx*.

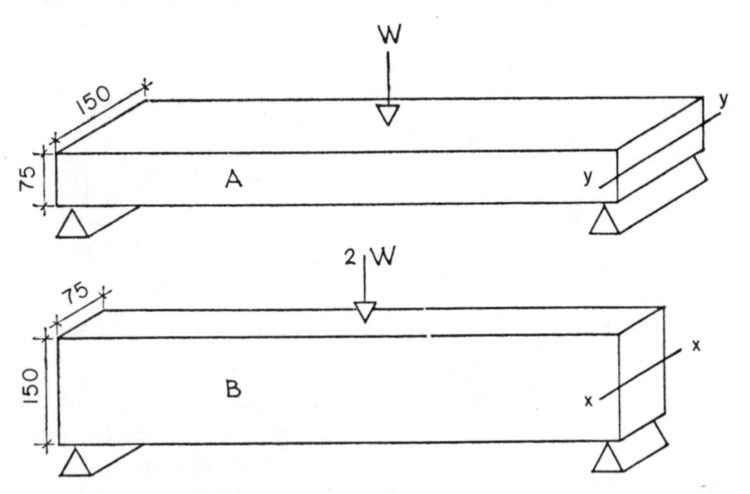

Fig. 5.6

A further example is given in Fig. 5.7. Beam *B*, which has twice as much material as beam *A* and is twice as wide, is also twice as strong. Beam *C*, which has the same amount of material as beam *B*, is twice as strong as *B* and four times as strong as *A*. Beam *D* is three times as strong as *B*. (This beam, however, might not be practical because of danger of sideways buckling. *See* page 76.

To resist a given bending moment, therefore, with the least amount of material, the deepest possible beam should be used. For example, the beams of Fig. 5.8 are all of equal strength for resisting bending.

The beam of rectangular cross-section is very common in timber, but this shape does not utilize the material to the best advantage because of the large amount near the neutral axis which is stressed only to a small extent. Some timber beams are now being 'built up,' as shown in Fig. 5.9, and these are stronger than solid rectangular beams of the same depth using an equal amount of material.

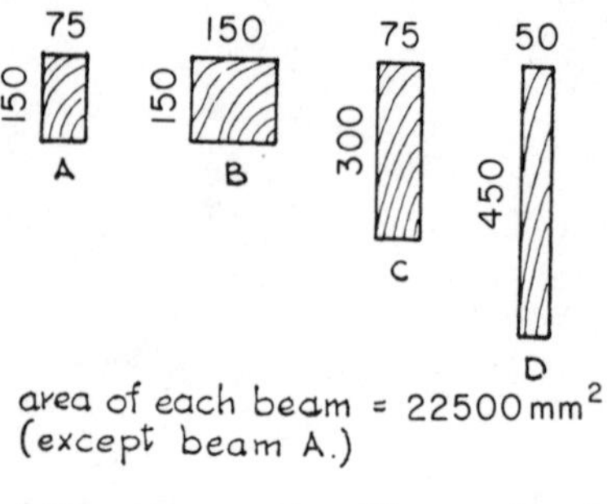

Fig. 5.7

Galileo Galilei (1564–1642), who invented the thermometer and was at one time imprisoned for asserting that the earth revolves around the sun, was one of the first scientists to make a theoretical investigation of the problem of bending. In a book published in 1638, he discussed the problem of the bending of a cantilever and attempted to determine the breaking load in terms of the

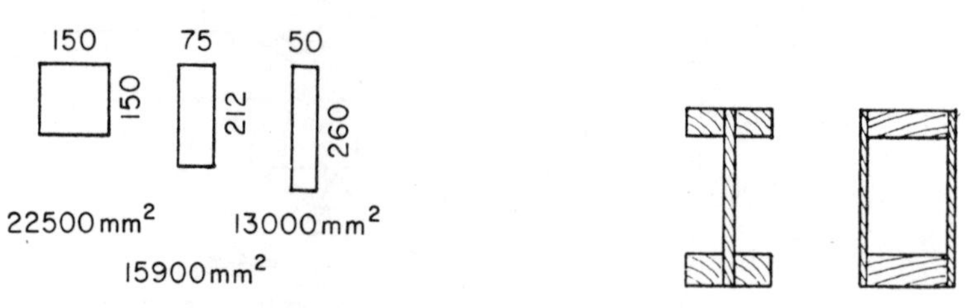

Fig. 5.8 Fig. 5.9

dimensions of the beam. He assumed, however, that all the resisting forces supplied by the beam fibres at various depths, indicated by arrows in Fig. 5.10 (*a*), were equal and all acting in one direction (the neutral axis being at the bottom of the beam). From the description on page 63 it will be seen that this assumption is wrong, the actual resisting forces being as shown in Fig. 5.10 (*b*). Galileo's method would give a safe load for a rectangular beam three times as great as the load obtained from the correct theory.

Edmé Marriott (1620–84), the inventor of the rain gauge, also assumed the forces in all the fibres of a cantilever beam to be tensile, but assumed that they ranged in intensity from zero at one extreme fibre to a maximum at the extreme fibre in tension (Fig. 5.11). A formula derived from these assumptions would give an apparent strength equal to twice the actual strength of a rectangular beam. These investigators, and others, failed in arriving at a correct result because they had no idea of the elasticity of materials, and a correct theory of bending was not possible until Robert Hooke discovered that 'stress is proportional to strain.'

After many other scientists and mathematicians had tackled the beam problem, it was Charles Augustin de Coulomb (1736–1806), a French engineer, who first enunciated the correct relationship between the bending moment and the moment of resistance of a beam, although he applied his beam theory only to rectangular

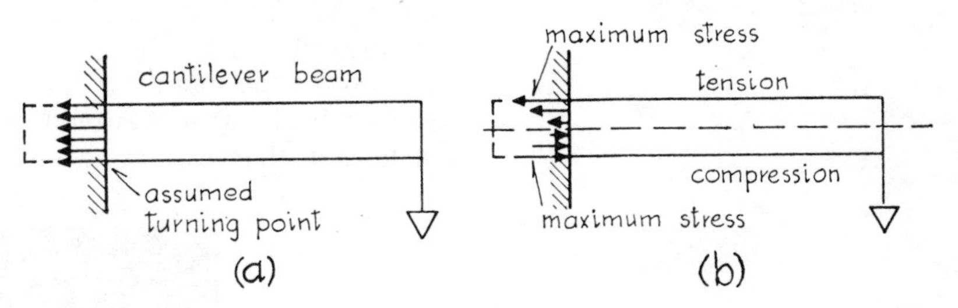

Fig. 5.10

sections. Later investigators developed the theory to cover any shape of cross-section.

Until the manufacture of cast iron on a large scale became possible, the only material available for beams was timber (apart from stone, which, because of its low tensile strength, is very uneconomical and impracticable). Cast-iron beams began to be used in buildings about the end of the 18th century. In 1801, James Watt specified beams of 4·25 m span for a Lancashire cotton mill. Cast-iron beams with spans up to about 12·5 m were used in the construction of the British Museum.

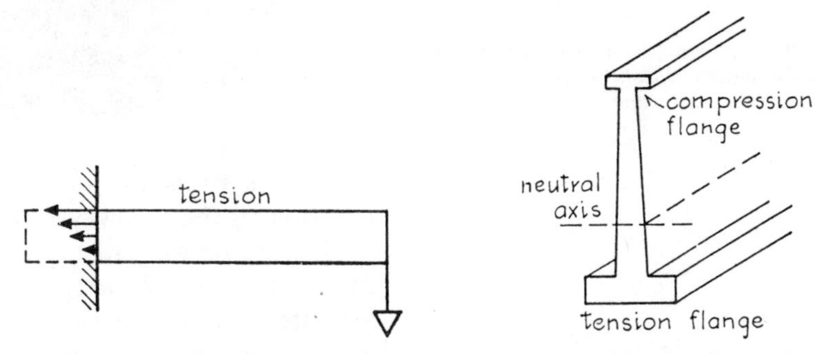

Fig. 5.11 Fig. 5.12

Early cast-iron beams were more the result of trial and error than of mathematical calculations, and before they left the foundry it was common practice to apply a test load. Many investigators, including Sir William Fairbairn and Eaton Hodgkinson (1789–1861) carried out tests to determine the most economical shape for cast-iron beams, and Hodgkinson's beam, as shown in Fig. 5.12, was of the shape generally used after about 1830.

Cast iron is brittle and is much stronger in compression than in tension. A smaller amount of material is therefore necessary to take the compressive stresses caused by bending than to take the tensile stresses, and this is the reason why the beam shown in Fig.

5.12 is more economical than a beam with top and bottom flanges of equal area.

It has been mentioned in Chapter 4 that, because of the brittleness of cast iron and inadequate knowledge of structural theory, failures of cast-iron structures sometimes occurred. Many structures, however, were successful, including the building designed by James Paxton and constructed in Hyde Park for the Great Exhibition of 1851. This building, which was one of the first examples of prefabrication on a large scale, was dismantled after the exhibition and re-erected at Penge as the Crystal Palace (destroyed by fire in 1936).

Towards the middle of the 19th century cast iron began to be superseded by wrought iron, which is ductile, has a higher strength in tension than cast iron, and is less susceptible to damage by shock. One of its first important uses in England was for the construction of the Conway Bridge of 500-ft span, and from about 1840, cast iron was practically superseded for bridge construction. The greater cost of wrought iron is probably the reason why it did not entirely supersede cast-iron floor beams in buildings, although from about 1840 the tendency to use wrought iron for beams increased whilst still retaining cast iron for columns.

Checking torque on friction-grip bolts in plate girders (*British Constructional Steelwork Association*)

Wrought iron, because of its method of manufacture, is not suitable for the rolling in one piece of a large beam. As a result of the Bessemer and Siemens–Martin processes for the manufacture of steel, and because steel is stronger than wrought iron, the latter material was eventually superseded, particularly when, in 1877, the Board of Trade raised its ban on the use of steel for bridge building. Probably the last big structure of wrought iron in Europe was the Eiffel Tower, built in 1879 from about 73 MN of metal.

Mild-steel joists were rolled by Dorman Long and Co. in 1885, the largest section being 406 mm deep. Today beam sizes range

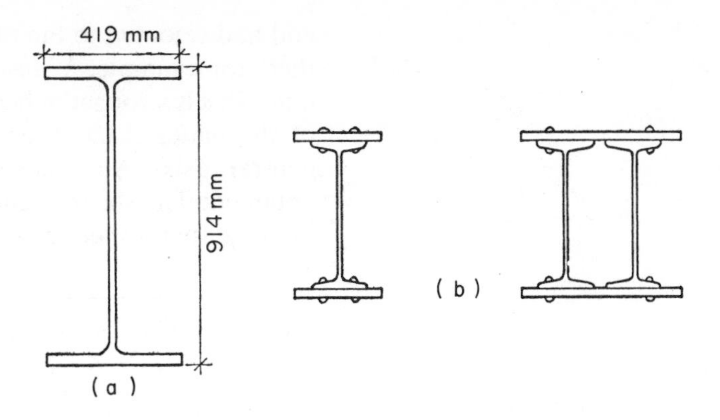

Fig. 5.13

from the largest in the Universal Beam series of 914 mm deep and 419 mm wide to the smallest in the Joist series which is 76 mm deep and 51 mm wide. If freely supported at each end of a 10 m span the largest beam [Fig. 5.13 (*a*)] will safely carry a load of 2 MN if it is uniformly distributed along the span. Greater loads can be carried by compound girders formed by riveting or welding steel plates to the flanges of beams [Fig. 5.13 (*b*)].

A useful beam is the *castellated beam*, which is produced by flame cutting an I-section (or other section) along the web as shown in Fig. 5.14 (*a*). One of the two pieces is then turned end for

Castellated beams in new Mechanical Engineering block, Imperial College, University of London (*British Constructional Steelwork Association*)

end and rejoined to the other piece by welding [Fig. 5.14 (*b*)], so that, for example, a joist originally 600 mm deep becomes 900 mm. This is stronger in bending than the original beam, since most of the material has been placed at a greater distance from the neutral axis. An increase in resistance to deflexion is also obtained. The shear strength is reduced, but this does not matter for large spans where bending is the prime consideration. It should, however, be considered carefully when the spans are short. This type of beam is very suitable when there are light loads on long spans, where normally deflexion is an important factor. Plate 2 (*a*) shows castellated beams with studs welded on for bonding to concrete.

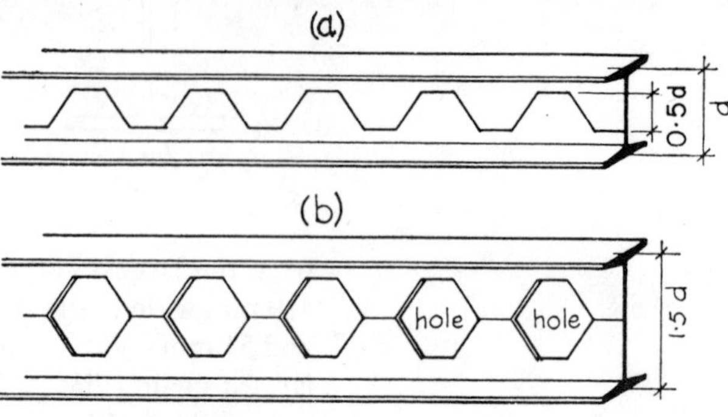

Fig. 5.14

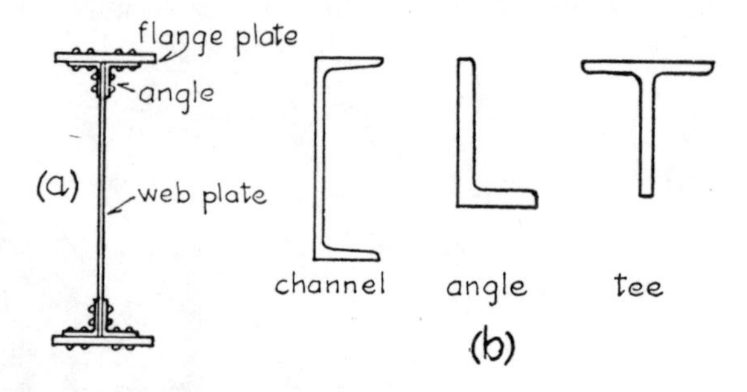

Fig. 5.15

When very deep girders are required and it is not possible to use the ordinary rolled sections, *plate girders* can be used which are built up of angles and plates. Components of the girder may be riveted, bolted or welded together. Figure 5.15 (*a*) is a cross-section of a riveted girder.

Since steel is as strong in tension as in compression, a section with equal flanges top and bottom is very suitable for resisting bending, but in addition to the I-shaped sections, others, as indicated in Fig. 5.15 (*b*), may sometimes be used.

Shear in beams

If timber and steel beams of spans normally used in buildings are made big enough to resist the tensile and compressive stresses caused by bending, they are usually strong enough also to resist the shear stresses caused by their tendency to fail, as shown in Fig.

5.16. (This type of failure, known as *vertical shear*, is unlikely to occur in beams, for reasons which will be explained later.)

Vertical shear stresses in beams are accompanied by horizontal shear stresses. This can be demonstrated by laying a number of strips of smooth wood one on the other, as shown in Fig. 5.17 (*a*). When the beam bends the strips will slide over one another. Horizontal shear can also be demonstrated by using a thick wad of A4 paper or cardboard strips or a pack of playing cards, as the

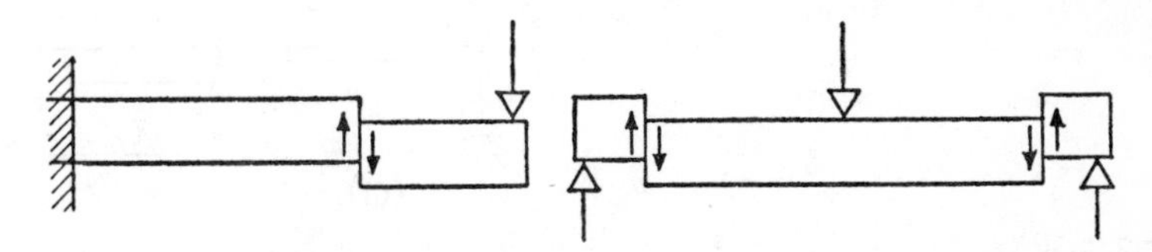

Fig. 5.16

beam. If the beam is solid [Fig. 5.17 (*b*)] instead of being built up of strips, the 'layers' cannot slide relative to one another, but the tendency to slide is present and shear stresses are developed because sliding is prevented. It is possible under certain circumstances for a timber beam to fail in horizontal shear, as shown in Fig. 5.18, which is a sketch of an actual failure, because timber can be weak in shear parallel to the grain.

Fig. 5.17

Shear forces in beams depend on load only, whereas bending moments depend on load and distance. To resist the bending moment, beam (*b*) of Fig. 5.19 must be bigger than beam (*a*), although the conditions of loading are identical. Ignoring the small weights of the beams the maximum shear force in beam (*a*) is, however, equal to that of beam (*b*), i.e. 2·5 kN. Thus, a beam is unlikely to fail in horizontal shear unless the span is very small; then the small amount of material required to resist the bending moment may be insufficient to resist the horizontal shear.

Another way of visualizing vertical and horizontal shear is demonstrated in Fig. 5.20. Imagine a small cube in the interior of the beam completely surrounded by the rest of the beam material. The cube could be imagined to have been embedded in the tree

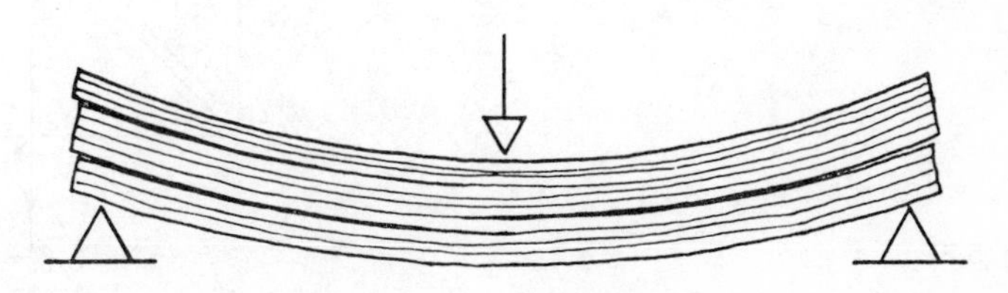

Fig. 5.18

during its growth. When the load is applied to the beam, vertical shear forces *a* are caused on the sides of the cube as shown. These two forces would cause the cube to turn in a clockwise direction, as can be verified if a cube is held between the hands and one hand is moved upwards whilst the other is moved downwards. Since the cube in the beam cannot turn it must be acted upon by forces *b*

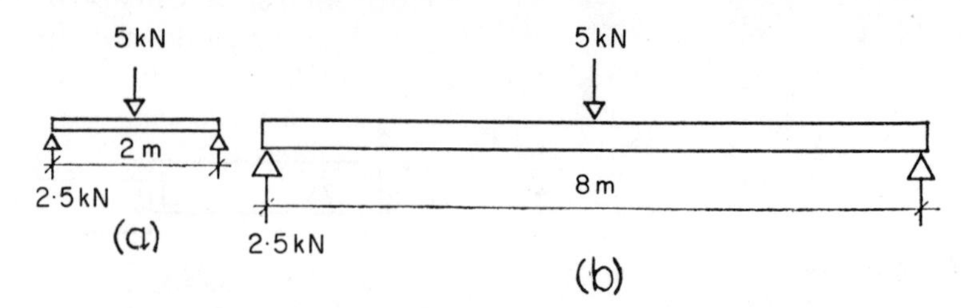

Fig. 5.19

which set up a counter-clockwise rotation. It is these forces *b* which tend to cause horizontal shear failure in a timber beam.

Forces *a* and *b* can be combined by the Parallelogram of Forces law to give a resultant *R* which tends to break the cube in tension, as shown in Fig. 5.21 (*a*). Also, as the corners *A* and *C* tend to get further apart (tension), corners *B* and *D* tend to come closer together (compression).

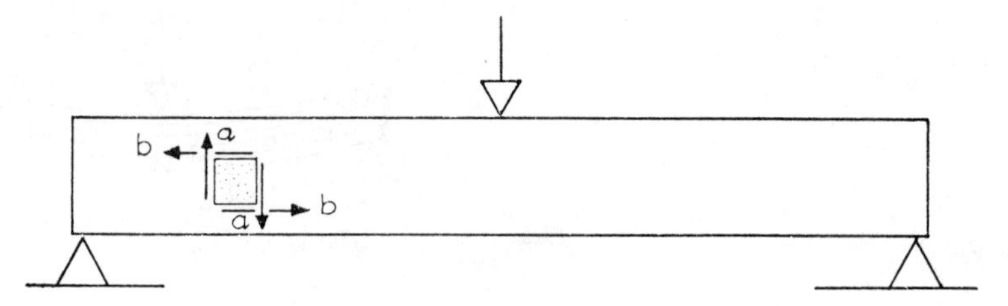

Fig. 5.20

It is thus demonstrated that vertical shear stresses in beams are accompanied by horizontal shear stresses, and these give rise to tensile and compressive stresses on diagonal planes. If the material is weak in compression it could fail as shown in Fig. 5.22 (*a*), whilst if it is weak in tension it could fail as shown at (*b*).

Steel is strong in compression and tension, but in deep girders with thin webs, such as plate girders, precautions have to be taken to prevent buckling of the webs due to diagonal compression

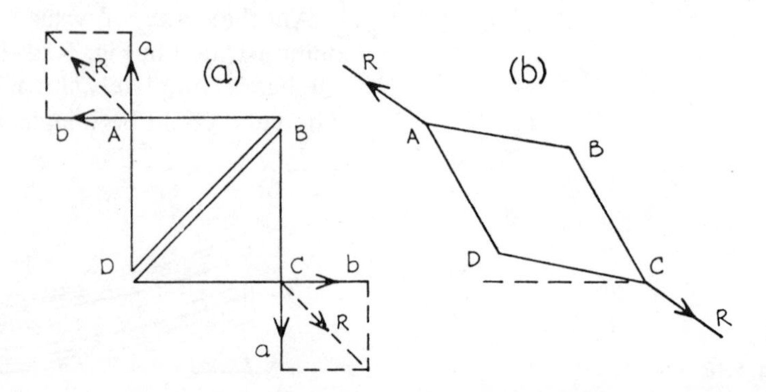

Fig. 5.21

caused by shear. Bridge girders have the webs strengthened by riveting angle sections to them, as shown in Fig. 5.23.

Concrete is much stronger in compression than in tension, and a shear failure in a reinforced-concrete beam would be due to diagonal tension. This will be discussed in greater detail in Chapter 6.

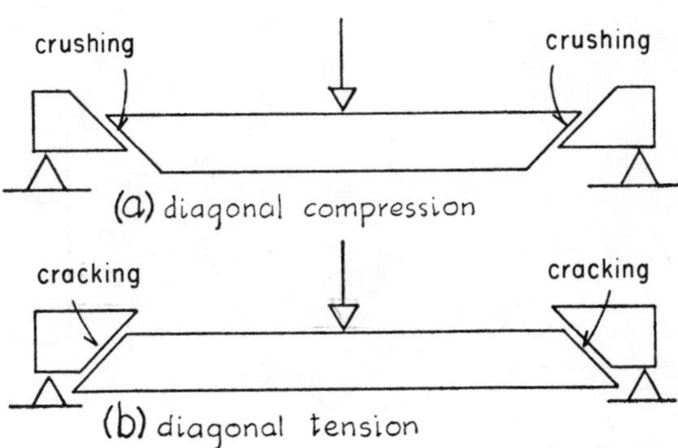

Fig. 5.22

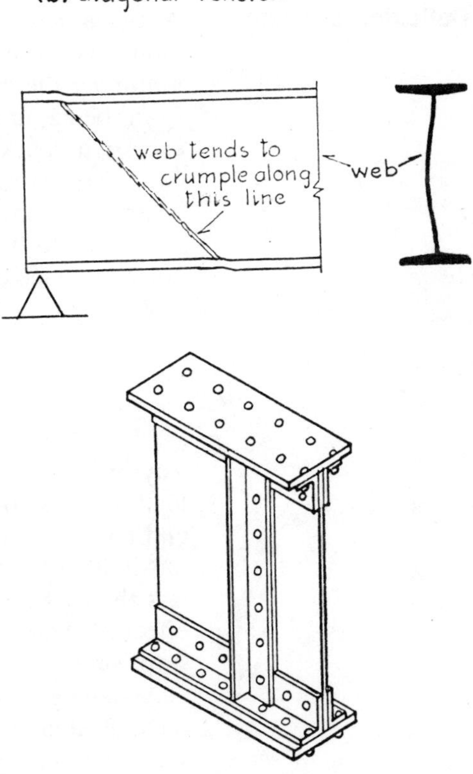

Fig. 5.23

Before leaving the subject of shear it is instructive to consider the behaviour of materials when tested in compression. A short length of mild steel when tested in a compression machine, being ductile, behaves as shown in Fig. 5.24 (*a*). A similar specimen of cast iron fails, however, as at (*b*) because of its weakness in shear. Timber specimens also sometimes fail in shear when subjected to compressive stresses [Fig. 5.24 (*c*)].

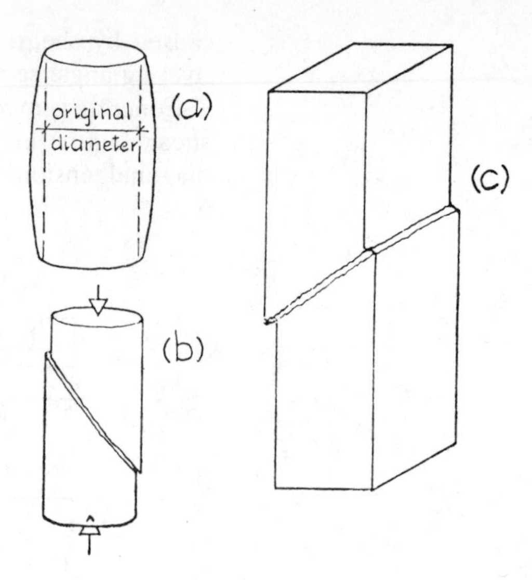

Fig. 5.24

Deflexion of beams

A beam may be strong enough to resist safely the bending moments and shear forces and yet be unsuitable because its deflexion under the calculated safe load is excessive (Fig. 5.25). Apart from being unsightly and giving an impression of insecurity, excessive deflexion can cause cracking of plaster ceilings and of partitions. Most codes of practice state that the deflexion of a beam must not exceed 1/325 of its span. This, for example, is equivalent to 10 mm deflexion in a beam of 3·25 m span when supported at each end. The amount a beam deflects depends on the

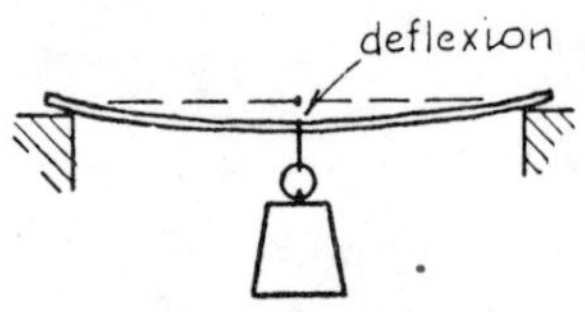

Fig. 5.25

way in which it is supported, the amount and disposition of the load, the span of the beam, the size and shape of its cross-section, and the nature of the material.

All other factors being equal, a beam with its ends fixed deflects less than a beam with its ends freely supported (Fig. 5.26). The effect of span on deflexion is very important and, in a way, very surprising.

Assuming that beams *a* and *b* of Fig. 5.27 are identical in size, the deflexion of beam *b* will be 8 times (not twice) that of beam *a*, since the deflexion of a beam is proportional to the cube of the span. If the span of beam *b* is 3 times that of beam *a*, other conditions being identical, the deflexion will be 27 times as great.

Deflexion becomes more of a problem as the span of a beam increases, because deflexion increases at a greater rate with increase

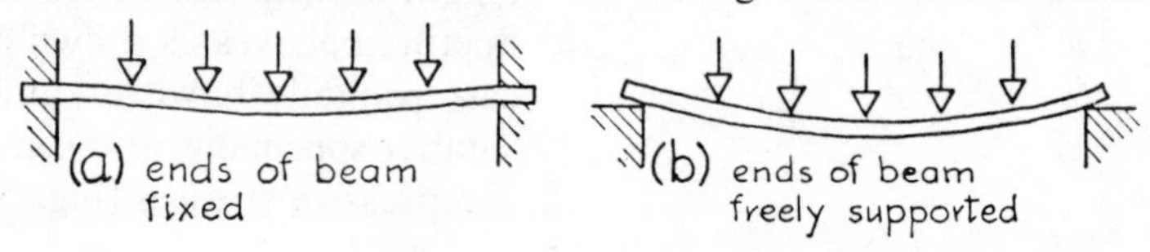

Fig. 5.26

of span than the size of a beam required to resist bending increases. For example, a timber beam 75 mm wide and 150 mm deep is suitable under certain conditions (bending stress of 7 N/mm^2 and elastic modulus of 8 kN/mm^2) for carrying a uniformly distributed load of 3·6 kN/m (total load, 7·2 kN) on a span of 2 m (Fig. 5.28). The calculated deflexion for such a beam is 4 mm, which is 1/500 of the span and much less than the normal permitted value.

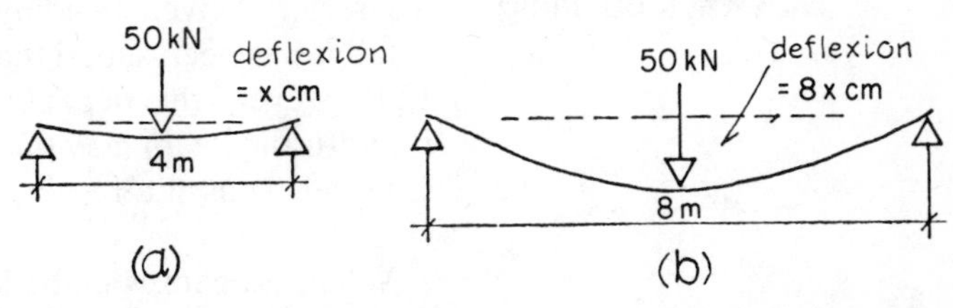

Fig. 5.27

If the span is increased to 4 m (total load, 14·4 kN at 3·6 kN/m), the beam dimensions required for resisting the bending moment, still keeping the depth of the beam equal to twice its breadth, are 115 mm by 230 mm. The calculated deflexion is 13 mm, which is 1/308 of the span (a little more than the normal permitted value).

When the span is increased still further to 6 m (total load, 21·6 kN at 3·6 kN/m), the beam dimensions required for resisting the bending are breadth 150 mm and depth 300 mm. The calculated deflexion is 22 mm, which is 1/273 of the span. The deflexion could be decreased to 1/500 of the span, the beam still being suitable for resisting the bending moment, by making the breadth of the beam 75 mm and the depth 465 mm, but this beam being so slender might be liable to sideways buckling (*see* p. 76).

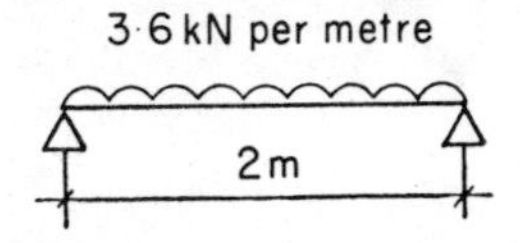

Fig. 5.28

It is, however, worth remembering that, for a given amount of material, the deepest beam is the best for limiting deflexion as well as being the most economical for resisting the bending moments due to the loads on the beam. Also, for large spans, beams may have to be of larger sizes than are necessary for resisting bending moments in order that the deflexions may be kept to reasonable limits.

One important factor which has obviously to be taken into account when calculating deflexions is the nature of the material. We know that force cannot be applied to a material without altering its dimensions. We also know (Hooke's law) that, for an elastic material, stress is proportional to strain, and that stress divided by strain is called modulus of elasticity, E. The greater the value of E, the stiffer is the beam, i.e. the greater is its resistance to being bent. It is possible to obtain aluminium alloys which have strengths approaching those of ordinary mild steel, but their E values are only about one-third of that of steel. It therefore follows that, for identical conditions, aluminium beams bigger than steel beams would be necessary in order to limit deflexions. Thus aluminium alloys are not suitable as alternatives to steel for the construction of beams and columns in multi-storey buildings, but for certain types of structure where deformations are small or not important (*see* Chapters 14 to 16) they may be used to advantage.

Sideways buckling

To resist a given bending moment and also to keep deflexions small it has been stated that the deepest beam is the most suitable. If, however, the depth is made too great in proportion to the breadth the beam may buckle sideways due to a *column effect* (*see* Chapter 9) as the result of compressive stresses in the top fibres (Fig. 5.29).

When a beam is embedded in a reinforced-concrete floor or is restrained by cross-beams as shown in Fig. 5.30, there is no need to consider the buckling tendency, since the floor slab (or beams) provides sufficient lateral restraint; but when a beam is not restrained laterally, the permissible bending stress for a slender beam must be reduced in accordance with code requirements.

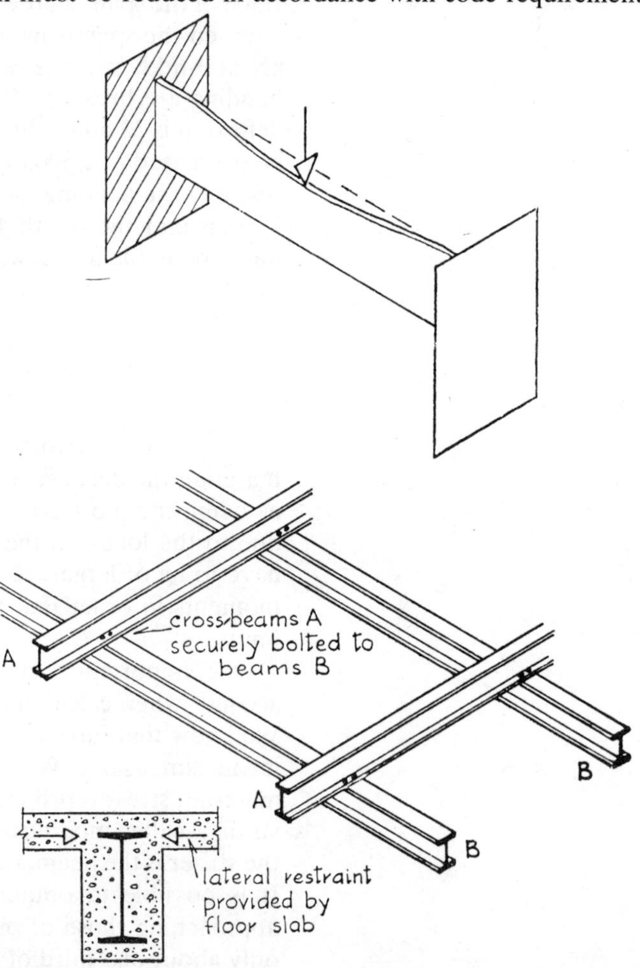

Fig. 5.29

Fig. 5.30

6 The Beam and the Slab: Reinforced Concrete

Concrete is an artificial stone, and in common with natural stone is a brittle material, strong in compression but weak in tension. From very early times, metal has occasionally been used to strengthen, or *reinforce*, masonry and concrete. The roof of one Roman bath was constructed of a coarse concrete reinforced with bronze and iron rods. Wren used an iron chain in the dome of St Paul's to help resist lateral thrust, and J. G. Soufflot (1713–81) used wrought-iron rods embedded in masonry in the construction of the church of St Sulpice, Paris. Until the 19th century, however, there was no knowledge of reinforcing principles as they are understood today, and satisfactory methods of calculation were only developed towards the end of the century.

An early example of reinforced concrete was displayed in the Paris International Exhibition of 1855. This was a rowing boat designed by J. L. Lambot, a French contractor, and made of hydraulic lime concrete reinforced with a skeleton of iron rods.

In 1854, a Newcastle plasterer, William B. Wilkinson, patented a method of constructing concrete slab floors reinforced either with a network of flat iron rods or with second-hand wire ropes. His manner of arranging the steel to take tensile stresses demonstrates that he appreciated the constructional principles of reinforced concrete.

Many patents were taken out during the 19th century, but only a few of them can be mentioned in this brief historical review. François Coignet patented a floor in 1855, and in following years built, among other structures, a lighthouse at Port Said and retaining walls in Paris. His son, Edmond Coignet (1850–1915), also took out various patents for the use of reinforced concrete.

Joseph Monier (1823–1906) was originally a gardener and a manufacturer of gardening tools. He conceived the idea of using tubs of concrete instead of wooden tubs for holding small trees, but finding that concrete was too brittle he made an iron network and constructed the tub by enveloping this network with mortar or concrete. This method he patented about 1867 and followed it with other uses of reinforced concrete, as for example a reservoir built in 1872. On the Continent at this time, reinforced concrete construction was commonly referred to as the Monier system.

An American, T. Hyatt, carried out experiments on beams and was among the first to appreciate the basic principles of reinforced concrete construction as we understand them today. He published his discoveries in London in 1877, but his patents did not receive the attention they deserved. Experiments were carried out in Germany and elsewhere towards the end of the century which enabled K. Koenen, Mörsch, Considère and others to produce mathematical design theories.

The first English textbook on reinforced concrete, by Marsh and Dunne, was published in 1904.

Another famous name associated with the development of reinforced concrete is that of François Hennibique (1824–1921), who constructed the first building with a complete reinforced concrete frame. He set up a branch office in London in 1897, in the charge of L. G. Mouchel, to design structures on the Hennibique system of patents. At the beginning of the 20th century, most reinforced concrete work was done by specialist firms of this sort, since the average architect or consulting engineer had not yet acquired sufficient theoretical and practical knowledge of the new method of construction to enable him to prepare designs with confidence. Although certain types of floor construction are still patented, all normal reinforced concrete work can now be designed and constructed without reference to patents.

The simple reinforced-concrete beam

The strength of concrete in compression is at least ten times as great as its strength in tension, and by using a material such as steel to take the tensile stresses induced by bending, full advantage can be taken of the high compressive strength of the concrete.

Bending reinforcement

A simple beam as shown in Fig. 6.1 can be constructed by placing steel bars in a wooden or steel mould and ramming in the wet concrete so that it completely surrounds the bars.

On setting and hardening, the concrete shrinks slightly and grips the steel bars firmly, so that when the beam is loaded the steel and concrete bend as one unit, there being no slip (in correct design) between them. (Tendency to slip is due to horizontal shear.) Ordinary mild steel is allowed to take a safe tensile stress of 140 N/mm^2 and the accompanying elongation of the steel, although very small, is sufficient to crack the concrete in the tension zone of the beam. (Since the elongation of the steel is so small, these minute cracks are normally not detectable by eye.)

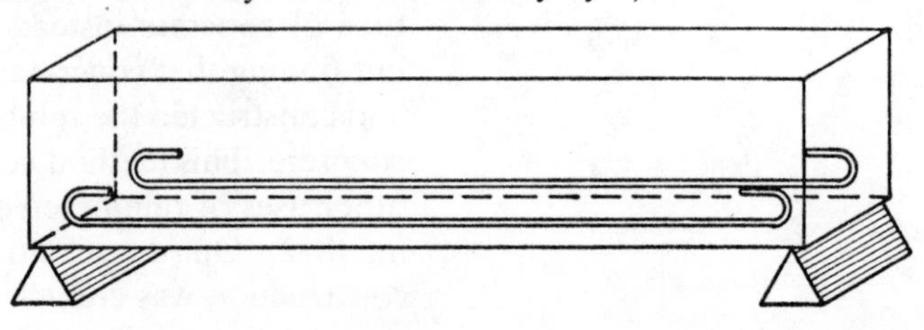

Fig. 6.1

The cracking of the concrete, which causes the steel to take all the tension below the neutral layer, may be explained in this way. The elastic modulus (E) of steel is about 15 times that of concrete, which means that if the steel is stressed to 140 N/mm^2, the concrete surrounding and gripping the steel, which stretches an equal amount, is called upon to transmit a stress of 1/15 of 140, i.e. 9·3 N/mm^2. But

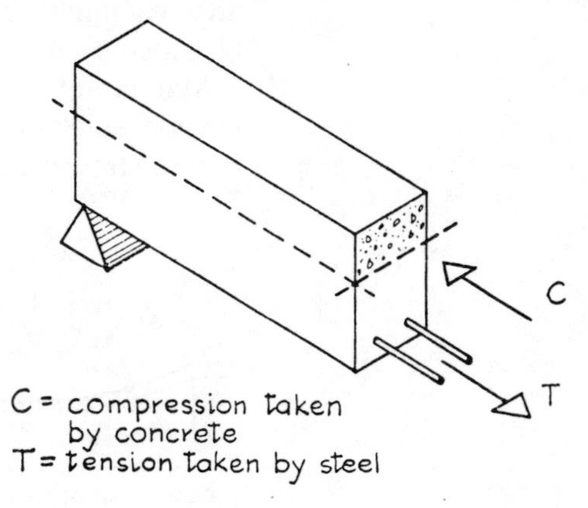

Fig. 6.2

concrete fails in tension at a stress of about 3 N/mm^2, the exact figure depending on the quality of the concrete. It follows, therefore, that the concrete below the neutral axis fails in tension and is ineffective for resisting bending stress (although it is still capable of resisting shear forces). In a simple reinforced-concrete beam as shown in Fig. 6.2, the compressive stresses are taken by the concrete and the tensile stresses are taken by the steel.

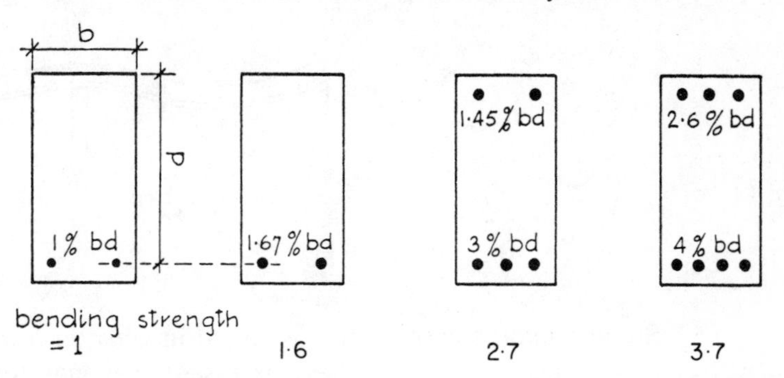

Fig. 6.3

The position of the neutral layer depends among other factors on the amount of the reinforcement. The resistance of a beam to bending can be varied by varying the amount of steel (and also the strength of the concrete). For example, with about 1 per cent of steel (such as 1250 mm^2 of steel in a beam 250 mm wide by 500 mm deep) and a concrete strength of 20 N/mm^2, the moment of resistance of the beam based on ultimate strength design methods is 115 kN-m. With 2·75 per cent of steel, the moment of resistance of a beam of equal overall dimensions is 270 kN-m. If for this mix the percentage of steel is increased further, no advantage results because the concrete in compression will be unable to supply enough resistance to balance the high tension the steel can supply.

The strength of the beam can be increased further, however, by placing steel in the compression zone of the beam to help the concrete, although this is not normally economical. It is a method adopted when it is required to keep the beam dimensions as small as possible. A comparison of the bending strengths of beams of equal overall dimensions using a 20 N/mm^2 concrete is approximately as shown in Fig. 6.3. Still higher strengths can be obtained by using stronger concrete.

Many reinforced-concrete beams and the majority of slabs have tension steel only, and this must, of course, be placed where tensile stresses due to bending occur (Fig. 6.4).

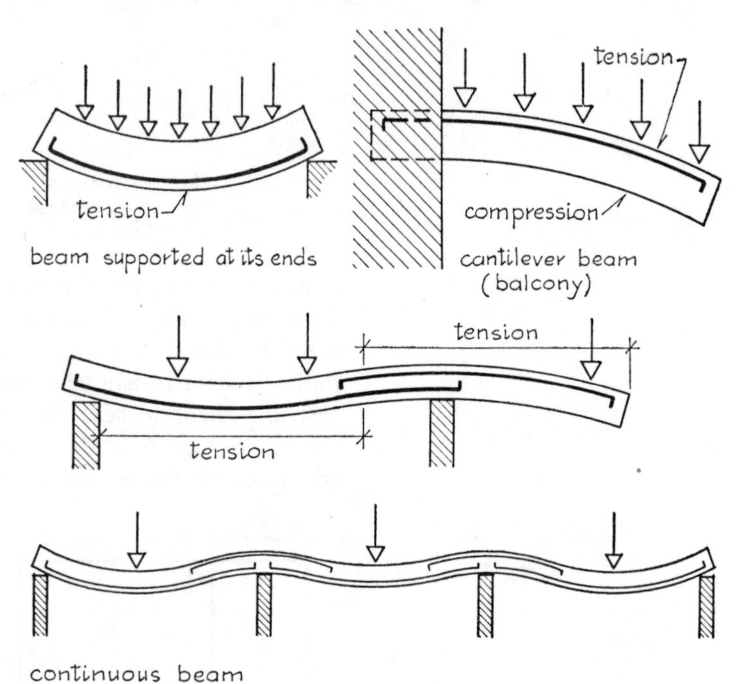

Fig. 6.4

Shear reinforcement

It was stated in Chapter 5 that vertical shear stresses, horizontal shear stresses, diagonal tensile and compressive stresses are caused in the material of a beam. Concrete being weak in tension will fail as indicated in Fig. 6.5 if the shear forces are big enough. The steel put in to resist tension due to bending (*see* Fig. 6.4) does not help the concrete to resist the diagonal tensile stresses. In some beams and in most slabs, the shear stresses are small enough to be taken by the concrete unaided, but if the calculated shear

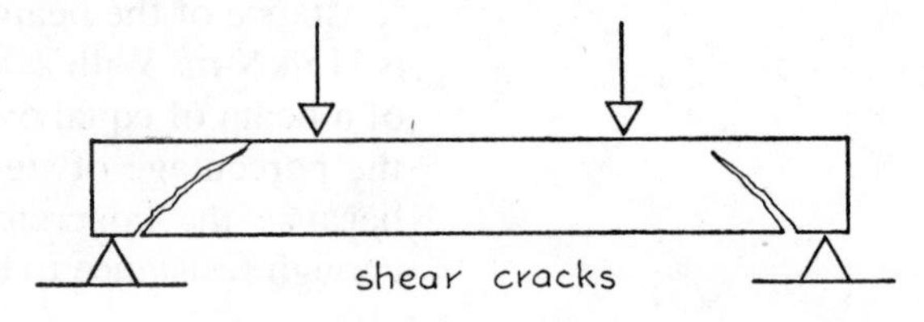

Fig. 6.5

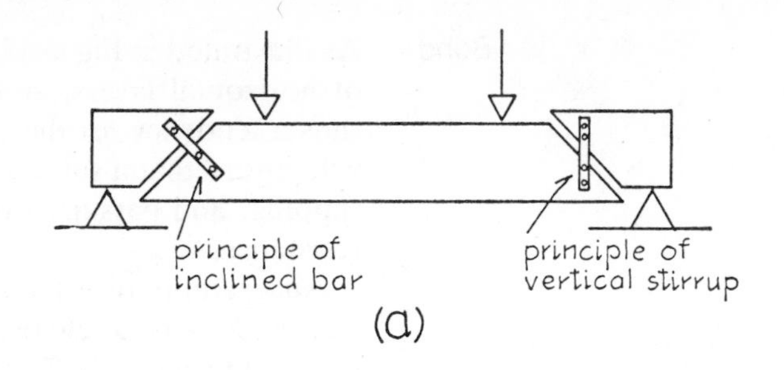

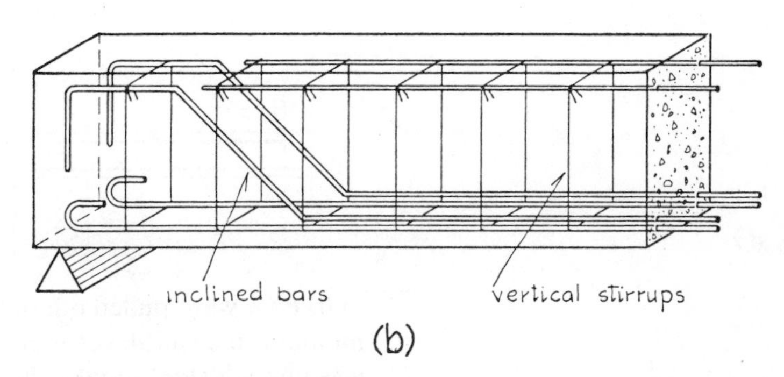

Fig. 6.6

stresses are greater than the permissible shear stresses, steel (called *shear steel*) must be supplied to take the diagonal tension. The reinforcement can take the form of inclined bars or vertical stirrups or a combination of both.

Figure 6.6 (*a*) illustrates the action of the shear reinforcement in holding the beam together and preventing failure, whilst Fig. 6.6 (*b*) shows shear reinforcement in a simple beam.

Reinforcement at Lead Works, Millwall (*Cement & Concrete Association*)

Bond As illustrated in Fig. 5.17, there is a tendency in a beam for sliding of horizontal layers, and in a reinforced-concrete beam there is thus a tendency for the steel to slip in the concrete. Good design will ensure that a sufficient length of steel is embedded to prevent slipping, and bars are often hooked at their ends to increase the resistance.

An example of a failure due to inadequate bond between the steel and the concrete occurred in a block of flats in England some years ago (Fig. 6.7). The balcony collapsed because the steel rods

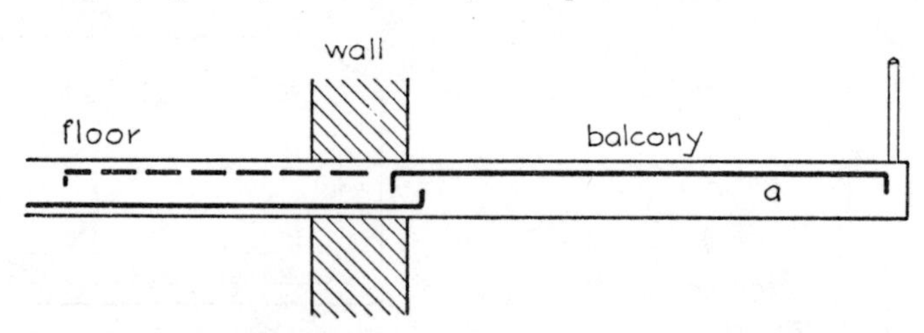

Fig. 6.7

marked *a* were pulled out of the concrete. The maximum bending moment in a cantilever is at the fixed support, and although there was enough steel to take the bending moment, the balcony failed because someone forgot to provide sufficient embedment for the bars. The steel should have extended for a greater distance backwards into the interior floor slab, as shown by the dotted line.

The reinforced-concrete slab A simple floor can be constructed by laying a number of precast beams, say 300 mm wide, side by side as shown in Fig. 6.8, and many floors are constructed in this manner although the precast units are usually hollow in cross-section. The steel runs in one direction only—parallel to the span. However, some floors are cast in place (*in situ*) on the building site and since they are often made

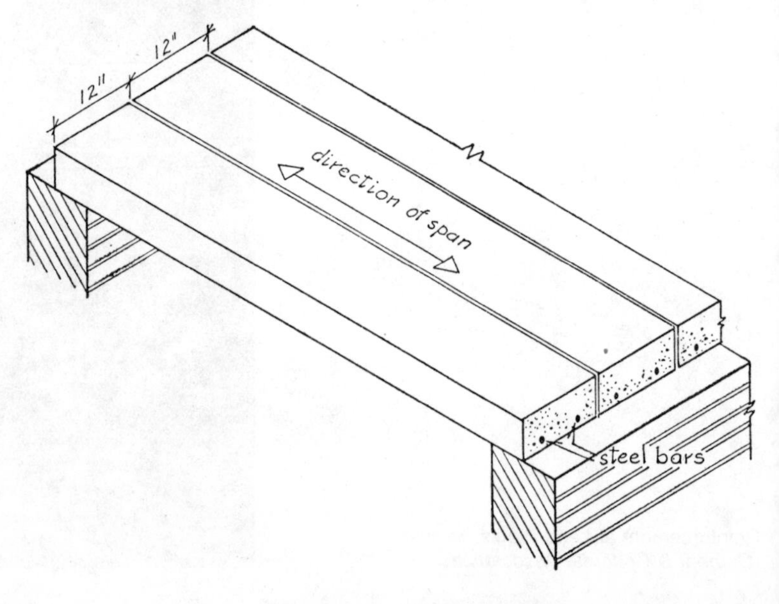

Fig. 6.8

in one continuous pour, cracks are likely to occur due to shrinkage and temperature changes in the concrete.

To distribute these secondary stresses throughout the mass of the concrete (primary stresses are due to the bending moments caused by the load), *distribution* or *temperature steel* is supplied, so that a mesh is formed as illustrated in Fig. 6.9. This steel is not assumed to take any of the bending moment, which is carried entirely by the main steel in conjunction with the concrete. (The distribution steel is smaller in quantity than the main steel.) Even if the slab is supported along its four edges instead of two, this assumption can still be made, and this type of construction is called a *one-way slab*, or a slab spanning in one direction. Some slabs, however, are constructed as *two-way slabs*, or slabs spanning in

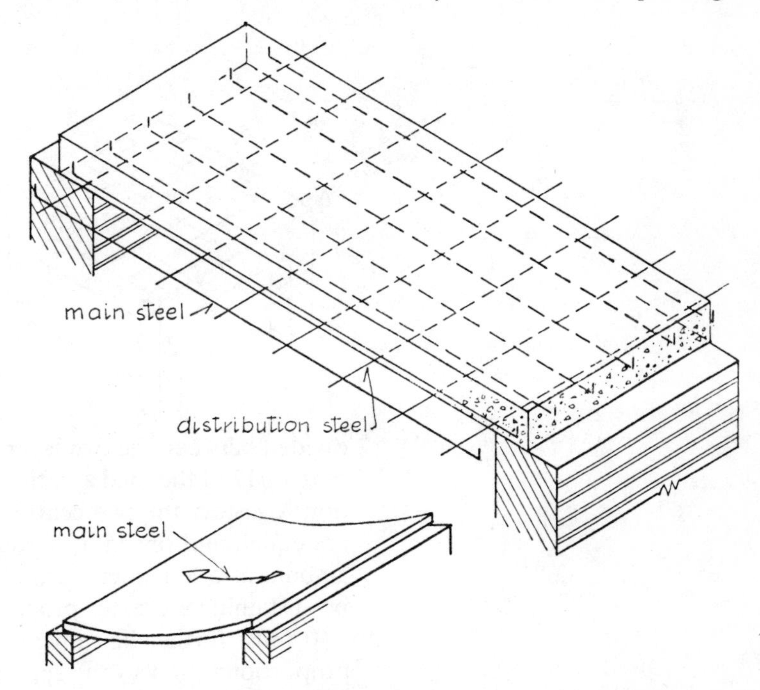

Fig. 6.9

two directions, and for such slabs both layers of steel are calculated to take bending moments.

The difference between one-way and two-way slabs can be illustrated by the following simple analogy. Imagine a stool with strips of webbing connected to the two sides *AB* and *CD* only [Fig. 6.10 (*a*)]. When one sits on such a stool, all the load is taken by these strips; this is comparable to a slab spanning in one direction. Now imagine an additional set of webbing strips connected to the sides *AC* and *BD* as at (*b*). If the stool is square, the load is now shared equally by the two layers of strips, one layer spanning from *AB* to *CD*, and the other from *AC* to *BD*. In a solid reinforced-concrete slab, the two layers of steel act similarly to the two layers of webbing, and this method of construction usually means a smaller thickness of slab than if a one-way system is used, more especially when the slab is square or nearly so.

When the slab is oblong, instead of the load being divided equally both ways, the shorter span takes a higher proportion of the load. For example, if in Fig. 6.11 the side *BD* is twice the side *AB*, the total load per square metre of floor is approximately

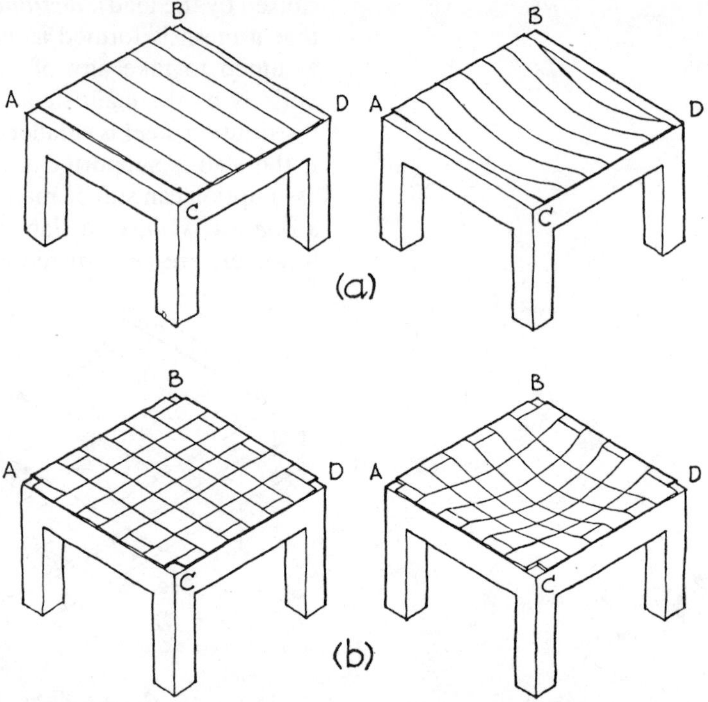

Fig. 6.10

divided between the two layers of steel so that the short strips support 16/17 of the load and the long strips 1/17. To find an explanation, consider the two centre strips. Their deflexions at mid-span are equal, and since it is harder to bend a short member the same amount as a long member, it follows that the short member must be sustaining a greater proportion of the total load.

It must further be remembered (*see* page 74) that deflexion is proportional, not to the span alone, but to the cube of the span;

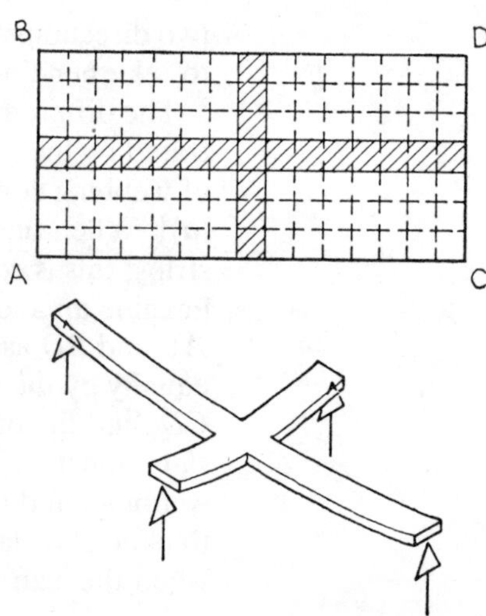

Fig. 6.11

moreover, when we are dealing with load per metre run on a beam, deflexion is proportional to the fourth power of the span. Thus if the total load per square metre of floor is 765 N and the long side *BD* is twice the short side *AB*, the division of the load per square metre is not 2 to 1 but 16 to 1 (2^4 to 1). Therefore the short-span strips are designed for 720 N/m² and the long-span strips for 45 N/m². It follows that, when a slab is very long and narrow, little is gained by designing it as a two-way system, since the short-span steel takes most of the load and the long-span steel will be much smaller in amount. When, however, the slab is square or nearly so, the load is divided equally and both layers of steel will be approximately of equal amounts. The bending of a solid slab is actually more complex than has been described above, but for a slab simply supported along its four edges this approximate treatment is permitted by most codes of practice.

Shear stresses in slabs are usually small enough not to require the use of shear reinforcement.

Hollow tile and lightweight floors

It has been mentioned earlier that for purposes of resisting bending moments the concrete below the neutral axis is useless (except, of course, that it connects the steel with the compression zone of

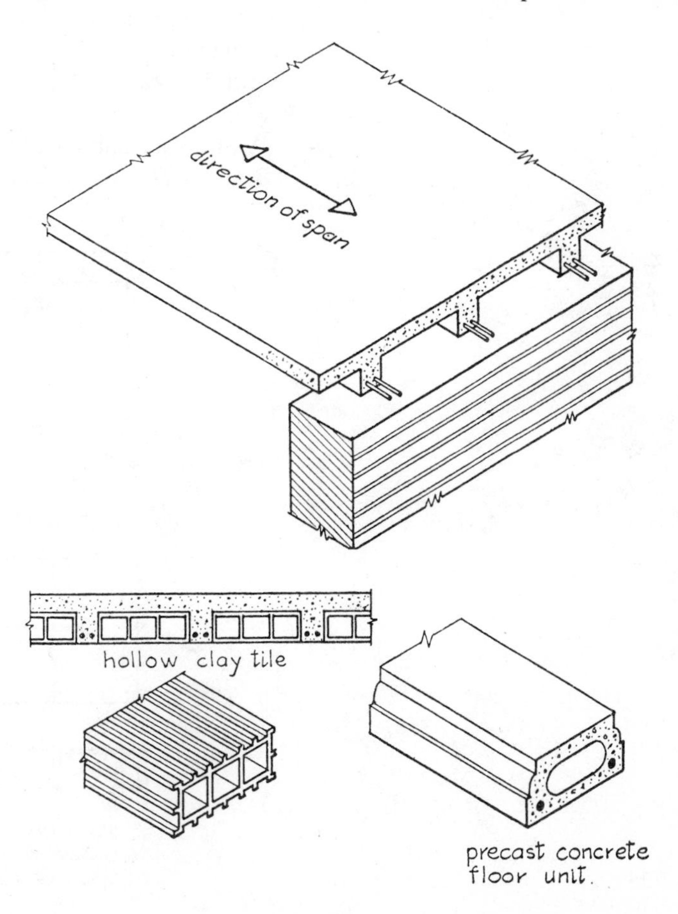

Fig. 6.12

the beam), so the bending strength of a beam or slab would be unaltered by leaving out some of this redundant concrete. Sufficient concrete must be left to accommodate the steel and to supply shear resistance. Thus a slab could be constructed, as shown in Fig. 6.12, which is as strong in bending as a solid slab of equal depth. It is usual, however, in order to give a continuous under-surface (the ceiling for the room below), to fill the gaps with hollow burnt-clay blocks which are cast in with, and subsequently gripped by, the concrete. This technique is commonly used in multi-storey buildings in order to minimize the self weight of the floors and so reduce the loads carried by the beams and columns of the building. When compared with equivalent strength solid slabs substantial savings in materials and cost can be made particularly in the lower storeys and foundations of high rise buildings.

Alternatively precast hollow floor units can be used, for which there are many proprietary types. The use of lightweight concrete can also minimize column loads and foundation costs and the combination of special floor units with this lighter material can have significant cost advantages.

Beam and slab construction

Economy is usually a prime consideration in the design of all types of structure, and economy can often be achieved by a suitable arrangement of structural members. As a simple example, assume that a flat roof is required to cover an area of 24 m by 8 m (Fig. 6.13).

A one-way slab could be designed to span the 8 m between walls *AB* and *CD*, but because of the large span a thick slab would be

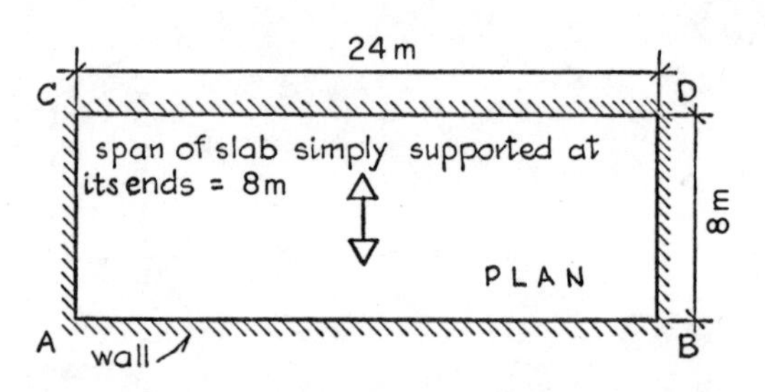

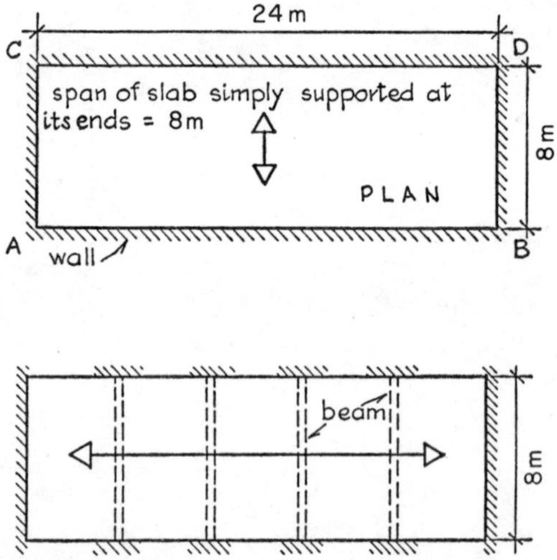

Fig. 6.13

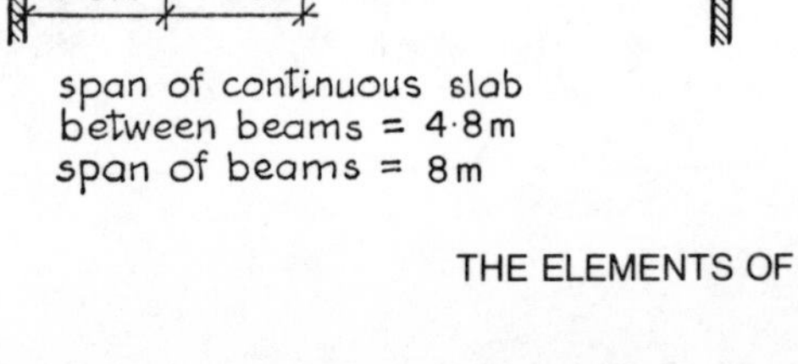

beam
8m
4·8m
4·8m
span of continuous slab between beams = 4·8m
span of beams = 8m

Fig. 6.14

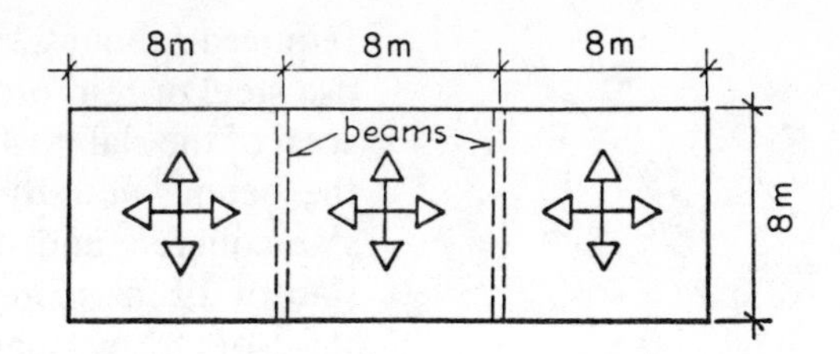

Fig. 6.15

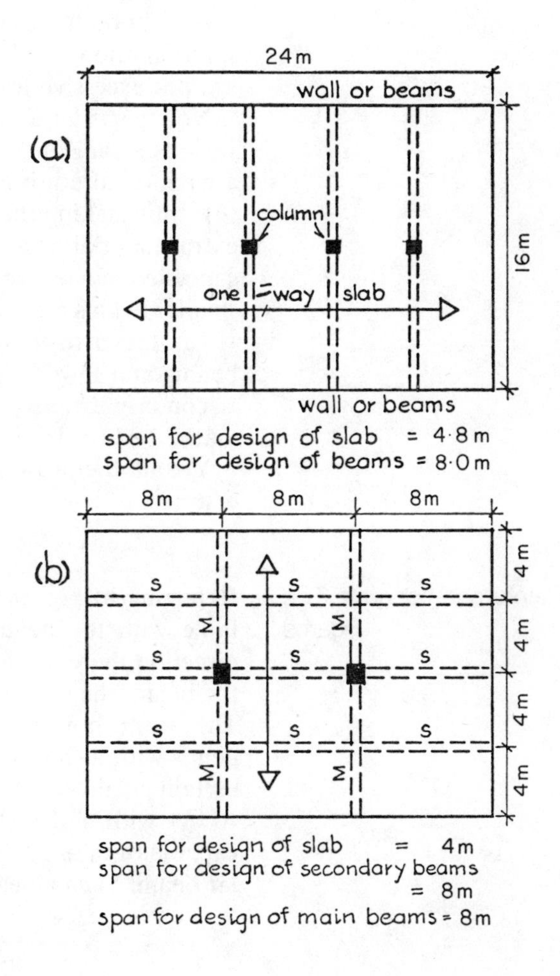

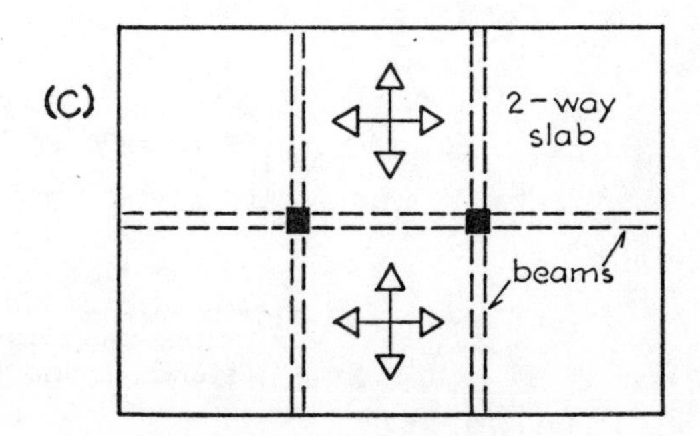

Fig. 6.16

required (about 225 or 250 mm). It might be more economical to use steel or reinforced beams, as shown in Fig. 6.14, to reduce the span of the slab to 4·8 m. The extra concrete and steel required for the beams would probably be more than offset by the saving of slab concrete and steel, since much less concrete is required for a slab of 4·8 m span (thickness approximately 125 mm) than for a slab of 8 m span. Furthermore, the bending moments in continuous slabs are less than those in simply supported slabs of equal span.

An alternative arrangement is to use a two-way slab of larger span, as shown in Fig. 6.15. Since the slabs are square the spans are not excessive for this type of construction.

Now consider an area of 24 m by 16 m, as shown in Fig. 6.16; various arrangements are possible. Unless an uninterrupted floor area is required, it may be more economical to use columns as at (*a*), thus making the span of the beams 8 m instead of 16 m. The extra material as a result of using columns would be more than cancelled by the saving of material in the beams, since much smaller beams are required for spans of 8 m than for spans of 16 m.

Another arrangement is shown in Fig. 6.16 (*b*), which uses more beams and fewer columns than the system of Fig. 6.16 (*a*). The 'secondary beams' marked *S* are supported by 'main beams' marked *M* (or by columns).

Yet another arrangement, using a two-way slab, is shown in Fig. 6.16 (*c*).

Reinforced-concrete Tee Beams

Referring to Fig. 6.17, and assuming that the slab is cast monolithic with the beam, it is spanning from E to W, and in this direction there are tensile stresses in the slab where it passes over the beam; the main steel at this point will be in the top of the slab. The beam, however, is spanning from S to N, and since the slab bends with it the whole of it in this direction is in compression. A certain portion of the slab (breadth *B* in Fig. 6.17) can be assumed to act with and to help the beam, so that the area of concrete capable of taking compression is much greater than in a rectangular beam. Such a beam is therefore designed as a T-beam and can

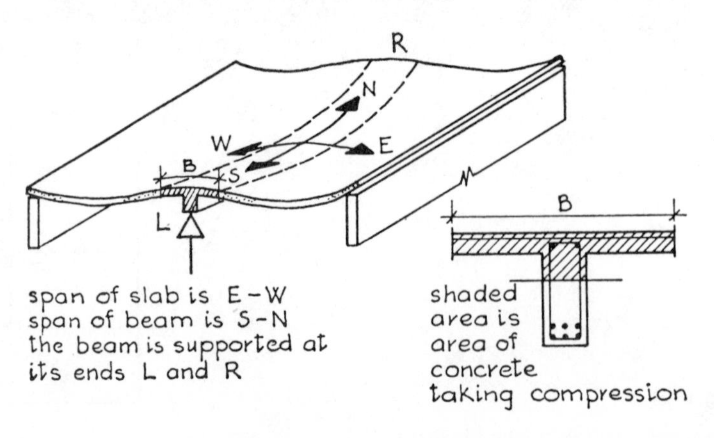

Fig. 6.17

be less deep than a comparable rectangular beam. The values of B to be used in design are given in various codes of practice.

Flat slab (beamless) floors

A type of construction which requires no beams is shown in Fig. 6.18. The slab is thicker than in beam-and-slab floors, but a great advantage is the extra head-room obtained, particularly in heavily loaded warehouse buildings where beams would be deep. The enlarged heads to the columns are to reduce the shear stresses, but if special reinforcement is provided, heads can be dispensed with.

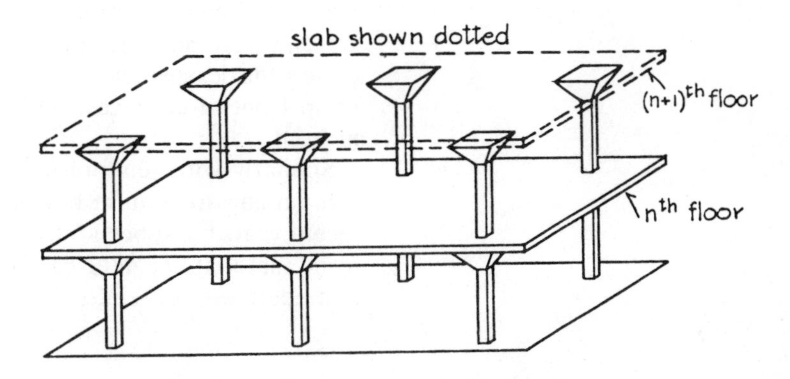

Fig. 6.18

Beamless floor in a factory at Brynmawr, South Wales (*Cement & Concrete Association*)

To fail in shear (diagonal tension) the concrete has to fail around the perimeter of the head in Fig. 6.19 (*a*). At (*b*), where the column has no head, the area resisting shear is much smaller, so a thicker slab is required than at (*a*) unless some other means to resist shear is used.

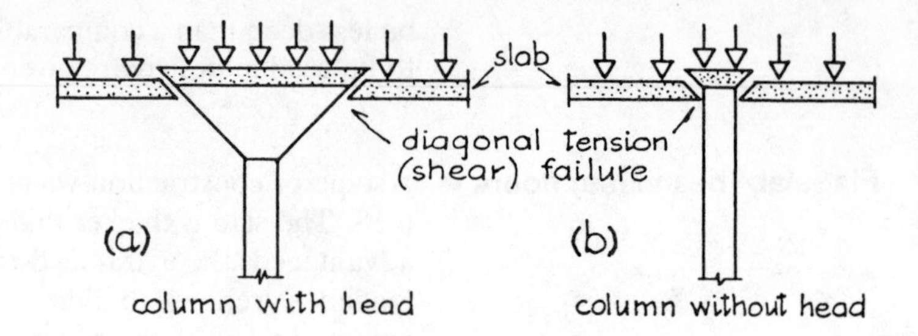

Fig. 6.19

Postscript New materials and new inventions nearly always follow older traditional methods of design. For example, the early motor-cars were in truth horseless carriages. The body design followed fairly closely the design of horse-drawn carriages, and since horses always go in front of carriages, the motor-car engines were also placed in front.

Similarly with reinforced concrete. The traditional beam-and-column construction in timber or steel was adopted, and only in recent years has it been fully realized that, since concrete is a 'free form' material, it can be cast into practically any shape. Examples of modern uses of reinforced concrete are given in later chapters.

7 The Beam: Prestressed Concrete

It was stated in Chapter 6 that in an ordinary reinforced concrete beam the concrete below the neutral axis (in the tension zone of the beam) is ineffective for resisting the tension due to bending. Also, because of these tensile stresses (which are taken by the steel) the concrete in the vicinity of the steel cracks.

Before the beginning of the 20th century attempts were made to improve reinforced concrete by using *prestressing* techniques to eliminate all tensile stresses, but these attempts were unsuccessful because the importance of using a high-strength steel and a high-quality concrete was not appreciated. Prestressed concrete as a practical method of construction dates from 1928, when the French engineer, Eugène Freyssinet used high-tensile steel. In 1936 he gave a lecture in Britain on prestressed concrete, and in the following year the Prestressed Concrete Co. was founded.

In 1938–39 the War Office in England roofed some underground munition stores with pretensioned beams, and during the war, when timber was scarce, prestressed-concrete railway sleepers and small prestressed floor beams for houses were used. A railway viaduct near Wigan in Lancashire, built in 1946, was the first prestressed-concrete bridge in Britain, and since that time the use of prestressed concrete has grown to the point where it is now a major industry, successfully competing with steel in many different structural applications from bridges to offshore platforms.

Principle of prestressing

The fundamental principle of prestressing may be described as follows. Forces are applied to the beam prior to its positioning in the building to place the concrete in compression; then, before tensile stresses can be produced due to the bending action of the final loads, the compressive stresses due to the prestress must be overcome. With competent design the compression initially imparted to the beam is of such an amount that, even when the beam is fully loaded, no tensile stresses are produced at any point, or at the worst there are only very small tensile stresses, which the concrete can sustain safely without cracking.

An illustration which has been frequently given in literature on prestressed concrete concerns a number of books or blocks. Since there is no adhesion between the books it is impossible for them to act as a beam, as shown in Fig. 7.1 (*a*); a row of books has no bending (tensile) strength or shear strength. If, however, the books are firmly pressed together, as at (*b*), they can act as a beam and can support loads, the magnitudes of the loads depending on

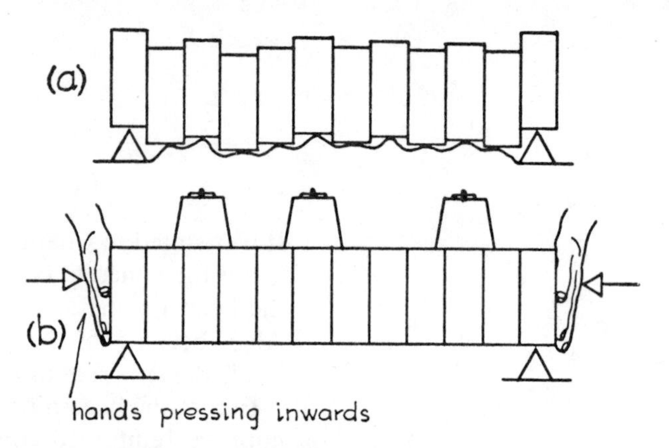

Fig. 7.1

the amount of compression applied as a result of the pressure exerted by the hands. The external pressure which squeezes the books together prevents shearing (i.e. prevents the books from falling vertically) and also prevents the joints from opening at the bottom, which they tend to do as a result of the tensile stresses caused by the weight of the books and the imposed load.

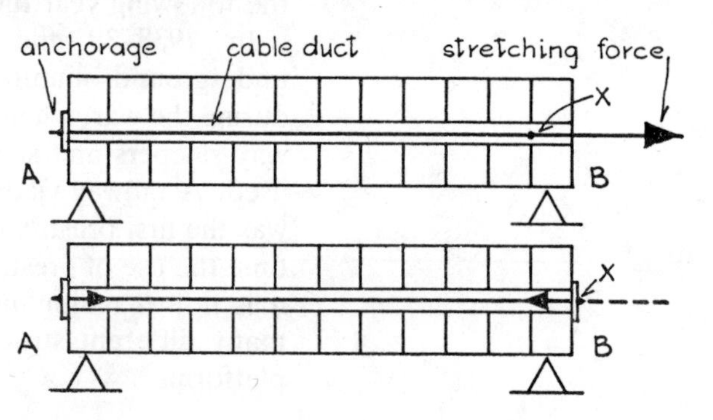

Fig. 7.2

Compression can be applied to a concrete beam made up either of separate blocks or of a single solid block, but normally it is not practical to apply pressure externally. It is more convenient to use internal steel wires or bars, and the principle of one method of prestressing is as follows.

In Fig. 7.2 there is a continuous hole right through the blocks (or solid beam) and the wire or cable is lying loosely in the duct. For convenience of illustration, X marks one particular spot on the wire (one end of which is securely anchored to end A of the beam). The wire or cable is now stretched by a considerable force applied by a jack until point X comes outside the beam. Whilst the wire is in this position it is firmly clamped to the beam at B and the jack is disconnected. The wire, being stretched, will try to return to its original length and in doing so will press the blocks tightly together, putting them into compression. The principle is much the same as stretching a strong rubber band to hold together several packets of cigarettes (Fig. 7.3), except that in this example the material (the rubber band) which causes the compressive prestress is outside the 'beam.'

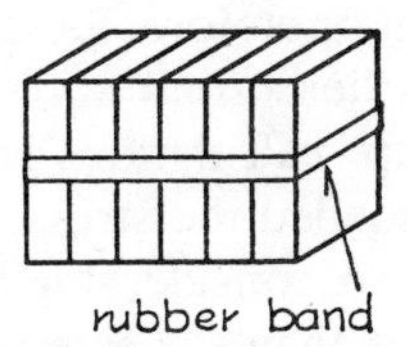

Fig. 7.3

Fig. 7.4

Note that, if the wire is placed at mid-depth as in Fig. 7.2, a uniform compression will be induced in the concrete. Thus there is considerable compression in the top fibre of the beam, and this compressive stress will be increased when the beam receives its external loads. In general, therefore, the wires are placed nearer the bottom of the beam, as in Fig. 7.4 (although, for a reason which will be given later, the wires are usually curved).

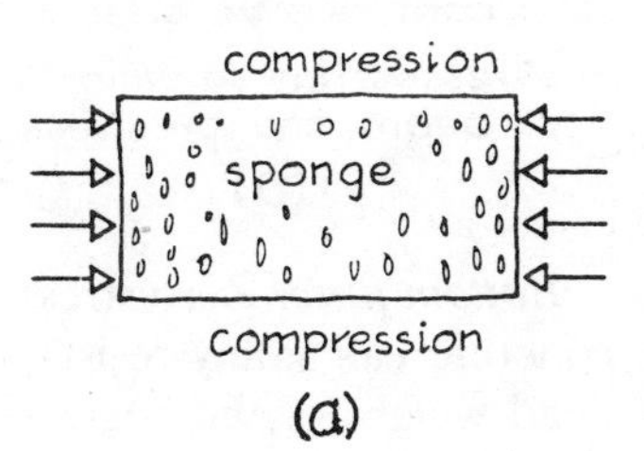

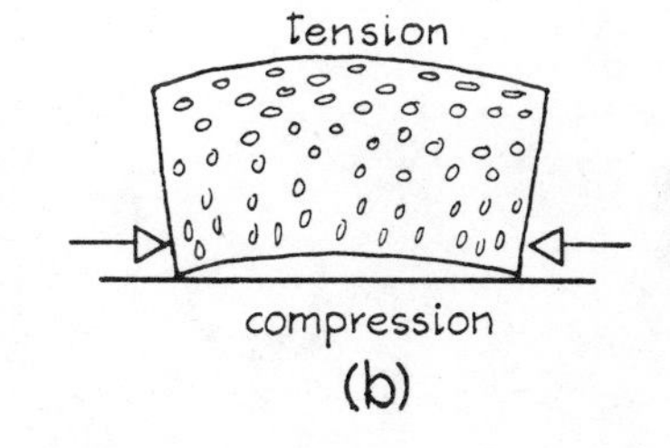

Fig. 7.5

The principle underlying this method of positioning the wires can be demonstrated by an ordinary household (or car) sponge. If a uniform external pressure is applied by the hands, as shown in Fig. 7.5 (*a*), uniform compressive stresses will be produced. If, however, the pressure is applied near the bottom, as at (*b*), tension is caused in the top fibre and compression in the bottom fibre.

Now, sponge is a light-weight material but concrete is heavy. If an unloaded concrete beam is supported, as shown in Fig. 7.6, the bending due to the weight of the concrete will produce compressive stresses in the top and tensile stresses in the bottom of the beam. With a prestressing force of sufficient amount and correctly applied near the bottom of the beam, the compressive stress in the

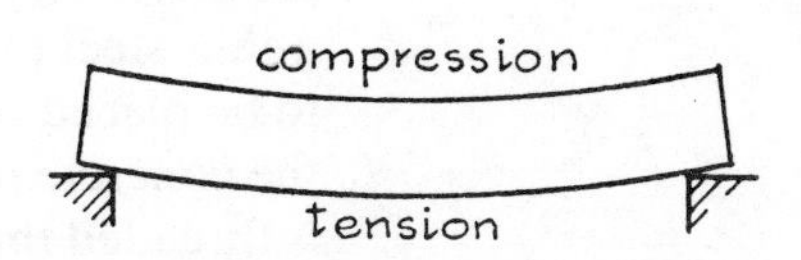

Fig. 7.6

top fibres due to the weight of the beam (Fig. 7.6) will be cancelled by the tensile stress caused by the prestress [Fig. 7.5 (*b*)]. Similarly, the tension in the bottom fibres caused by the beam weight (Fig. 7.6) will be more than cancelled by the compression induced by the prestressing force, so that the bottom of the beam will have a considerable compressive stress [Fig. 7.7 (*a*)].

The beam is now ready to receive its external imposed loads, which, as demonstrated in Chapter 5, cause compressive stresses in the top fibres and tensile stresses in the bottom fibres. The final loaded beam, therefore, has compressive stress in the top fibre and no stress in the bottom fibre [Fig. 7.7 (*b*)].

To sum up: calculations ensure that in a prestressed concrete beam fully loaded the stresses are as follows.

Top fibre. A considerable compressive stress (which, however, is not greater than the safe stress for the concrete) as a result of compressive stress due to dead weight of beam less tensile stress due to the prestressing force plus compressive stress due to the applied or imposed loads.

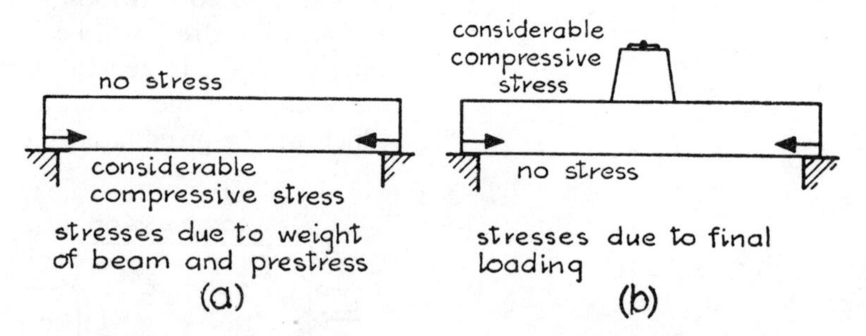

Fig. 7.7

Bottom fibre. Zero stress (or very small tensile stress which the concrete can safely resist) as a result of tensile stress due to the dead weight of the beam less compressive stress due to the prestressing force plus tensile stress due to the final imposed loads.

In a prestressed concrete beam, therefore, the concrete is in compression throughout the whole (or practically the whole) depth of the beam, and is therefore doing useful work in resisting bending moments. Compare with the ordinary reinforced-concrete beam, where about half of the concrete (in the tension zone) is ineffective for resisting bending.

Post-tensioning

The method of prestressing described above is known as the *post-tensioning* or *end anchorage* method because the wires are tensioned after the concrete has hardened. Two alternative methods are used. In the first, the cable (usually a bundle of wires twisted into a single strand) is placed in the beam mould and protected from the concrete during the casting process by means of a thin steel or rubber sheath. The second method requires a duct to be placed in the beam and the concrete poured around it. When the concrete has hardened and gained sufficient strength the cable is threaded through the duct, tensioned and finally anchored to the ends of the beam. It is then usual to force cement grout into the cable duct, not to transfer force from cable to concrete but to protect the cables against corrosion. The transfer of prestressing force takes place at the anchorages in the ends of the beam. Post-tensioning is used for large beams and for site work, and the cables are usually curved as in Fig. 7.8.

The maximum bending stresses due to the weight of the beam and the superimposed load are (for a simply supported beam) at

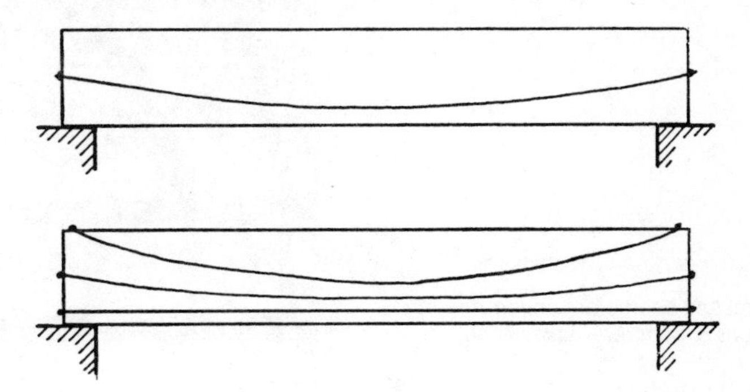

Fig. 7.8

midspan. Prestressing must counteract these bending stresses so that the final stresses are compressive in the top fibre and zero or near zero at the bottom fibre. If the cables were maintained straight near the bottom of the beam for their whole length, tensile stresses would develop in the top fibres near the beam supports since there are no bending stresses to counteract the prestress. (Bending stresses due to the beam weight and superimposed load

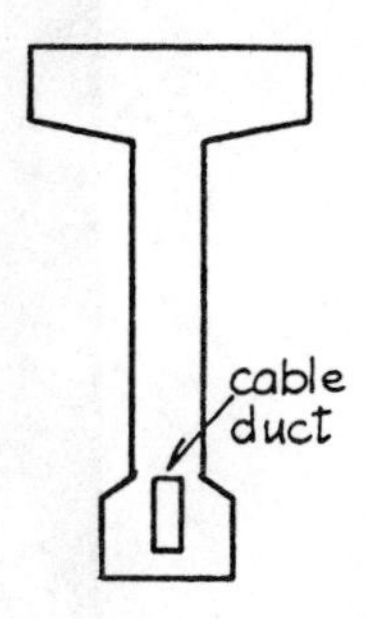

Fig. 7.9

become less towards the ends of the beam and are zero at the extreme ends.) The curving of the cables ensures that compressive and tensile stresses due to the prestressing force balance exactly the stresses due to the bending moments at any point of the beam.

Because prestressing increases the resistance to shear forces compared with ordinary reinforced-concrete beams, it is possible to use shapes (Fig. 7.9) in prestressed concrete which would normally not be practicable with ordinary reinforced concrete.

The adequate anchorage of the wires at the ends of the beam is of paramount importance in post-tensioned work and several methods are in use.

Prestressed beams at London Airport
(*Cement & Concrete Association*)

Freyssinet jack and anchorages
(*Cement & Concrete Association*)

Pretensioned or fully bonded method of prestressing

In this method the wires are stretched and thus put into tension before the concrete is cast into the mould. The concrete is in intimate contact with the steel for the whole length of the beam, and when the concrete is hard the stretching force is removed. The steel tries to revert to its original shorter length but cannot do so because of the friction or bond between the steel and the concrete. Prestressing forces are thus induced as shown by arrows in Fig. 7.10. This method is used principally for units such as concrete floor joists and railway sleepers, mass-produced in a factory.

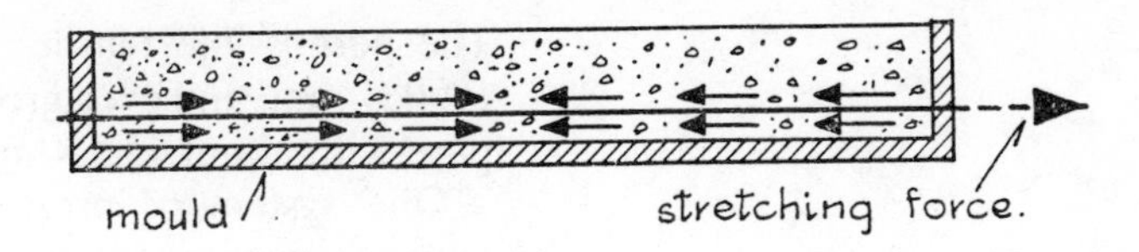

Fig. 7.10

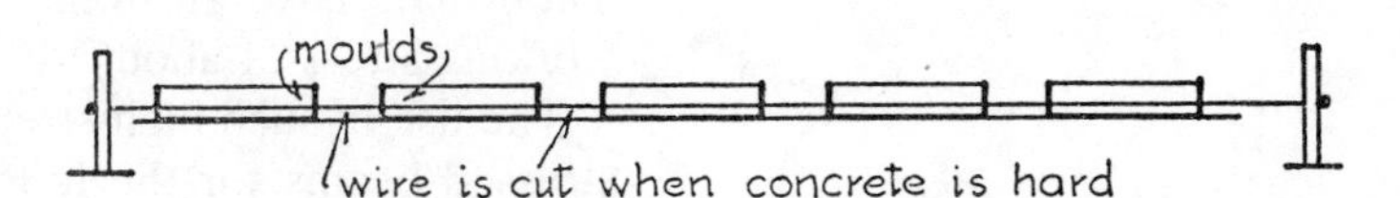

Fig. 7.11

For such small units the *long line* process is normally used, the wires being held in tension at two points a considerable distance apart (Fig. 7.11). A series of moulds are placed around the wires and the concrete is cast. When the concrete is hard the wire between the beams is cut and a number of prestressed beams are thus produced. Larger beams and overhead transmission poles, telegraph poles, etc., would be cast in their individual moulds.

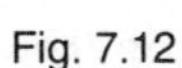

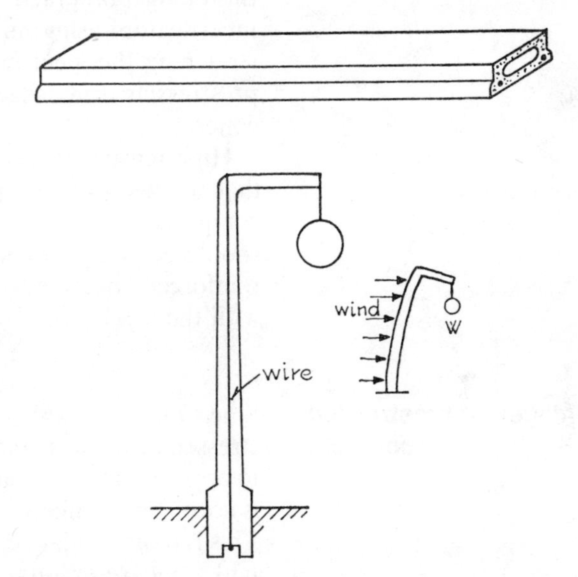

Fig. 7.12

Fig. 7.13

An example of a floor joist produced by the pretensioned process is given in Fig. 7.12.

The principle of prestressing as applied to transmission poles and lamp-posts is illustrated in Fig. 7.13. The action of the wind and the weight, *W*, would cause tension in the vertical cantilever portion. This tension can be counteracted by suitable prestressing, thus enabling smaller and more graceful members to be used than would otherwise be possible.

Prestressing steel

Cold-drawn steel wires with high tensile strengths are used, the diameter of the wire ranging from 2 to 7 mm. The smaller-diameter wires are used in the pretensioning process, and wires of 5 and 7 mm diameter are common in the post-tensioned method.

The permissible tensile stress for such steel is frequently about 1000 N/mm^2 and even greater for the smaller-diameter wires (compare with about 150 N/mm^2 for ordinary mild steel).

One system of prestressing uses special bars of alloy steel (silicon–manganese) from 12·5 mm up to 28·5 mm diameter. These bars have an ultimate stress of about 1100 N/mm^2 and a working stress of about 610 N/mm^2.

The use of such high-strength wires and bars is essential in prestressed beams for the following reasons. Ordinary mild steel is unsuitable because its relatively low strength limits the amount of extension that can be obtained. If this were used in a prestressed beam, part of the small elongation would be lost because of shrinkage of the concrete, and most of the remainder of the elongation would disappear owing to creep in the concrete under load (for example, getting shorter under a sustained compressive load). The compressive prestress in concrete depends on the force exerted by the steel, which in turn depends on the elongation of the steel. If this elongation practically disappears the beam will fail, and early investigators using mild steel found that, within a few weeks of stretching the steel in their beams, practically the whole of the prestressing force disappeared due principally to creep of the concrete.

High-tensile steel can, however, be extended to such an extent that the loss in elongation due to shrinkage and creep is small by comparison and can be allowed for in calculations. The concrete should be of higher quality and strength than is normal in ordinary reinforced concrete to reduce creep effects and improve bonding with the steel.

Advantages of prestressed concrete

Since there are no (or only very small) tensile stresses in a prestressed beam, it is practically impossible for small cracks or fissures to appear under normal loading conditions; thus corrosion of steel is normally less likely to occur than in ordinary reinforced concrete. This is of particular advantage in water-retaining structures such as tanks and reservoirs. Prestressed concrete is more suitable for large-span beams than ordinary reinforced concrete, because the weight of the beam itself has little or no effect when calculations are made for its size. In fact, the beam may be said to 'carry itself,' since its weight is 'cancelled out' by the prestressing. In ordinary reinforced concrete, a large span requires a large beam, and the weight of the concrete forms a high proportion of the total load used for calculating the size of the beam. More graceful structures are possible with prestressed concrete, and less concrete and steel are required than in ordinary reinforced-concrete construction, but the materials are more costly and expert supervision is essential. The deflexions of prestressed beams are usually small, and there is a high resistance to fatigue. If required, a beam can be made by precasting small units as illustrated in Fig. 7.2.

8 The Truss

In solid timber, steel and reinforced-concrete beams it has been shown that the maximum compressive stress occurs in the top fibres (for a beam simply supported at each end), and the maximum tensile stress at the bottom fibres. It is therefore an advantage to concentrate most of the material in the top and bottom fibres of the beam. It has also been demonstrated that there are vertical and horizontal shear stresses which give rise to diagonal tensile and compressive stresses.

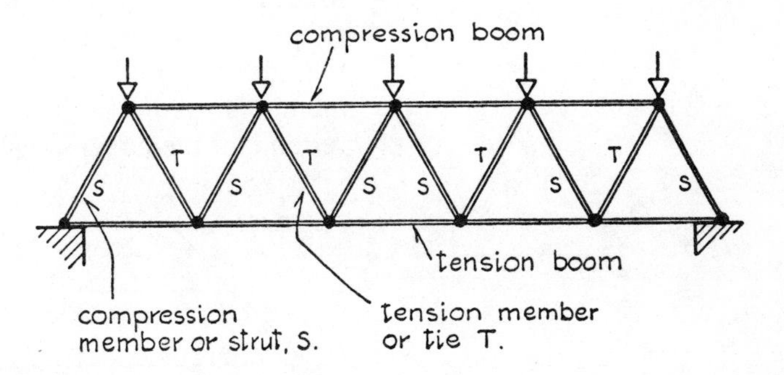

Fig. 8.1

Instead of having a solid continuous web, a beam built up of separate members jointed together can be used (Fig. 8.1). This type of structure is particularly suitable for large spans for which it would be impossible to obtain 'one-piece' girders of sufficient depth and length.

The members forming the compression boom act in a similar manner to the top flange of an I-section beam, and the tension boom acts as the bottom flange. The diagonal tensile and compressive stresses are taken by the internal members. Note that the ties run in the same direction as the inclined bars used to resist shear in a reinforced-concrete beam.

If all the joints are hinged and the loads are applied at the joints, as in Fig. 8.1, there is no bending in any of the individual members and the resulting structure is called a truss. When the truss is triangulated correctly (*see* below) the stresses in all members can be calculated by the three laws of statics given in Chapter 3 or obtained graphically by the polygon of forces. The same principle applies to roof trusses as shown in Fig. 8.2.

A correctly triangulated truss has just sufficient members to prevent the truss from being unstable. In such 'perfect' trusses the number of members equals twice the number of joints, minus three. For example, in Fig. 8.1 the number of members is 19 and the number of joints is 11. $(2 \times 11) - 3 = 19$. In Fig. 8.2 there are 27 members and 15 joints. $(2 \times 15) - 3 = 27$.

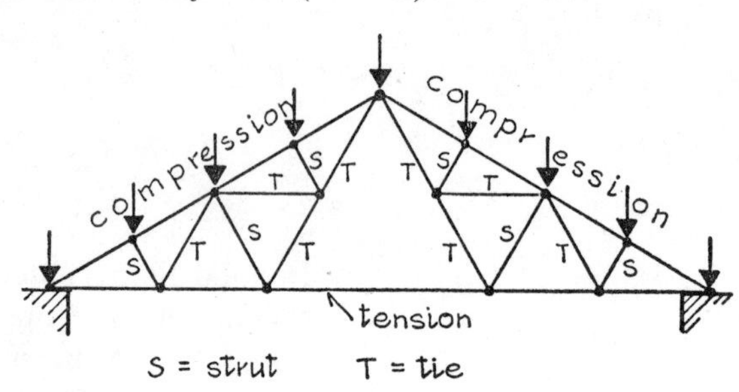

Fig. 8.2

It is possible to construct trusses with a smaller or greater number of members than is given by the above rule. The truss in Fig. 8.3 (*a*) is said to be 'deficient,' and with hinged joints it would not be stable and would distort under load. Such trusses, when used, must have rigid or semi-rigid joints and would then be more properly called a frame. When a truss contains more members than are required for a 'perfect' truss it is said to have redundant members. An example is given in Fig. 8.3 (*b*), where the redundant members are shown by dotted lines.

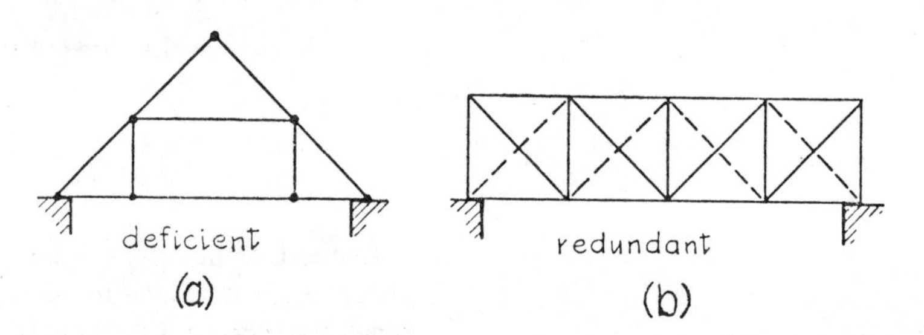

Fig. 8.3

'Redundant' does not mean that the extra members are of no value; it means that more members are provided than the minimum number necessary for stability. Such a truss is actually stronger and stiffer than a 'perfect' truss but more difficult to calculate. It is said to be 'statically indeterminate,' since the forces in the members cannot be determined merely by the application of the three laws of equilibrium.

Many timber bridges and roofs made up of jointed members have been constructed over the ages, but until the 19th century little was known about the nature and direction of the forces in the various members, although Andrea Palladio (1518–80) gives an illustration of a truss which is correctly triangulated (Fig. 8.4). It appears that he understood which members were in compression and which in tension, but there is no evidence that he had any idea of the magnitude of the forces.

The metal truss or lattice girder preceded the rolled-steel beam in Britain. Open-webbed cast-iron beams were used in the construction of the 1851 Exhibition building. A wrought-iron roof truss of 12·2 m span was used for a railway station in London in 1839.

During the 19th century timber- and metal-truss bridges were developed on a wide scale in America, and failures were not uncommon. The American, Whipple, was one of the first investigators to deal with the analysis of forces in the members of a trussed girder.

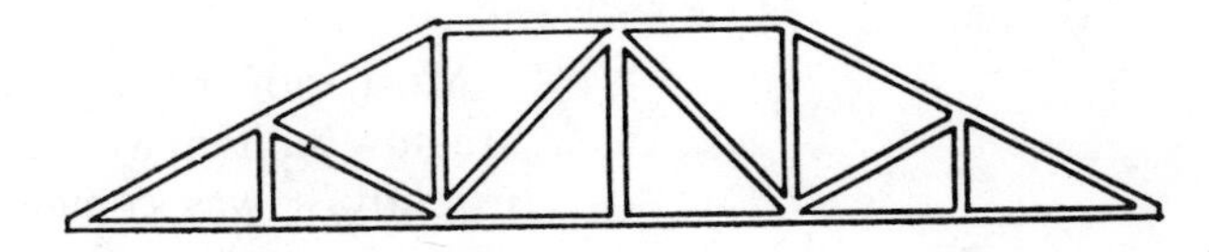

Fig. 8.4

Several well-known types of lattice girder were first developed in America, such as the N or Pratt truss (designed by T. W. Pratt), the Whipple–Murphy truss and the Howe truss (Fig. 8.5). Warren girders of wrought iron were used in the Crumlin (Monmouthshire) viaduct, opened in 1857.

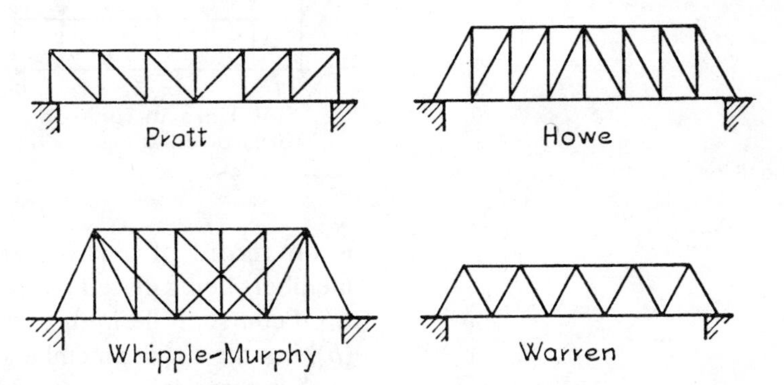

Fig. 8.5

W. J. M. Rankine (1820–72), Professor at Glasgow University for many years, made use of graphical methods (*see* Chapter 3) for determining the forces in the members of trussed structures. Graphical methods for dealing with trusses were also devised by Carl Culman (1821–81), who published his work on graphic statics in 1864.

Robert Bow was responsible in 1873 for a simple system of notation whereby the various loads acting on a truss and the members of the truss itself could be easily lettered and identified on a scale drawing made to determine the stresses in the truss members. This method of lettering is known to students of architecture and engineering as *Bow's notation.*

The early designers' inadequate knowledge of the directions of the forces or stresses in the members of a trussed girder is shown by a test made on a wrought-iron lattice girder designed in connexion with the Dublin Exhibition of 1853. In this test, flat bars used for members sloping towards the supports (similar to members 1 and 2 in Fig. 8.6) buckled, and had it not been for the test the engineers would probably not have realized that these members were in compression.

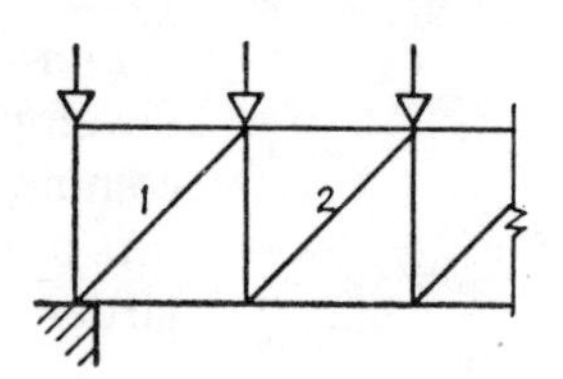

Fig. 8.6

Most roof trusses and light lattice girders for supporting roofs are now built up with lengths of steel angle sections, but until quite recently it was common practice to use flat steel bars for the tension members. Such flat bars can be very strong in tension but weak in compression because of the tendency of long slender members to fail by buckling (*see* Chapter 9). At least one example

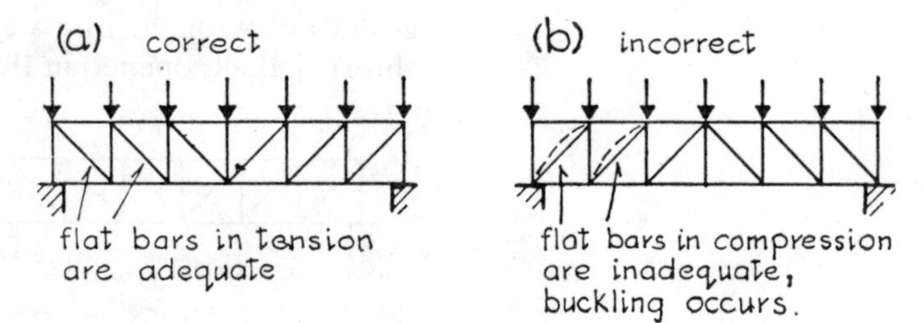

Fig. 8.7

is on record of a lattice roof girder with flat bars as tension members being placed upside down in the building [Fig. 8.7 (*b*)].

Of course, if desired, a truss could be constructed as in Fig. 8.7 (*b*), but the sloping members would have to be stouter to resist the buckling tendency due to compression. Even so, the truss shown in Fig. 8.7 (*a*) is better, because the shorter members are in compression (length being important in compression) and the longer web members (the diagonals) are in tension (length not being important in tension).

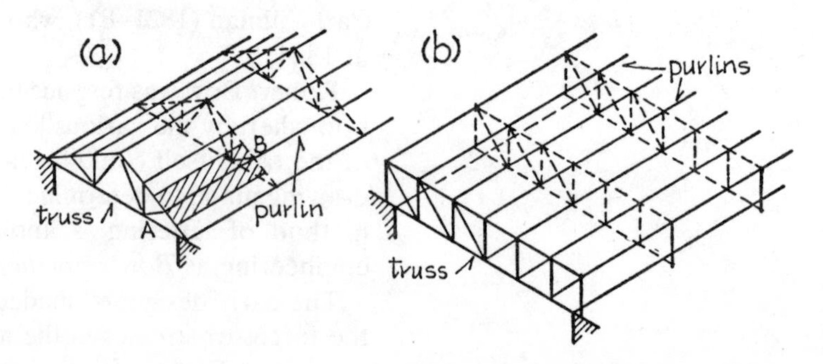

Fig. 8.8

When designing roof trusses and lattice girders they are treated as 'one-plane' (i.e. two-dimensional) structures, whereas many types of structures should be treated as three-dimensional (or space) structures. Space structures are dealt with in Chapters 14–16, where comparisons will be made between two-dimensional and three-dimensional trusses.

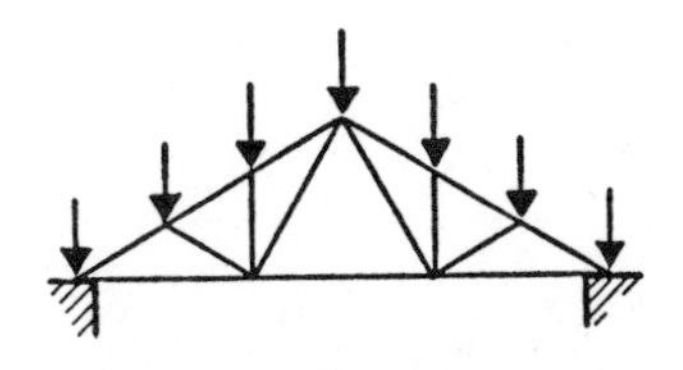

Fig. 8.9

Whether the roof has sloping or horizontal top members (Fig. 8.8), the method of design is as follows. First, the *purlins* (timber or steel beams) are designed either as simply supported beams spanning from truss to truss, or as continuous beams supported by a number of trusses. For example, purlin *AB* in Fig. 8.8. (*a*) would be designed to support the load from an area of roof covering shown shaded (and from pressure due to wind). The load from each purlin is then transmitted to the trusses to give point loading as shown in Fig. 8.9, and the stresses in the ties and struts are determined. Any stiffening effect on the structure the roof covering may have is ignored. Thus, the purlin is first designed as a one-plane structure and then the truss is also designed as a one-plane structure.

Lattice girders in factory buildings (*British Constructional Steelwork Association*)

9 The Column

The column is essentially a compression member, but the manner in which a column tends to fail and the amount of the load which causes failure depend on—

(1) The material of which the column is made.
(2) The slenderness of the column.

The first point is obvious. The example, other conditions (size, height, etc.) being identical, a steel column is capable of supporting a much greater load than a timber column.

Slenderness

If a short block of material, as indicated in Fig. 9.1, is gradually loaded, failure will not occur until the ultimate resistance to crushing of the material is reached. The shape of the cross-section is not important. If all the blocks of Fig. 9.1 have equal areas of cross-section, and if the loads are applied in line with the centre-of-gravity axis, the crushing loads will be practically equal (assuming all blocks to be made from the same material). If, however, a column is very long and slender, failure will occur due to buckling at a much lower load than would cause failure in a short column of equal cross-section.

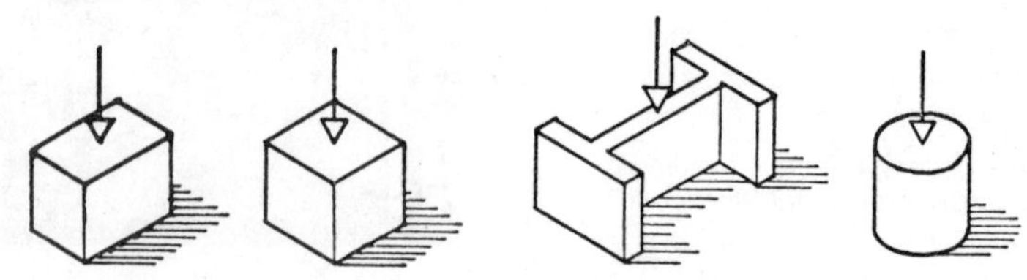

Fig. 9.1

This can be demonstrated quite easily with a light cane (Fig. 9.2). At a very small load the cane will buckle as shown, and if the load is increased slightly the cane will snap. Buckling, of course, is the same as bending, and it was stated in Chapter 5 that the shape of a beam (i.e. the arrangement of the material with respect to the axis of bending) is very important.

The shape of a column is also very important. For example, a sheet of cardboard has practically no strength as a column, but if bent to form an angle section or other shapes, as shown in Fig. 9.3, it is capable of supporting a load.

It follows that by intelligent use of available material economical columns can be constructed.

Although the buckling of a column can be compared with the bending of a beam, there is an important difference. The designer can choose the axis about which a beam bends but normally has not a similar choice with columns.

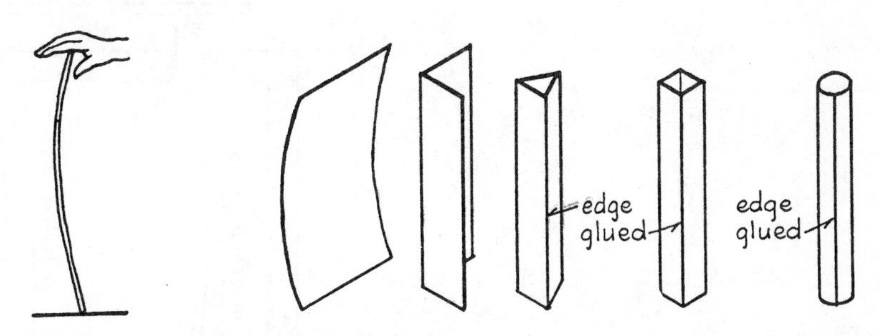

Fig. 9.2 Fig. 9.3

As stated in Chapter 5, beams are stronger when bending about axis *XX* (beams (*a*) and (*b*) of Fig. 9.4) than when bending about axis *YY (beams* (*c*) and (*d*)), and the designer would usually arrange beams to obtain their greatest resistances to bending. When, however, these sections are used as columns the designer has no choice and cannot dictate the manner of bending. The columns will take the line of least resistance and under axial load will buckle about axis *YY*.

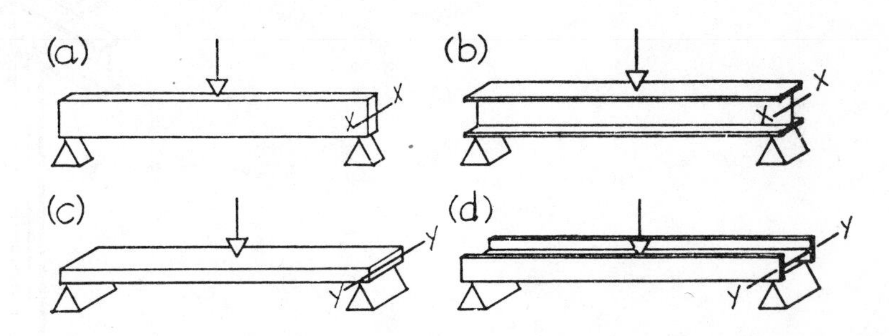

Fig. 9.4

Effective height of columns

It was stated in Chapter 3 (Fig. 3.18) that beams fixed at both ends behave differently from beams freely supported at both ends and, other conditions being identical, are capable of supporting more load. In a similar manner, the way in which the top and bottom of a column are held or secured affects considerably its load-carrying capacity. Figure 9.5 shows three identical strips of steel and the loads required to cause buckling. Column formula and tables of permissible stresses for various slenderness ratios are based on column type (*a*) which is hinged at both ends. Column (*b*) buckles in a different manner from column (*a*), and in fact behaves as a shorter column of type (*a*), the length of curve *CD* corresponding to the full curve AB of column (a). In calculations relating to column (b), for example, the slenderness ratio can be expressed as

$$\frac{\text{Effective height}}{\text{Least radius of gyration}} = \frac{CD}{r_{min}}$$

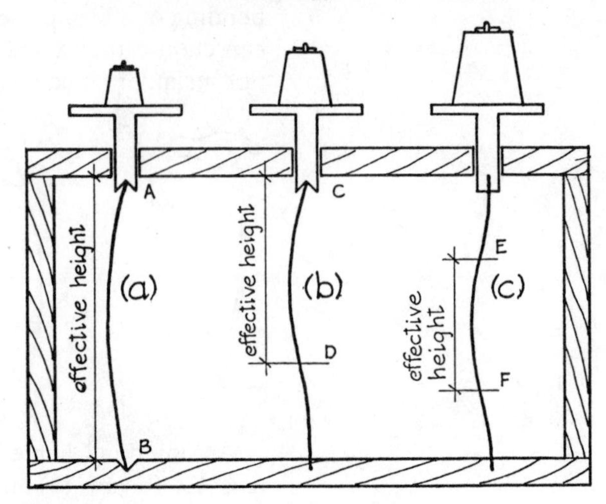

Fig. 9.5

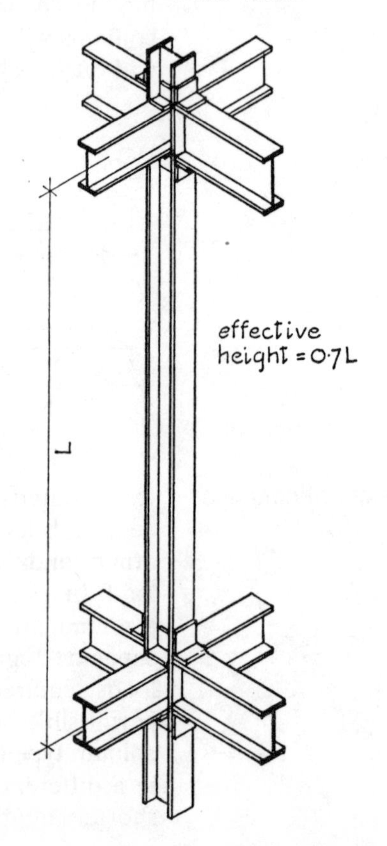

Fig. 9.6

where the least radius of gyration is a property of the column section that represents its resistance to bending about the *YY* axis. See section on steel columns for further explanation of r_{min}. Similarly, the length of curve *EF* of column (*c*) compares with the full curve *AB* of column (*a*); therefore the slenderness ratio is EF/r_{min}.

Note that the better the column is secured at both ends the greater is the load it can carry. Perfect end support conditions, such as those in Fig. 9.5, cannot be assumed in practice, so the designer has to use his discretion in estimating the effective height of a column. (As stated in Chapter 1, structural engineering is not an exact science, and mathematics is merely a tool which has to be used with knowledge, skill and judgment.) B.S. 449 gives a guide to the effective lengths to be assumed for various ways in which the ends of columns are secured. For example, an internal column supporting four beams (Fig. 9.6) can be assumed to have an effective height of 0·7 times its actual height.

Timber columns

In Fig. 9.7 (*a*) the dimension *b* is important in defining slenderness, and for timber posts, which are generally rectangular in cross-section, the slenderness is often expressed as the ratio

$$\frac{\text{Effective height of post}}{\text{Least width of post}} \text{ i.e. } \frac{l}{b}$$

Codes of practice and building by-laws give tables of permissible stresses for various values of *slenderness ratio*. For example, if the effective height is 2 m (2000 mm) and the post is 225 mm by 100

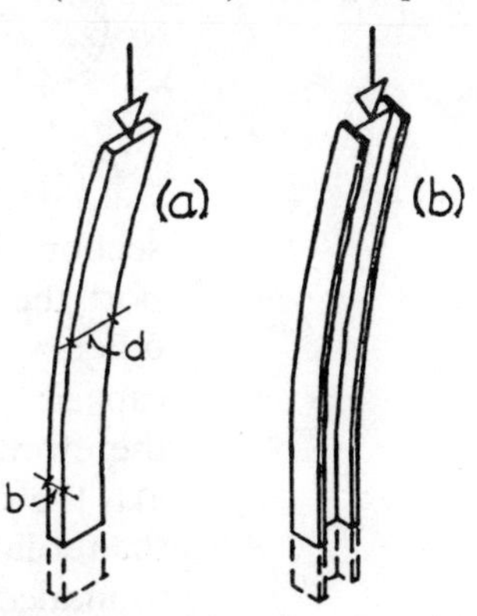

Fig. 9.7

mm in cross-section, the slenderness ratio is 2000/100 = 20. From code tables the permissible stress for this ratio for pine or spruce is 4·35 N/mm², so the safe axial load for the post is 97·875 kN (stress multiplied by area). If the same cross-section post has an effective height of 4 m (4000 mm), l/b = 4000/100 = 40, and the permissible stress for this slenderness ratio is 1·75 N/mm², giving a safe axial load of 39·375 kN.

It should now be obvious that, for rectangular cross-section columns or posts, a section with equal sides (i.e. a square cross-section) makes the best use of the material, since the resistances to bending about axes XX and YY are equal. A post 150 mm square has the same amount of material as a 225 mm by 100 mm post as quoted above. For a height of 2000 mm, l/b = 2000/150 = 13·33. The permissible stress from the same tables as used for the 225 mm by 100 mm post is 4·85 N/mm^2 giving a safe load of 22500 × 4·85, i.e. 109·125 kN. Similarly, when l = 4000 mm, l/b = 4000/150 = 26·66, the permissible stress is 3·30 N/mm^2, and the safe axial load (22500 × 3·30) is 74·25 kN (compare with the safe load for the 225 mm by 100 mm post of equal height, i.e. 39·375 kN).

For a timber column, therefore, a square section is more economical than a rectangular column having an equal amount of material, whereas for a beam (provided that bending is about axis XX) the rectangular section is the more economical.

Steel columns

A similar argument applies to I-sections. A broad-flange section in which the width of the flange is equal to the depth is more efficient as a column than a narrow-flange deep section having an equal amount of material. (A narrow deep section is more efficient as a beam.)

Accepting the fact that the square column is more efficient, it is, however, still true (as in beams) that the further away the material is from the neutral axis, the greater is the resistance offered to buckling. All the shapes of Fig. 9.8 have equal areas of cross-

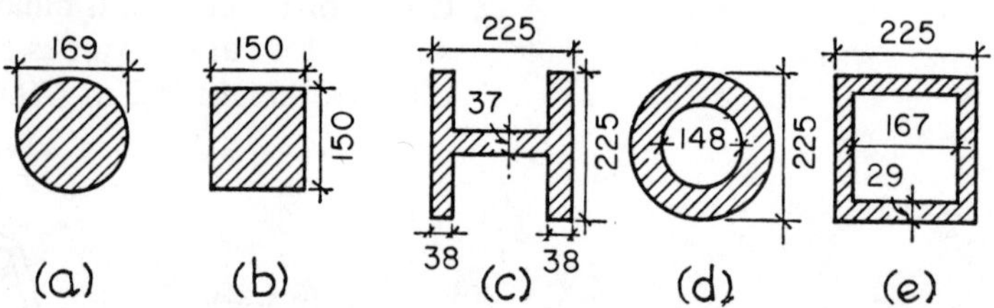

Fig. 9.8

section (22,500 mm^2). The column which would be allowed to support the greatest axial load (for any height) is column (e). The other sections, arranged in descending order of load-carrying capacity, are (d), (c), (b) and (a). For a given amount of material, therefore, the greater the distance it is placed from the buckling axis the stronger is the column. Hollow sections are more efficient than solid sections, provided that the walls are not made so thin as to induce local crumpling.

In comparing the slendernesses of steel columns of different shapes, and therefore their load-bearing capacities, it should now be obvious that some quantity must be used which takes into account, not only the size of the cross-section (the amount of material), but also its shape (the way in which the material is arranged with respect to the axis about which buckling is most likely to occur). Therefore, slenderness is defined as

$$\frac{\text{Effective height of column}}{\text{Size-shape factor}} \text{ i.e. } \frac{l}{r}$$

The *size-shape factor*, which it is found most convenient to use is called the *radius of gyration*. This is the distance to an imaginary point or axis (i.e. line) at which all the material of the column is assumed to be concentrated with respect to the axis of buckling. At this radius the material has the same resistance to buckling as the actual section, in which the various particles which make up the column are at varying distances from the buckling axis.

The term 'radius of gyration' has been borrowed from the science of *dynamics*, which deals with moving bodies. In rotating flywheels, for example, it is convenient, in order to calculate the total energy, to assume that all the material of the wheel is concentrated into one point at a certain radius from the axis of rotation.

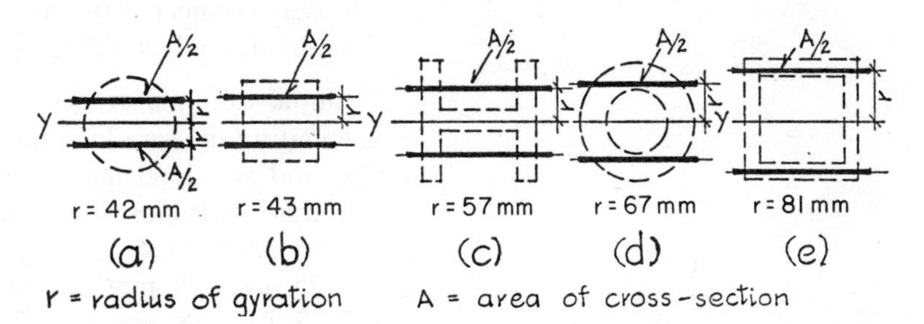

Fig. 9.9

Values of the radii of gyration for different shapes can be obtained by mathematics, and for the columns shown in Fig. 9.8, are given in Fig. 9.9. *YY* is in every case the axis about which buckling would occur (although in (*a*), (*b*), (*d*) and (*e*) the buckling resistances about axes *XX* and *YY* are equal). Assuming that the columns are of ordinary mild steel and that each has an effective height of 3 m, the safe axial loads, using tables of permissible stresses from B.S. 449, 'The Use of Structural Steel in Building,' are given in Table 9.1. It should be remembered that each column has an area of cross-section of 22,500 mm².

Table 9.1

Column	*r*	$\frac{l}{r}$	*Permissible stress*	*Safe axial load*
	mm		*N/mm²*	*MN*
(*a*)	42	71	114	2·57
(*b*)	43	70	115	2·59
(*c*)	57	53	131	2·95
(*d*)	67	45	136	3·06
(*e*)	81	37	140	3·15

Leonardo da Vinci (1452–1519), the great Italian painter, sculptor, architect, scientist, engineer and inventor, was one of the first experimenters on columns. The following quotation from his manuscripts was contained in a lecture on the history of theory of structures by A. A. Fordham, reported in the *Structural Engineer* of May, 1938:

> 'Many little supports held together (in a bundle) are capable of bearing a greater load than if they are separated from each other. Of 1000 such rushes of the same thickness and length which are separated from one another, each one will bend if you stick it upright and load it with a common weight. And if you bind them together with cords so that they touch each other they will be able to carry a weight such that each single rush is in the position of supporting 12 times more weight than formerly. The increase of carrying capacity is entirely dependent upon the firmness of the binding and if loosely connected the total load becomes merely the sum of the loads each can carry separately.'

A simple example of Leonardo's experiment can be obtained by calculating the safe load for four columns of mild steel 25 mm square and say 1000 mm effective height (Fig. 9.8). It can be proved that the radius of gyration of a square cross-section is 0·29 times the length of side; therefore each column has an r value of 0·29 × 25, i.e. 7·25 mm. The slenderness ratio for each column is 1000/7·25, i.e. 138, and from B.S. 449 the permissible stress is 47 N/mm^2. Since each column has an area of 625 mm^2 the safe load for one column is 29·375 kN, and the total safe load for the four columns is 4 × 29·375 kN = 117·5 kN.

Figure 9.10 (*b*) shows one column 50 mm square (or it could be considered as 4 columns 25 mm square tightly bound together). The radius of gyration is 0·29 × 50 = 14·5 mm, and l/r = 1000/14·5 = 69. The permissible stress for a slenderness ratio of 69 is 117 N/mm^2, and since the area of cross-section is 2500 mm^2, the total safe load is 2500 × 117 = 292·5 kN.

As a further example, the safe load for 16 columns 25 mm square acting as shown in Fig. 9.10 (*a*) is 16 times the safe load for one column, i.e. 16 × 29·375 kN = 470 kN.

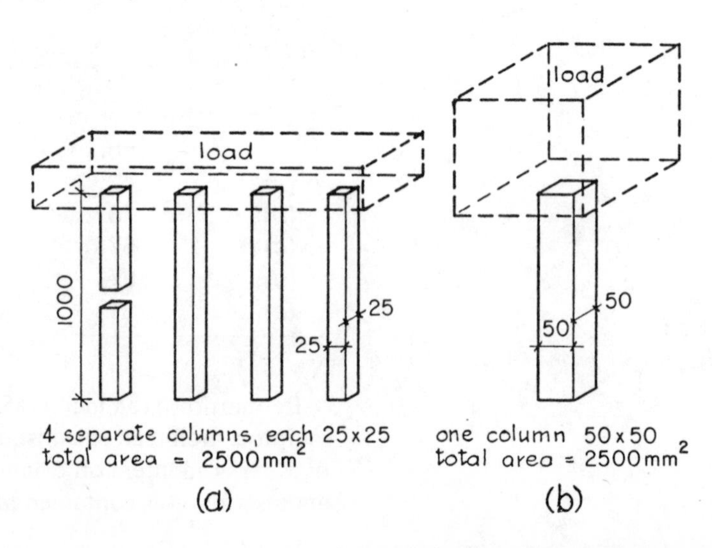

Fig. 9.10

For one column 100 mm square, $l/r = 1000/(0{\cdot}29 \times 100) = 34{\cdot}5$; the permissible stress is 141 N/mm^2; and the safe axial load (area × stress) is 10,000 × 141 = 1410 kN.

For 36 columns 25 mm square and of 1000 mm effective height, the safe load is 36 × 29·375 kN = 1057·5 kN. For a solid column 150 mm square, $l/r = 1000/(0{\cdot}29 \times 150) = 23$; the permissible stress is 146 N/mm^2; and the safe load is 22,500 × 146 = 3285 kN.

These results are summarized in Table 9.2.

Table 9.2

(*a*) No. of steel columns 25 mm square 1000 mm high	(*b*) Total safe load for (*a*)	(*c*) Solid steel columns 1000 mm high	(*d*) Total safe load for (*c*)	Load (*d*) / Load (*b*)
	kN		*kN*	
4	117·5	50 mm square	292·5	2·49
16	470·0	100 mm square	1410·0	3·00
36	1057·5	150 mm square	3285·0	3·11

A famous name in connexion with columns is Leonhard Euler (1707–83), a Swiss mathematician who derived mathematically a formula for the collapsing (or buckling) load of long columns. Unfortunately, Euler's formula depends on assumptions of perfection (perfect straightness, perfect homogeneity) which cannot be realized in an imperfect world, and is therefore not a practical formula for column design. It gives values of collapsing loads which are reasonably accurate for very long slender columns in which the imperfections are not significant, when compared with the high slenderness ratio, but it predicts values which are unsafe for short and intermediate length columns. Failure in this range is affected not only by slenderness of the column but also by the presence of imperfections in the column and the yield or crushing strength of the column material.

Eaton Hodgkinson (1789–1861) derived in 1840 a formula which appears to have been the basis of the well-known *Rankine formula* —still given in textbooks. This formula is derived using the following argument. In a very short stout column in which buckling is impossible, the failing load depends on the crushing strength of the material and the simple formula is: failing load is equal to failing stress for the material multiplied by the area of cross-section of the member. A very long slender column fails by buckling and Euler's formula gives reasonable results for elastic materials. Why not combine these two formulae in such a manner that the resulting single formula can be used for a column of any slenderness ratio? The way in which this was done can be studied in books on strength of materials and theory of structures. Rankine's formula is no longer used for the practical design of columns, having been superseded by more exact formulae based on mathematical and experimental research. It is an interesting fact, however, that Euler's work has been used even in the derivation of the more exact design rules which are the basis of the tables of permissible stresses that are given in various codes of practice. Permissible stresses from B.S. 449 were used in the examples given above.

Steel columns cased in concrete

Most steel columns in buildings have to be protected from a possible fire by a casing of concrete (Fig. 9.11). Unprotected steel, under heat, becomes soft and plastic at a comparatively low temperature, well below the temperature which can be attained in a

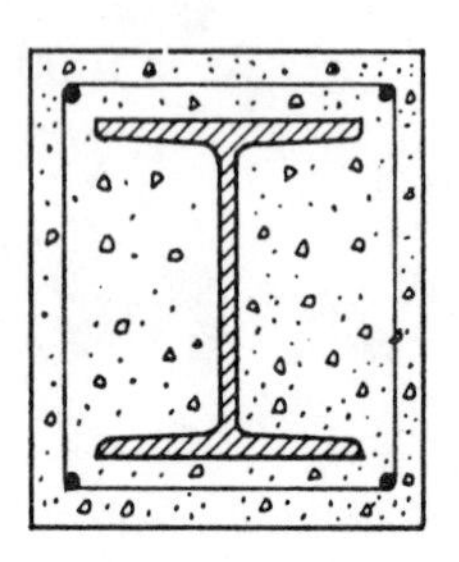

Fig. 9.11

big fire. If certain conditions are complied with, cased columns or stanchions of I-sections can be allowed to support more load than uncased sections because of the strengthening effect of the concrete. A bigger radius of gyration than for the uncased section can be assumed, and the concrete can be allowed to take a stress equal to one-thirtieth of the stress permissible on the steel. Under certain conditions, it is possible for a cased steel column of I-section to support a load up to twice the permissible load for an equivalent uncased section.

Reinforced-concrete columns

The safe load for a reinforced-concrete column of the type shown in Fig. 9.12 is equal to the safe load for the concrete plus the safe load for the longitudinal steel.

Reinforced concrete columns have cross-sections significantly larger than corresponding steel columns and failure due to buckling is not common because the slenderness ratio is usually low. Most codes do not require a reduction in permissible stresses unless the slenderness ratio is greater than 50. A column 300 mm square can therefore have an effective height of up to 4 m before reduction in axial stresses would be required. Most concrete columns in buildings are therefore classified as 'short' for buckling purposes.

The minimum permitted area of cross-section of the longitudinal steel is 0·8 per cent of the total area of cross-section of the column, and the maximum permitted area is 8 per cent. Because of overlapping of bars at certain points, usually at floor levels [*see* Fig.

10.11 (*a*)], the maximum percentage of steel in any one length of column would not in general exceed 4 or 5. Since steel is stronger than concrete, the greater the percentage of steel the more load a given column can support. Alternatively, a smaller section can be used to support a certain load by increasing the area of steel.

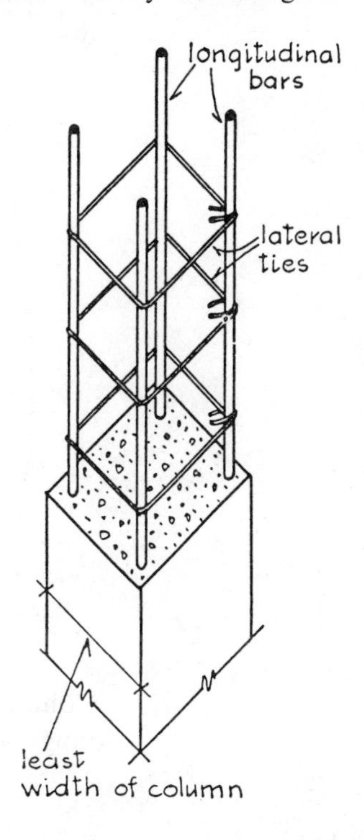

Fig. 9.12

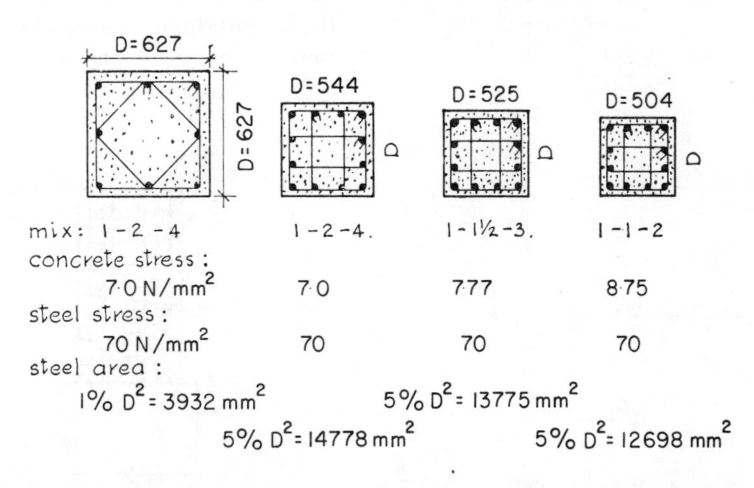

Fig. 9.13

Stronger concrete mixes also affect the load-carrying capacities of reinforced-concrete columns. All the columns (assumed short) of Fig. 9.13 are capable of supporting an axial load of 3 MN. If the smallest possible column of this type is required, the solution is to use the strongest permissible concrete with the greatest permissible amount of steel.

The lateral ties shown in Fig. 9.12 are not assumed to contribute to the strength of the column but are essential—

(1) To prevent the longitudinal bars from buckling and bursting off the comparatively thin layer of concrete surrounding them.

(2) To prevent shearing of the concrete on a diagonal plane (Fig. 9.14).

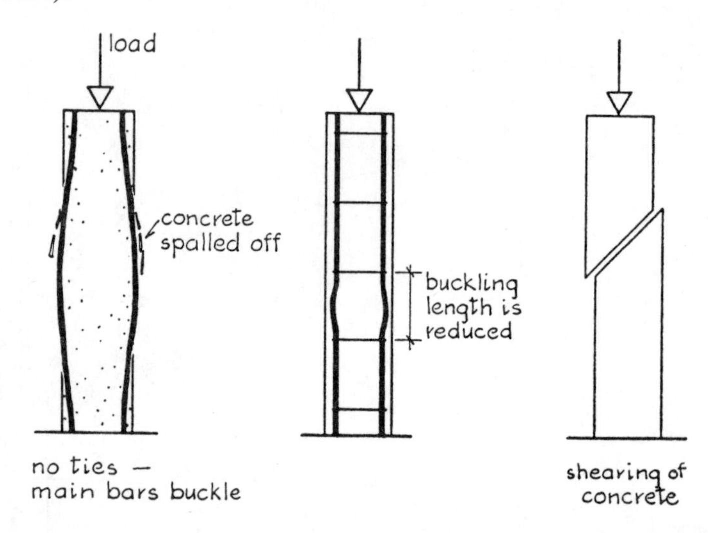

Fig. 9.14

Columns with lateral ties may be of any shape, e.g. square, rectangular, circular, octagonal, etc.

Columns with helical binding

Some columns (usually circular or octagonal in cross-section) have the longitudinal bars enveloped in a cage formed by winding a continuous length of bar round the main bars to form a helix (Fig. 9.15). This closely wound binding is allowed to contribute to the

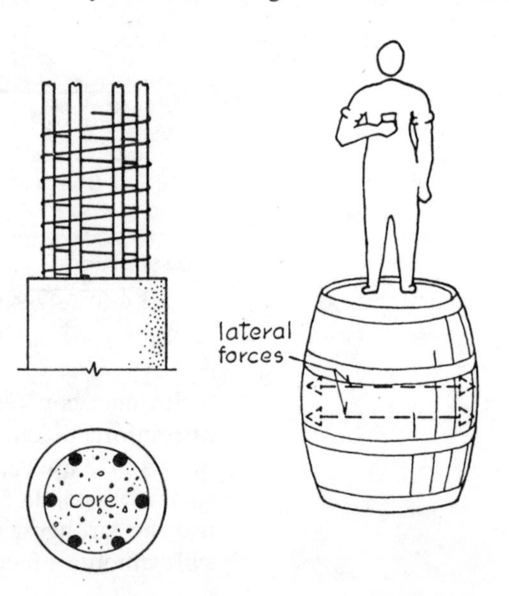

Fig. 9.15

load-carrying capacity of the column. The principle can be compared to the restraining action of steel hoops on a barrel. Assuming the barrel of Fig. 9.15 to be full of earth, the bursting of the barrel due to lateral forces caused by the downward load is prevented by the steel hoops which are put into a state of tensile stress. Similarly, in a reinforced-concrete column, the downward load tends to make the column shorter and fatter. The binding acts like the steel hoops on a barrel or like a woman's corset, and prevents bulging, thus enabling the column to take more load. Like a corset, however, the binding cannot prevent bulging outside the area on which it acts; thus only the core area (Fig. 9.15) is allowed to support vertical load.

The safe load for a helically bound column is therefore equal to the safe load for the concrete in the core plus the safe load for the longitudinal steel bars, plus the safe load taken by the binding. In order that such a column may support more load than an identical column with lateral ties, the loss in load-bearing capacity due to ignoring the concrete outside the core must be more than compensated by the load taken by the binding. In fact, by using the largest possible diameter of bar for the binding and the closest permissible spacing of the 'rings,' a helically bound column can be designed to take an appreciably higher load than a laterally tied column with identical area of cross-section and identical amount of longitudinal steel.

Typical stanchion base for heavy universal column (*British Constructional Steelwork Association*)

It can be appreciated why lateral ties (Fig. 9.12) do not contribute directly to load-carrying capacity if it be remembered that these ties are never closer than about 150 mm and are usually spaced at 300 mm, which is the maximum allowed. It is like a woman trying to restrain the bulges by wrapping round her a few separate lengths of elastic. The turns or rings in a helically bound column are never more than 75 mm apart and the distance may be as small as 40 mm.

Eccentric loads on columns

Axial or concentric loading means that the line of action of the resultant load coincides with the centre-of-gravity line of the column. The stress caused in the material of the column is due to the direct downward action of the load and is called *direct compressive stress.* This stress is uniform at any cross-section of the column and is given by

$$\frac{\text{Load}}{\text{Area of cross-section}} = \frac{W}{A} \text{ newtons per square millimetre}$$

Eccentric means 'off-centre,' and the line of the resultant load does not coincide with the centre-of-gravity line of the column.

Before dealing with 'long' columns it is instructive to consider the behaviour of a short block resting on soft earth (Fig. 9.16). The sinking of the blocks into the soil has been exaggerated in the diagrams; in practice, the sinking would be very small so that for all practical purposes the blocks could be assumed to remain horizontal.

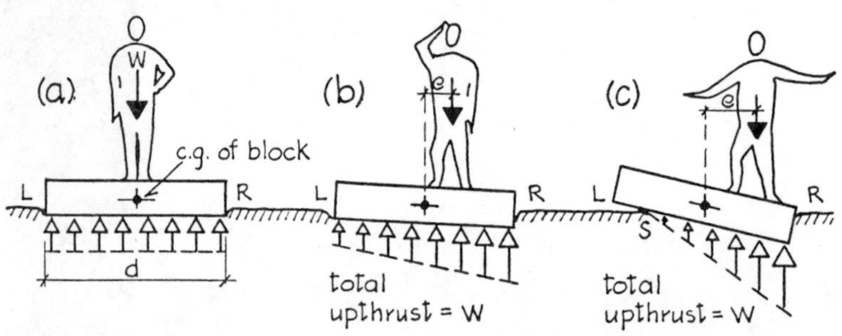

Fig. 9.16

In Fig. 9.16 (*a*) the load is axial and the pressure on the soil is uniform. At (*b*) the load is eccentric and the edge *R* sinks further into the soil than the edge *L*, indicating that there is a greater intensity of stress at *R* than at *L*. Since the total upward reaction of the soil must equal the total downward load, it follows that the stress at *R* is greater than W/A, where *A* is the area of the base of the block, and the stress at *L* is less than W/A. In fact, if the block is rectangular (and ignoring the weight of the block itself), when the *eccentricity* $e = d/6$, the stress at *L* will be zero and the stress at *R* will be $2 \times (W/A)$. If the eccentricity is increased to a greater value than $d/6$ [Fig. 9.16 (*c*)], edge *L* will lift so that all the pressure is concentrated on the length *SR*, giving a much greater stress at *R* than $2 \times W/A$, i.e. more than twice the stress which would be caused by an equal axial load. The increased stresses at edges *R* are due to the moment *We*, and the greater the value of this moment the greater will be the maximum stress in the soil. Since design is based on maximum stresses, eccentric loading is avoided whenever possible.

Figure 9.17 (*a*) shows two loads on a block. The load W_1 due to its moment about the *XX* axis of $W_1 \times e_x$ causes a maximum compressive stress on the soil along the edge *RS*, and the load W_2

causes a maximum compressive stress along edge *LR* (due to moment $W_2 \times e_y$). The point *R*, therefore, is the most highly stressed and the block will dig into the soil at this corner. In Fig. 9.17 (*b*), although there is only one load, it is eccentric about both axes (*XX* and *YY*). The maximum pressure on the soil will be at the corner *L*. This can be demonstrated by standing on a drawing board placed on soft earth or by standing on a raft floating on water.

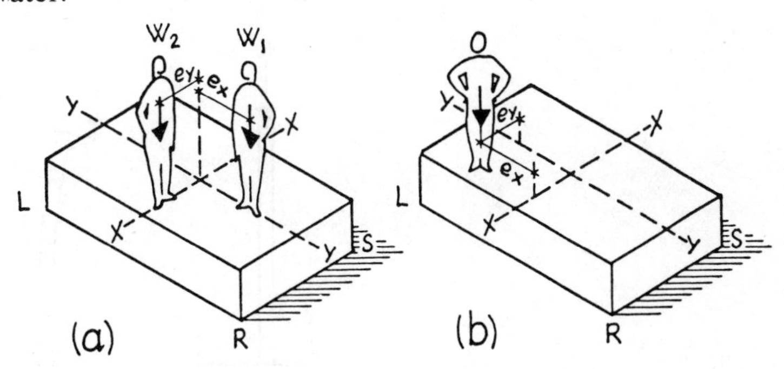

Fig. 9.17

Eccentric loads on 'long' columns

The same principle applies as explained above. An eccentric load, in addition to causing direct compressive stresses due to its downward action, sets up bending stresses. A member carrying an eccentric load is behaving both as a column and as a beam.

In Fig. 9.18, the bending action due to the moment *We* increases the compressive stress on one side of the column and decreases it on the other. If the eccentricity is large enough the tensile stresses caused on one side of the column due to the bending effect of the moment *We* will more than cancel out the compressive stresses due to the direct downward action of the load [Fig. 9.18 (*c*)], and this could be dangerous in piers (i.e. columns) of brittle material such as brick, stone and unreinforced concrete.

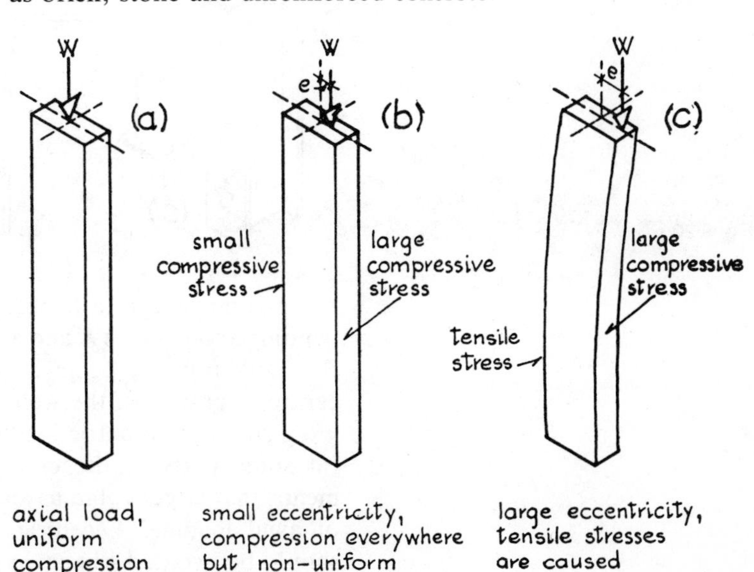

Fig. 9.18

Figure 9.19 gives further examples of eccentric loading and illustrates that, although a column may be supporting several eccentric loads, the resulting effect may be that of an axial load.

Figure 9.20 (*a*) shows one beam connected to a stanchion; there is, therefore, bending about axis *XX*. At (*b*), provided that $W_1 \times e_1 = W_2 \times e_2$, the bending tendency of one moment cancels out that of the other, and the column can be designed for an axial load. Figure 9.20 (*c*) could be a corner column in a building, one beam

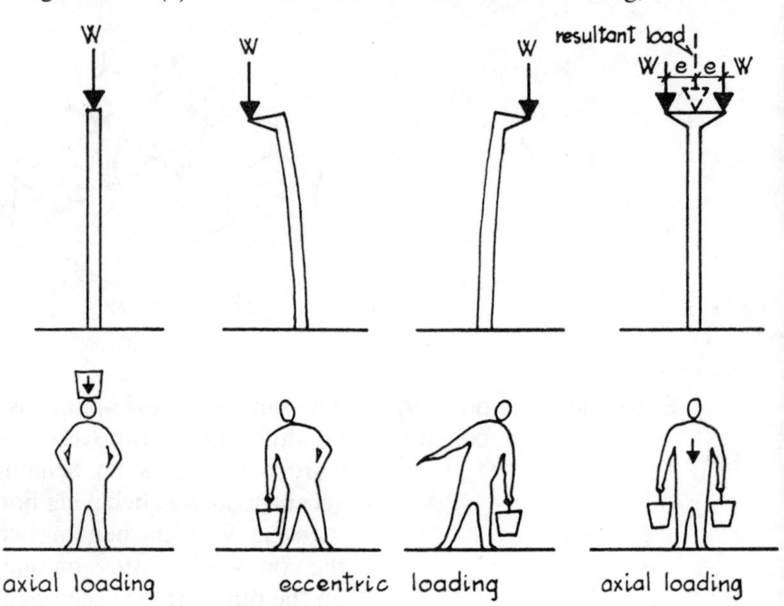

Fig. 9.19

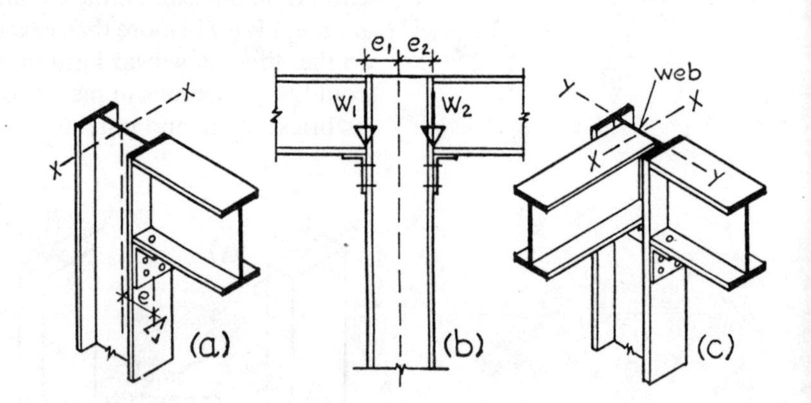

Fig. 9.20

being connected to a flange and one beam to the web. There is bending about axis *XX* and a small moment about axis *YY* because the load from the beam is applied at a small distance from the centre of gravity of the web.

To sum up: eccentric loading causes a higher intensity of stress (at some parts of the cross-section) than axial loading, which means that larger columns are necessary than for an equal amount of axial loading. Eccentric loading is therefore expensive and should be avoided whenever possible.

10 The Connexions

The various members (beams, columns, ties and struts in roof trusses, etc.) must be adequately connected together so that all the loads and forces acting on the structure are safely conducted to the final supporting medium—the ground. Many methods of connecting members together have developed (and are still developing) for different materials and for different requirements which the joints between members have to satisfy.

Timber

Many types of joint such as the dovetail, mortice and tenon, etc., are used for non-load-bearing members such as furniture, doors and window frames. Details are given in books on building construction. Here, however, we are concerned with methods of connecting timber to timber or timber to metal whereby load or stress is transmitted from one member to another.

Most traditional jointing methods require cutting or notching, thus considerably reducing the strength of one or both members. A few examples are given in Fig. 10.1, where (*a*) and (*b*) show the method of 'halving' for connecting two members in line, and (*c*) shows the same method when the two members to be connected are at right angles to each other. (*d*) is another method for connecting two members at right angles. All these joints should be made where the bending moment is zero or nearly so. (*e*) represents a floor joist A supported by another joist B. The bending moment at the end of A is zero but the shear force is a maximum. The depth X should therefore be sufficient to resist the shear stresses. Some traditional joints involve the use of metal splice plates and bolts as shown at (*f*).

In addition to jointing by notching, other methods of joining timber members are: adhesives; nails, screws and bolts; metal connectors in conjunction with bolts.

Adhesives

Adhesives of organic origin such as hide glue and casein have been used for many years, but because these are susceptible to moisture with consequent loss of strength and are also liable under damp

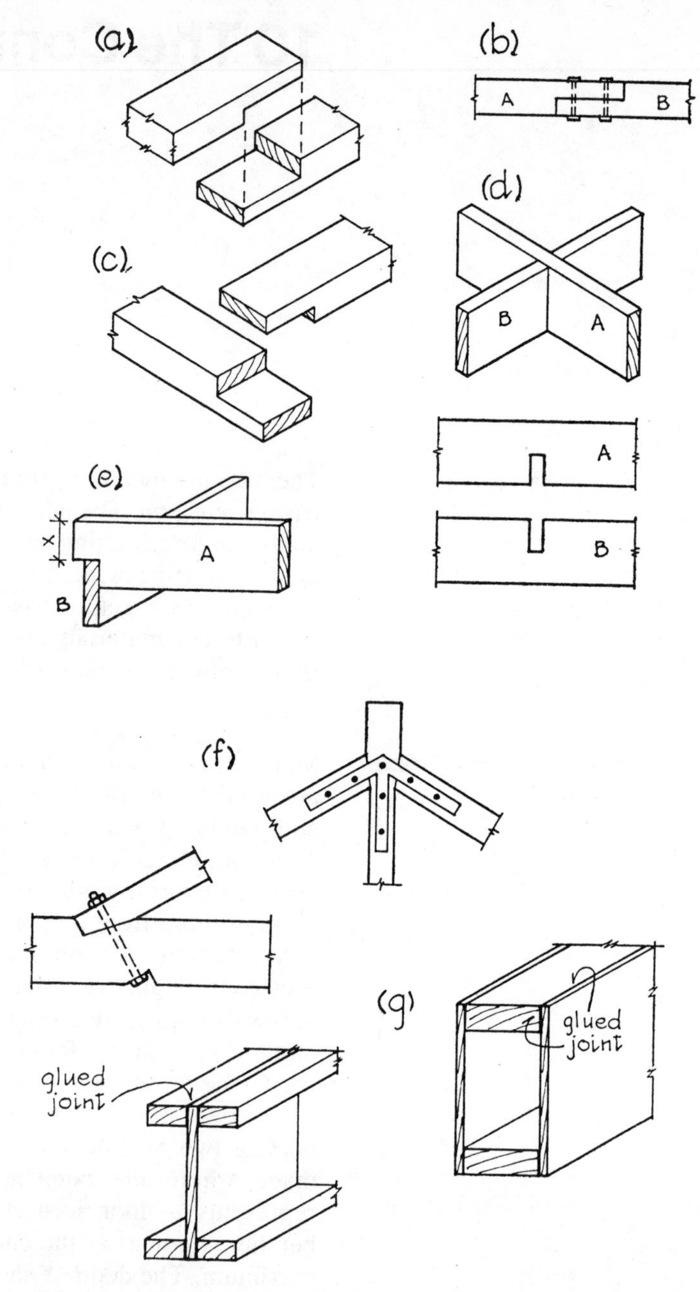

Fig. 10.1

conditions to attack by moulds or bacteria, they are generally suitable only for non-load-bearing and internal woodwork. Cold-setting adhesives of the synthetic-resin type are not susceptible to the type of attack mentioned above, and most codes now allow resin-type adhesives to be used for all conditions of exposure.

Glued joints should normally be allowed to be stressed only in shear along the plane of the glue line, and the permissible stress is equal to that for the timber in shear parallel to the grain. The jointing material is thus considered to be of equal strength to the timber. Glued joints are used for constructing beams of I or box section as shown in Fig. 10.1 (*g*). Laminated beams and arches are possible by squeezing together under considerable pressure layers of timber coated with glue. The availability of reliable glues and adhesives and the improvement of assembly techniques has enabled the recent construction of several large and impressive timber structures.

Nails, screws and bolts

Joints can be made very simply with these connectors. By driving nails into holes pre-bored to a diameter slightly less than the diameter of the nail a small increase in holding power results. The efficiency of an ordinary nailed, screwed or bolted joint may be as low as 15 per cent, i.e. the joint develops only about 15 per cent of the strength of the timber members connected together. This low efficiency is due mainly to the low shear strength of timber parallel to the grain [Fig. 10.2 (*a*)] and the manner in which the nail or bolt bears or bites into the timber [Fig. 10.2 (*b*)].

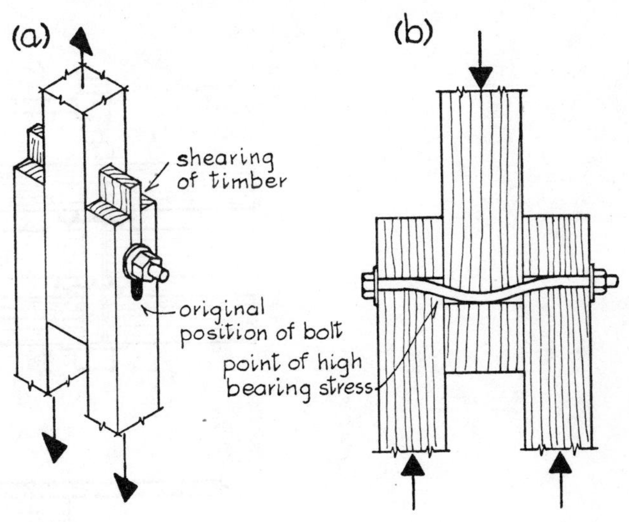

Fig. 10.2

Timber connectors

Timber connectors consisting of metal rings or discs were used in Europe before 1914 but were not developed and used extensively until the Second World War. The greatest development was in the United States, where, owing to the shortage of steel, timber structures were extensively used. Timber connectors increase the efficiency of bolted joints since their positions at the points of maximum stress (*see* Fig. 10.2) give a much larger area of metal in contact with the timber.

Many patents have been taken out and there are many types of connector now in use, but only one, the *Bulldog connector* (Fig. 10.3), will be illustrated. This should be sufficient to give an idea of the function of this comparatively modern method of constructing joints in timber.

For the connexion of several timber members, holes are bored through the assembled members and the toothed connectors put in place [Fig. 10.4 (*a*)]. A high-tensile steel rod threaded at both ends is connected up and tightened, thus forcing the teeth of the connectors to bite into the timber. When the teeth are fully embedded the high-tensile steel rod is withdrawn and replaced by the permanent fastening, which is usually a black bolt (described on

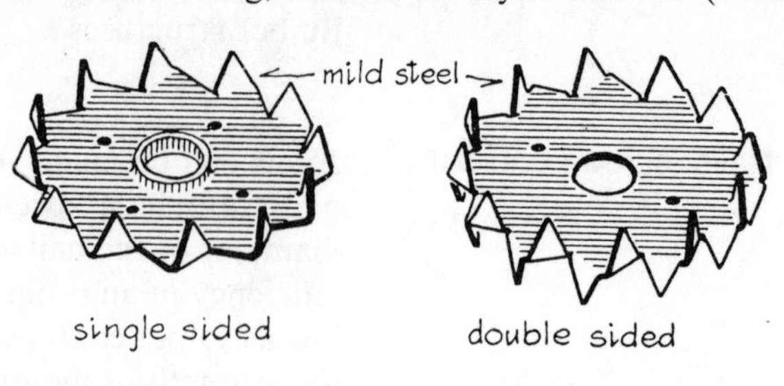

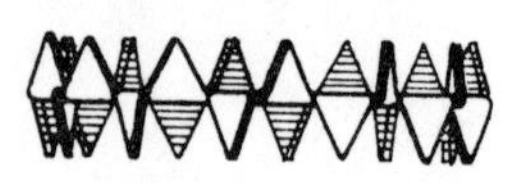

Fig. 10.3 Bulldog connectors

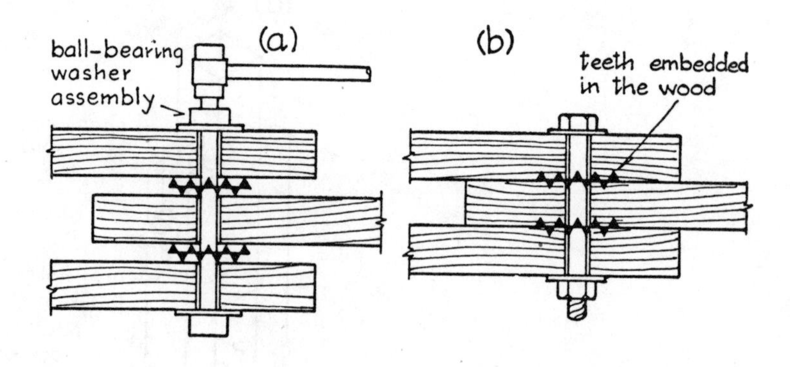

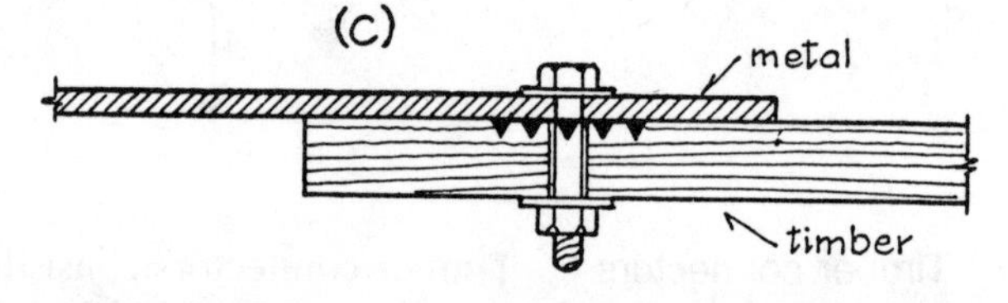

Fig. 10.4

p. 125) as shown in Fig. 10.4 (*b*). In simple joints such as this, the high-tensile rod may not be necessary. The tightening-up process required to embed the connectors is carried out by the black bolt, which is then left as the permanent fastening. For connecting a metal member to timber a single-sided connector can be used [Fig. 10.4 (*c*)].

Steel structures

Connexion of steel to steel is usually accomplished by—

(1) Hinged or pinned joints and rocker bearings.
(2) Rivets and bolts.
(3) Welding.

Hinged joints

Many roof trusses and bridges constructed in the 19th century had pin joints. These allow relative movement of the members, and their one great advantage (when the advantage is required) is that they cannot resist bending moments. For example, an ordinary door is fixed to the door frame by its hinges but is capable of free rotation. In a roof truss or framed structure, some of the members are in direct tension only and the remainder are in direct compression only, provided that the loads are applied at the joints and that the joints are incapable of resisting bending moments.

In Fig. 10.5, (*a*) is a bar with an enlarged end to receive the pin, and (*b*) is a bar with a forked end. There is, of course, a certain amount of friction between members connected together in this manner, but, ignoring friction, no moment can be resisted. Hinged or pinned joints were once common in roofs, but have been superseded by riveted, bolted or welded joints. When loads (roof covering and wind pressure) are applied to the joints of a roof truss, it

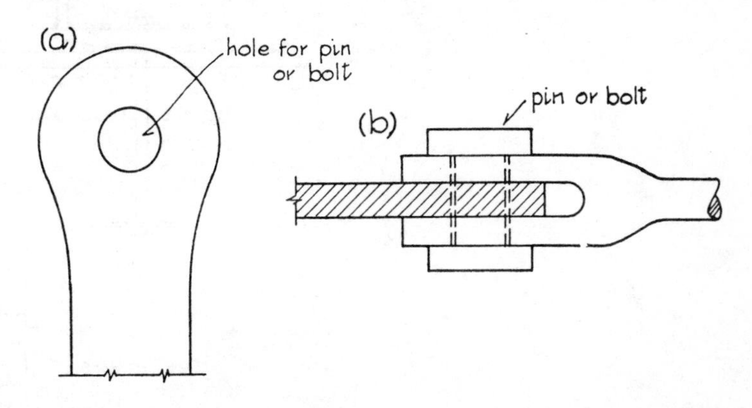

Fig. 10.5

deforms slightly. If the members are connected at their junctions by pin joints, they can move relatively to one another to enable the truss to take up its slightly different shape. The stresses (primary stresses) caused in the members are either tensile or compressive according to whether the members are acting as ties or struts.

If, however, the members are riveted, bolted, or welded together, they cannot move relative to one another at the joints to take up the slightly different shape of the truss (although there is a certain amount of play in bolted joints). There is consequently a slight distortion at the joints causing bending moments and consequently bending stresses (secondary stresses) in the members, in addition to the primary stresses of tension or compression. In small roof trusses these secondary stresses are usually ignored, and calculations are made on the assumption that the joints are hinged. In important structures, particularly trussed bridges of large span, the secondary stresses are taken into account when calculating the sizes of the struts and ties.

Although hinge joints have practically disappeared from roof truss construction they are frequently used for *portal frames*, which structures will be discussed in Chapter 13. Some portal frames have three hinges, whilst others have two, and the hinges are employed for a deliberate reason—to avoid bending moments at certain points of the structure. For example, with hinges at the feet of portal frames the foundations have to resist only vertical and horizontal forces.

Figure 10.6 (*a*) shows a hinge or rocker base. A curved steel plate *A* is welded to the steel stanchion and is seated on a flat steel base-plate *B*. Figure 10.6 (*b*) shows the principle of another type of rocker hinge base, and (*c*) is a simplified sketch of a roller-rocker bearing. This bearing allows for horizontal movement (as shown by the arrow) due to changes in temperature, and is suitable for large-span bridges.

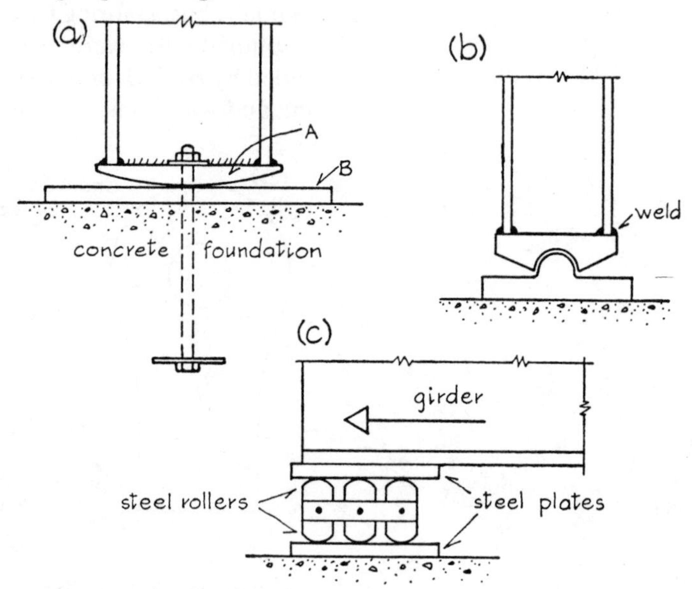

Fig. 10.6

Rivets and bolts

Rivets and bolts of mild steel are used extensively for steelwork connexions. The normal practice is to use rivets for those connexions or parts of connexions which can be fabricated in the workshop, and bolts for the final assembly on the building site. Site or field riveting is not so common today as it was before the Second World War.

The rivet usually employed in buildings has a round head, as shown in Fig. 10.7. The holes in the plates to be joined together are drilled 2 mm larger in diameter than the nominal (cold) diameter of the rivet. The rivet is heated to make the metal soft and plastic, pushed through the hole, and the head *A* held firmly against the plate. The other head *B* is formed by pressure, which squeezes the metal into the finished shape and also forces the metal to fill the hole completely. Calculations can be based on the finished diameter of the rivet: for example, a 20 mm diameter rivet (nominal diameter) becomes 22 mm diameter when it is finally in position.

The bolts most commonly used are *black bolts* and *turned and fitted bolts.*

Black bolts are manufactured direct from round mild-steel bars which have been formed by rolls in the steel mill. It is not possible to obtain precise dimensions by rolling in this way, so holes to receive black bolts are usually drilled 2 mm larger in diameter than the stated size of the bolt. Since bolts are always inserted 'cold,' a black bolt [Fig. 10.7 (*c*)] does not fill the hole in the same way that a rivet does, and there is likely to be a certain amount of play. (The lack of fit is exaggerated in the diagram.) Black bolts are allowed to be used only for certain types of connexion, as for example in completing beam-to-stanchion connexions in which the rivets have been calculated to take the whole of the load. Black bolts are used for roof truss connexions and for certain types of beam-to-beam connexions, but the permissible stresses are less than those for rivets and turned bolts.

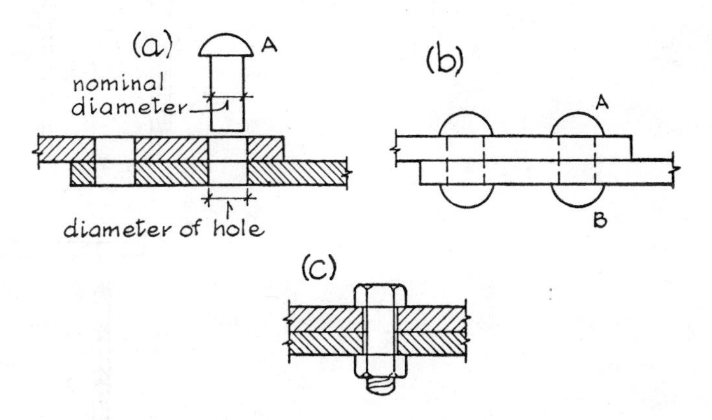

Fig. 10.7

Turned bolts are more accurately made than black bolts, since they are turned down to the required diameter from rolled steel bars of larger size. They are fitted bolts in the sense that they fit the hole better than black bolts and are allowed to take stresses almost equal to those for rivets driven accurately in the shop.

Although rivets and bolts may sometimes have to resist tensile and compressive stresses, their usual function is to resist shear stresses.

Figure 10.8 (*a*) shows simple connexions between beams and the flanges and web of a stanchion. (The other ends of the beams are assumed to be supported in a similar manner.) Only the rivets in the vertical legs are assumed to support the loads from the beams (the black bolts in the top cleat are 'loose' in their holes). If, for example, we imagine that the rivets fail in shear, beam *A* will slide down the stanchion exposing the cross-section of the four cut rivets, as shown at (*c*). The permissible shear stress for shop rivets is 100 N/mm^2, so if the rivets are 22 mm finished diameter (area, 380 mm^2) the safe load for the connexion (4 rivets) in shear is $4 \times 380 \times 100 = 152$ kN.

Another way in which a riveted or bolted connexion can fail is by 'bearing,' i.e. by the rivet or bolt bearing against or biting into the metal plates. Bearing failures are more likely to occur than shear failures when the plates or other parts joined are thin. The holes become elongated as a result of the crushing of the plate

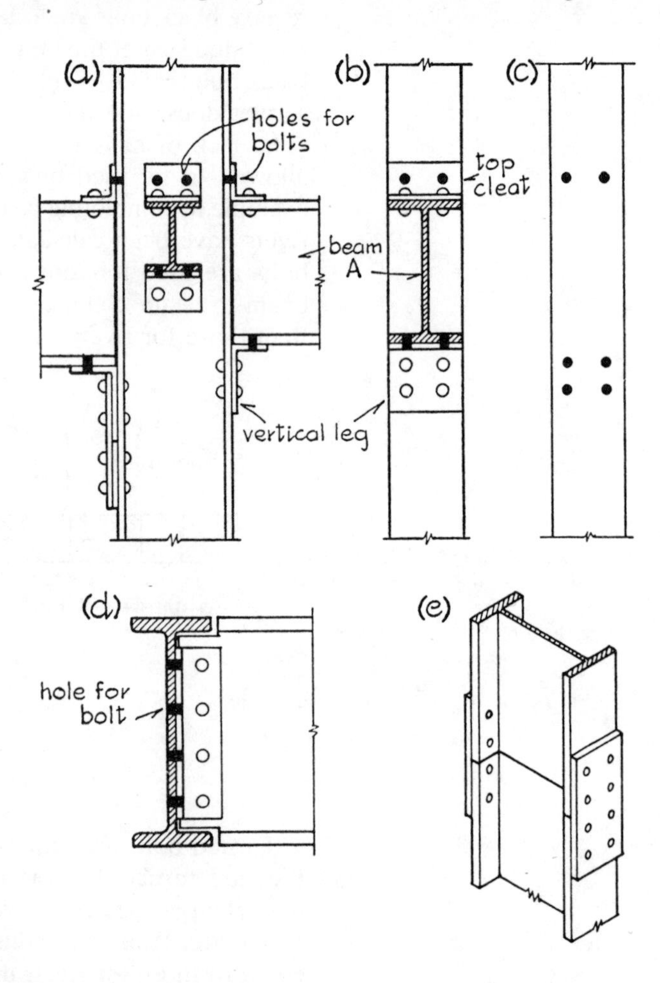

metal, and the joint becomes loose and inefficient. Figure 10.8 (*d*) is an example of a beam-to-beam connexion, and (*e*) shows a simple splice or join in a stanchion.

High-strength friction-grip bolts

During recent years an important development in the joining of steel members has been the introduction of *high-strength friction-grip bolts.* These are bolts of high-tensile-strength steel with high-tensile nuts and hardened washers. These are tightened up by special spanners to a precalculated extent so as to give a definite clamping force, compressing the members together. Loads in the connected members are then transferred from one to the other because of friction between the several parts, and not by relying on the shear and bearing strengths of the bolts.

Welded connexions

The two main methods of welding are oxy-acetylene and electric arc. The electric-arc method is the one most commonly used for structural steelwork.

Oxy-acetylene

Acetylene, which is a compound of carbon and hydrogen, when mixed with oxygen and ignited gives a flame of high temperature (about 2000° to 3000°C). The flame is used for melting a rod of

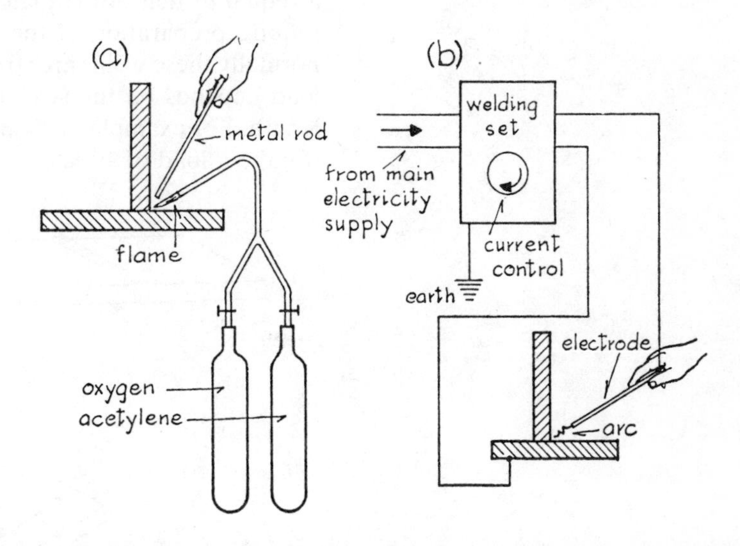

Fig. 10.9

metal and causing it to be deposited at the junction of two parts to be joined together [Fig. 10.9 (*a*)]. The 'parent' metal at the junction also becomes molten so that the weld metal fuses with it. Oxy-acetylene welding is used principally for work on light-gauge sheet metal. The flame is also used for cutting steel.

Electric-arc welding

The heat required to melt the welding rod is caused by an electric arc which requires a voltage of 70 to 100 volts [Fig. 10.9 (*b*)]. The alternating voltage of the mains is reduced to this figure by means of a transformer. If direct current is used the mains voltage is reduced by a motor-generator set. The rod of metal which is melted to form the weld is called the *electrode* and forms part of the electric circuit. To form a weld the electric circuit is completed by touching the parent metal with the electrode near the proposed joint. The electrode is immediately drawn away a short distance and the current jumps the gap in the form of an arc. The heat of the arc melts the electrode and also the parent metal near the join, so that the deposited metal fuses with the parts to be joined together. When welding structural (mild) steel parts, the electrode is also of mild steel coated with a material which acts as a flux and prevents chemical combination between the oxygen of the atmosphere and the molten steel.

Butt welds and fillet welds

Square butt welds, as shown in Fig. 10.10 (*a*), are only used for joining thin plates up to about 5 mm thick. For thicker plates the edges of the plates are prepared for welding by bevelling or gouging. A single-V butt weld is one example [Fig. 10.10 (*b*)], and there are many other types of butt welds known as double-V, single-U, double-U, etc., according to the shapes of the ends of the plates.

If properly constructed the strength of a butt weld can be taken as equal to that of the plates it joins together. Fillet welds do not require preparation of the plates or members to be joined, and normally these welds are stressed in shear [Fig. 10.10 (*c*)]. The safe load depends on the size (i.e. the thickness) of the weld and its length. For example, a 25 mm length of 5 mm weld can be allowed to take a load of 10 kN.

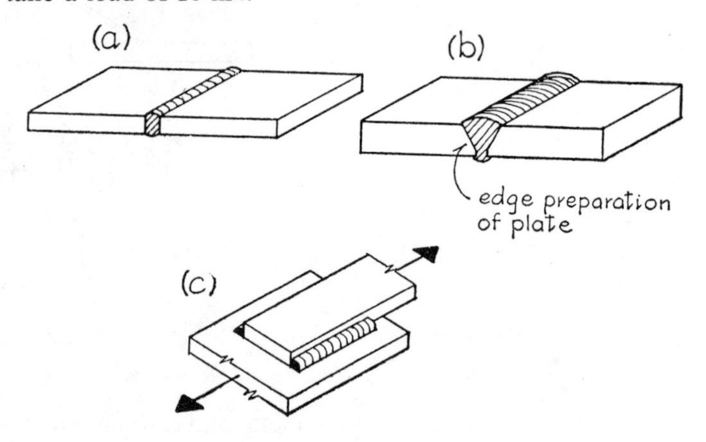

Fig. 10.10

Reinforced concrete

When concrete is placed *in situ* (on the building site) joints are automatically formed by the pouring of the concrete and its subsequent hardening. Figure 10.11 (*a*) illustrates a joint between a beam and a column. Normally, the procedure is to extend the bars (*a*) from the lower column a certain distance above the floor level. The beam (and slab) and the lower length of column are then concreted and the concrete is allowed to become hard. Bars (*b*) are then placed in position and concreting of the upper column commenced. The purpose of the overlapping of bars (*a*) and (*b*) is to transfer the load gradually from one set of bars to the other. Bars (*b*) cease to carry load at the level $A–A$, and bars (*a*) begin to carry the load at level $B–B$, so that by the time level $A–A$ is reached bars (*a*) have taken over completely from bars (*b*).

The same principle applies when joins have to be made in bars in beams or slabs. Bars are overlapped a sufficient distance to enable the stresses to be transferred from one set to the other. Joints, as shown in Fig. 10.11 (*a*), are monolithic (one-stone) and rigid, and capable of resisting bending moments. Sometimes, hinge joints incapable of resisting bending moments are required. Figure 10.11 (*b*) shows approximately the main reinforcement at a hinge joint on the Medway Bridge, U.K. The main bars in each member terminate at the end of the member so that there is no continuity of reinforcement from member to member. The hinge bars marked (*h*) are capable of taking horizontal shear, but there is no, or very little, resistance to bending at the hinge position.

Other types of hinge joints can be similar to those shown in Fig. 10.6, the steel plates being cast into the concrete members and secured to the concrete by anchor bars. Another type of hinge bearing is shown in Fig. 10.11 (*c*).

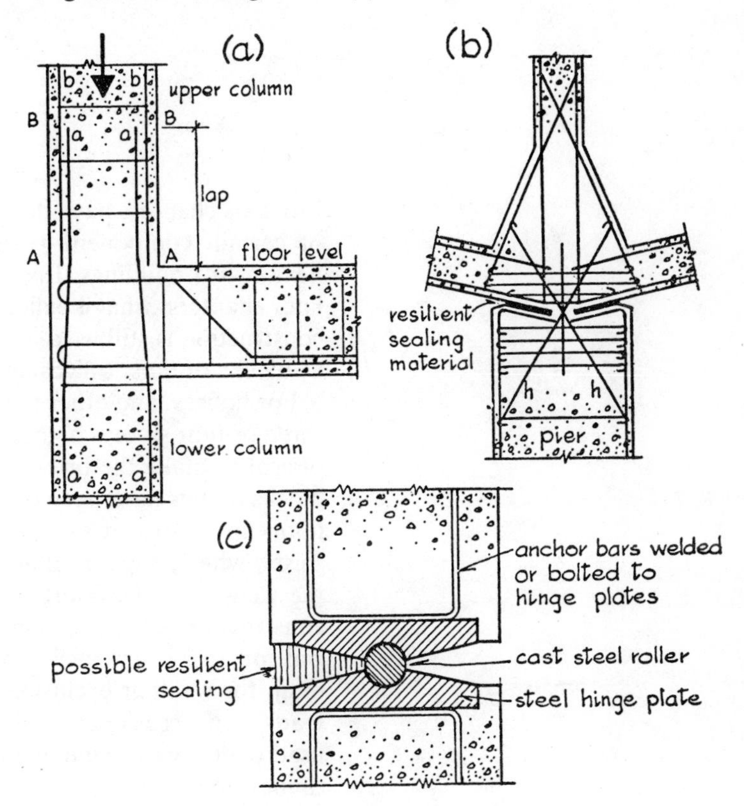

Fig. 10.11

11 The Building

Previous chapters have dealt with principles underlying the design of certain component parts (e.g. slabs, beams, columns, connexions) of buildings. In spite of modern techniques described in later chapters, what is called *post and lintel* (i.e. column and beam) construction is still used extensively in multi-storey buildings such as blocks of flats, offices and warehouses.

For houses, and other types of building which are not more than three or four storeys high, it is still usual and generally more economical to make the brick walls load-bearing. (Recently, even tall blocks of flats have been constructed with load-bearing walls.) The floors in houses of two storeys are usually supported by timber joists, which, supported at their ends by the brick walls, transfer the weights of the floors to them. Since the loads (the weights of construction, etc.) are comparatively small, a 225 mm thick solid wall or a 275 mm thick cavity wall, which is the practical minimum for weather exclusion, is capable of easily supporting these loads. An occasional steel or reinforced-concrete beam may be used over a wide window or door opening to support the ends of timber floor joists.

Other buildings, such as factory and office buildings of not more than a few storeys, may employ steel or reinforced-concrete beams to support solid reinforced-concrete or hollow-tile floors. In turn, the beams would be supported by brick walls thickened into piers (if necessary) at those points where the ends of beams are supported. This type of construction was used extensively in the 19th century, first with cast-iron beams and cast-iron internal columns (the external walls being load-bearing stone or brickwork), followed by wrought-iron beams, and eventually by steel beams.

As buildings became higher the brick or stone walls had to be made of considerable thickness to support the weight of the many floors. A building 14 storeys high, completed in America in 1890, had walls on the ground floor 2·75 m thick. The first building of any importance which had a complete steel skeleton to carry all the loads through beams and columns to the foundations was built in Chicago about 1890. This was followed, in 1892, also in Chicago, by the 20-storey Masonic Temple. New York followed suit, and by the beginning of the 20th century had, among other skyscrapers, one with a height of 118 m.

Britain was slower than the United States in developing steel-frame construction. The first steel frame in England was probably the one designed in 1896 by W. Basil Scott for a warehouse in West Hartlepool. The Ritz Hotel, built in 1904, was the first multi-storey building in London which had a complete steel skeleton to carry the weights of all floors and walls. Steel frames were used in 1906 in the construction of Selfridge's store in Oxford Street, and one of the earliest buildings in England with a reinforced-concrete frame was the General Post Office, London, built in 1910.

Castrol House, Marylebone Road, London (*Cement & Concrete Association*)

Although, with steel beams and columns to carry the entire weight of the building, the functions of the wall were merely weather exclusion, heat insulation and fire resistance, Building Acts in operation in Britain during the construction of the early steel-framed buildings still required walls to be of considerable thickness. For example, in the Ritz Hotel, although the external walls were supported on steel beams at each floor level, the London Building Act of 1894 required them to be of equivalent thickness to load-bearing walls. A first step towards thinner walls was when the Royal Institute of British Architects recommended in 1904 a wall thickness of 215 mm for the top 6 m of a framed building and 325 mm for the remainder of the height.

Simshill School, Glasgow (*Cement & Concrete Association*)

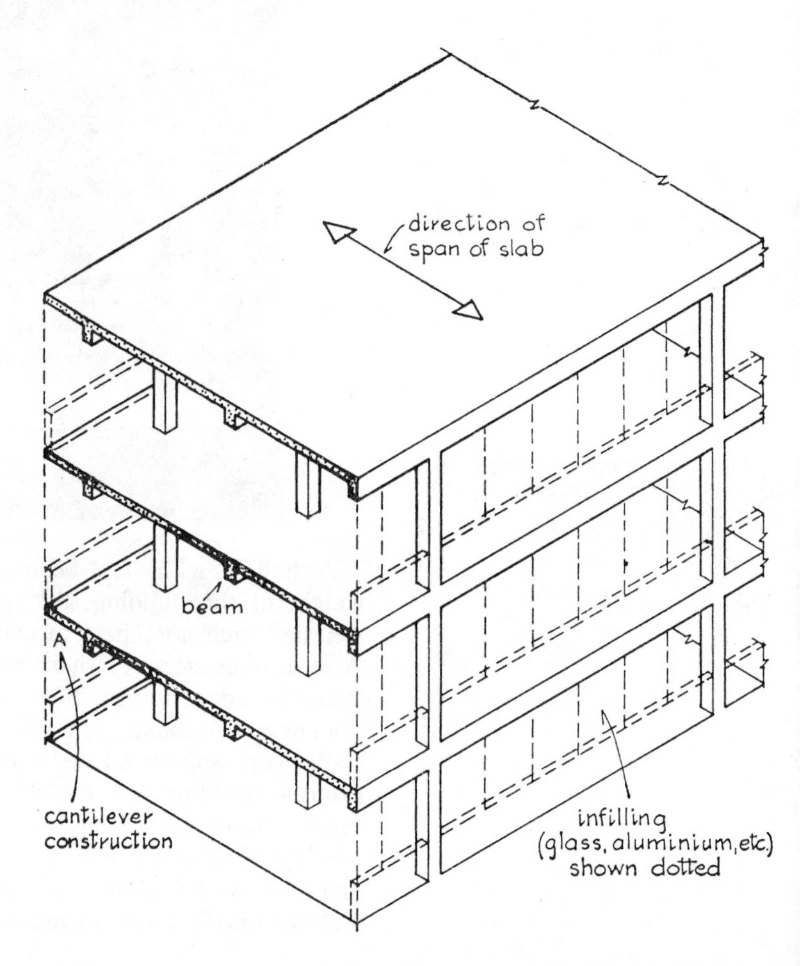

Fig. 11.1

Building Acts still govern thicknesses of walls for framed buildings, but other materials apart from masonry are now allowed, so that in many modern multi-storey buildings some of the walls may be completely devoid of brickwork and present a front consisting entirely of glass and aluminium or plastic.

In many of the early buildings the steel or reinforced-concrete frame (like the human skeleton covered with flesh) was completely hidden by a skin of brickwork or stonework to simulate the older traditional masonry buildings. This gave to the onlooker the false impression that all the loads of the building were carried by these

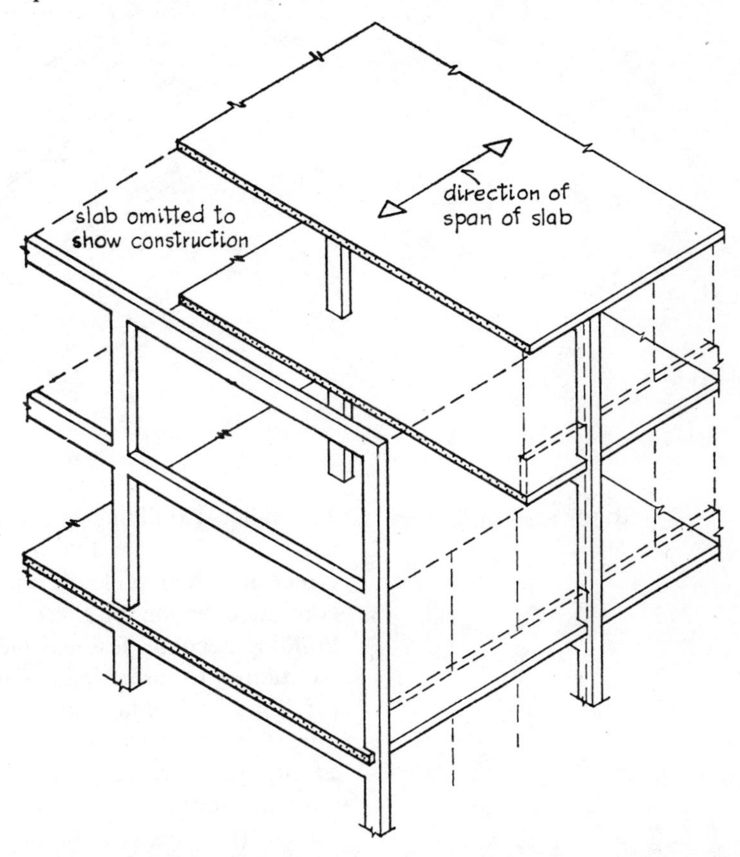

Fig. 11.2

walls. Now, however, the columns and beams on the outside elevations are more often than not exposed to view, although when the frame is of steel it is encased in a fire-resisting material such as concrete.

The use of steel and reinforced-concrete skeletons or frames to carry all loads, and the ease with which beams can be cantilevered beyond the external columns (particularly in reinforced-concrete construction) have allowed great freedom in the treatment of outside elevations of buildings. Figures 11.1 and 11.2 are examples of structures with columns and beams (or slabs) showing on the outside elevation. If desired, cantilever construction can be adopted as shown at *A* in Fig. 11.1, where the columns are some little distance away from the external walls.

Another example of cantilever construction is shown in Fig. 11.3. Beams cantilever from central pairs of columns and support the floor slabs. No loads are taken by the external walls, which can therefore be of any material provided that it satisfies the relevant building by-laws.

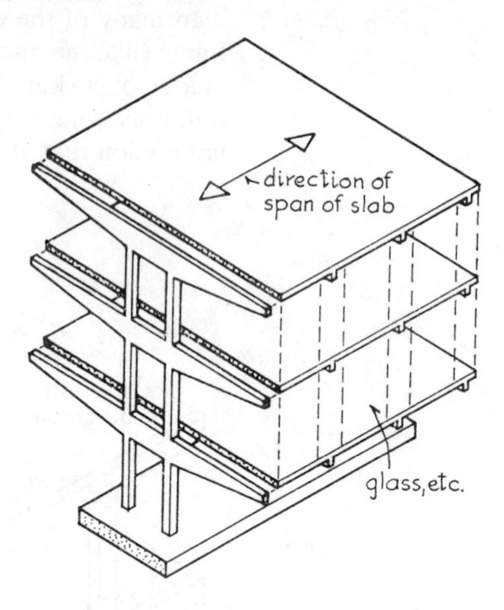

Fig. 11.3

Shear wall buildings

The taller a building becomes the greater are the vertical loads carried by the ground floor columns and the more important is the effect of horizontal loads due to wind and earthquake. Due to the combined action of these loads, the traditional frame type of building becomes less satisfactory as the height increases and one solution is to insert load-bearing shear walls into the building to carry the horizontal loads. The frame is then designed to support only the vertical loads. Concrete shear wall buildings are very common for structures in the 10 to 20 storey range. Some designs omit the frame altogether and the building is then comprised of intersecting walls and floor slabs. However, single walls are inefficient for lateral load when the height exceeds about 20 storeys and as a result the tube building has become popular, particularly for very tall buildings (above 40 stories).

Tube buildings

A tube building is one where the outside frames of the building act together like the walls of a giant tube which is fixed in the ground. To achieve this action in practice, the external walls must be stiffened either by diagonal bracing or by numerous closely spaced columns. In the latter case, the building resembles a perforated tube with window openings framed by adjacent columns and heavy floor beams.

Apart from the Empire State Building (102 stories) in New York, the world's tallest buildings are of the tube type. The John Hancock building (100 stories) in Chicago is a diagonally braced tube; the Sears Tower (110 stories, 443 m high) also in Chicago, is a bundled tube (several multiple-column tubes grouped together but of varying height); and each of the twin towers of the World Trade Center (110 stories; 412 m high) in New York is a tube-in-tube (two multiple-column tubes, one inside the other).

Structural design of a building

The architect is responsible for the architectural design of the building, i.e. he decides its shape, the planning of the accommodation to suit his client's requirements, the positioning of staircases, lift shafts, heating and ventilating equipment, etc., and he specifies the materials to be used internally and externally. He also decides the spacing of the beams and columns, although modifications may subsequently be made after consultation with the structural engineer. In the design of a small building requiring only a few beams and columns the architect may even make his own calculations for the sizes of the structural members. Normally, however, especially for multi-storey buildings, the architect passes his preliminary drawings to the structural engineer, who is then responsible for determining the sizes of all beams, columns, floor slabs and foundations.

A building is constructed from the foundations upwards, but calculations for the structure are made from the roof downwards, taking into account all loads: *dead loads, live* or *imposed loads, wind pressure* and *earthquake loads* if appropriate.

The dead load, which is always present, is due to the weight of all material with which the building is constructed, e.g. floors, beams, columns, walls, partitions, etc. Tables which give the weights per cubic metre of building materials are available; for example, the weight of reinforced concrete is usually taken as 24 kN/m^3, and the weight of ordinary brickwork as 19 kN/m^3.

The live or imposed load which may sometimes be present in full, sometimes in part, and sometimes completely absent, is due to: the occupants and furniture of the building; the weight of merchandise in warehouses; the weight of snow on roofs, etc. Codes of practice and building by-laws specify minimum loads which must be assumed to be imposed on every square metre of floor area to allow for the different types of occupancy. These loads are based in part on investigations of existing buildings, but it cannot be said that there is any mathematical accuracy in them. All that can be tentatively claimed is that, if floors are designed to support the specified loads, a sound safe structure should result. As experience has accumulated over the years the tendency has been to reduce the specified imposed loads. For example, the London Building Act of 1909 required office floors to be designed for an imposed load of 5 kN/m^2, whereas today the figure is 2·5 kN/m^2.

Other examples of imposed loads are given below—

Floors	*kN/m²*
Dwelling-houses of not more than two storeys . .	1·5
Dwelling-houses of more than two storeys; flats; hospital wards	2·0
Office floors above entrance floor	2·5
Office entrance floors; classrooms in schools . .	3·0
Retail shops; garages for vehicles not exceeding 25 kN gross weight; churches; restaurants	4·0
Office floors used for storage and filing; places of assembly without fixed seating (dance halls, etc.) .	5·0
Warehouses	5·0–10·0
Flat Roofs	
Where there is no access except for maintenance purposes	0·75
Where access is provided	1·5

Most building codes require that (except for floors for residential purposes) a notice should be permanently displayed on every floor stating the imposed load for which the floor has been designed. Floors have collapsed because they have been used for a purpose for which they were not originally designed, as, for example, placing heavy machinery on a floor designed for office-type loading. The public usually display lamentable ignorance in this respect, as is evidenced by crowding on to roofs or light sheds and canopies to watch processions and other public spectacles.

A simple example will now be given to illustrate how the structural design of a framed building is proceeded with. Figure 11.4 represents part of the roof plan (and also the floor plan) of an office building with solid reinforced-concrete floors, beams and columns. To avoid complications in the descriptions which follow, no staircases or lift shafts are assumed to occur in this part of the building. The slabs are spanning in one direction, supported at 3 m intervals by secondary beams, *S*1, *S*2, *S*3, etc. Some secondary beams such as *S*1 and *S*2 are supported by main beams such as *M*1, and others such as *S*3 and *S*4 are supported directly by columns. (An alternative design would be to dispense with the secondary beams and design the slab as spanning in two directions as described in Chapter 6.)

Design of the roof slab

The roof slab has to support—

(1) The weight of the *screeding* (a layer of mortar or fine concrete) placed on top of the structural slab and tapered in thickness to allow water to run off the roof into the gutters provided.

(2) The weight of the asphalt required for waterproofing the roof.

(3) The weight of the plaster on the underside of the slab (the ceiling of the room below).

(4) The weight of the slab itself.

(5) The imposed load specified in codes of practice, by-laws, etc.

Item 4 is a simple example of a problem which is always confronting the structural engineer. He does not know the weight of the slab until he has calculated its thickness, but he cannot calculate its thickness without allowing for its weight which has to be added to the other loads! In a simple slab design, experience of previous constructions will enable the engineer to make an accurate first guess of the thickness (100 mm in this example). In more complicated constructions, preliminary assumptions of sizes and weights may have to be made and calculations adjusted until the final correct weight is arrived at.

The loading for which the slab has to be designed will be somewhat as follows—

	kN/m²
20 mm thick asphalt	0·45
50 mm thick screed	1·15
100 mm thick slab	2·40
Ceiling finish	0·30
Dead load	4·30
Imposed load	1·50
Total load	5·80

The bending moments are calculated (the span of the continuous slab being 3 m), the assumed thickness of 100 mm is checked by calculation, and the slab reinforcement is designed.

Design of the secondary beams

Beam *S*1 is assumed to support an area of floor (shown shaded in Fig. 11.4) of 8 m × 3 m, i.e. 24 m². The load on the beam is therefore 5·80 × 24 = 140 kN (by slide-rule, which is accurate enough for structural design) plus the weight of the beam, which will be taken here as about 18 kN. The total load on the beam is therefore 158 kN. The design of the beam and its reinforcement will take into account the fact that the beam is continuous over the main beams. (An external secondary beam such as *S*5 will support only half the slab load carried by beam *S*1, plus the weight of any wall or parapet.)

Design of the main beams

A beam such as *M*1 supports the ends of two secondary beams and therefore receives 158 kN as a point load at mid-span and has also to carry its own weight (say 18 kN) as a uniformly distributed load.

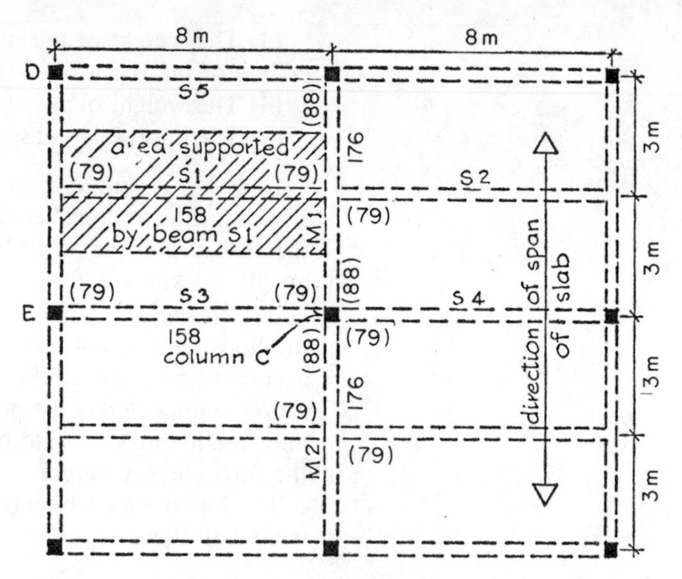

Fig. 11.4

The total load on beam $M1$ is therefore about 176 kN, and 88 kN is transferred to column C.

If the frame were steel instead of reinforced concrete the connexions between the various members would be designed for the loads shown in brackets in Fig. 11.4.

Design of column C

The load carried by column C at roof level is $2(88) + 2(79) = 334$ kN. To this must be added the weight of the column length (say 5 kN) from the roof to the floor below. Column C can therefore be designed for an axial load of 339 kN. Columns such as D and E may be less lightly loaded than column C, but the loading is not axial and allowance must be made for bending moments due to eccentricities.

Floor slab and beams

Again, calculations show that a 100 mm thick slab is adequate to carry its own weight plus the weight of floor finishes (which depends on the type, e.g. whether wood block or composition flooring, etc.) plus a minimum allowance (in office buildings) of $1{\cdot}0$ kN/m² for light partitions which may not be fixed in the preliminary design, plus the imposed load of $2{\cdot}5$ kN/m² for office floors. A typical loading is as follows—

	kN/m²
Floor finish	0·67
100 mm slab	2·40
Ceiling finish	0·30
Light partitions	1·00
Dead load	4·37
Imposed load	2·50
Total	6·87

Load supported by beam $S1 = 6{\cdot}87 \times 24 = 165$ kN plus, say, 18 kN for the beam = 183 kN.

Load supported by beam $M1 = 183 +$ (say) $18 = 201$ kN.

Load supported by column C from floor alone

$$= \frac{183}{2} + \frac{183}{2} + \frac{201}{2} + \frac{201}{2} = 384 \text{ kN},$$

plus, say, 10 kN for weight of column = 394 kN. Therefore, the column length below this floor which is supporting one floor and the roof has to carry 339 + 394 = 733 kN. All the floors below will impose a load on column C of approximately 394 kN; therefore a ground-floor column which has to support 10 floors plus the flat roof will be designed for 339 + (10 × 394), i.e. 4279 kN.

The calculations given above are approximate; for example, some reduction in imposed load is allowed for certain of the floors in multi-storey buildings, and refinement has not been displayed in the estimations of weights of beams and columns. It should be noted that if the columns were placed closer together the beams would be smaller and each column would carry less load (but the number of columns would be increased). Less weight on columns would also result if hollow-tile floors were used.

Wind pressure

The wind pressure acting on the vertical faces of a building depends on the wind speed, the condition of exposure and the size of the building. It also varies with height up the face of the building, although for buildings less than 5 m high, the pressure is usually assumed to be uniformly distributed.

For example a single-storeyed portal frame that is 4·5 m high at the eaves would be designed for a wind pressure of about 400 N/m² which would be assumed constant over the vertical faces of the building. This pressure corresponds to a basic wind speed of 40 m/s (144 km/h) and normal exposure. Of course higher pressures would be used if the exposure was severe (along a sea coast for example) or if the structure was of particular importance. A ten-storey building in the same location as the portal frame would be designed for pressures varying from about 600 N/m² near

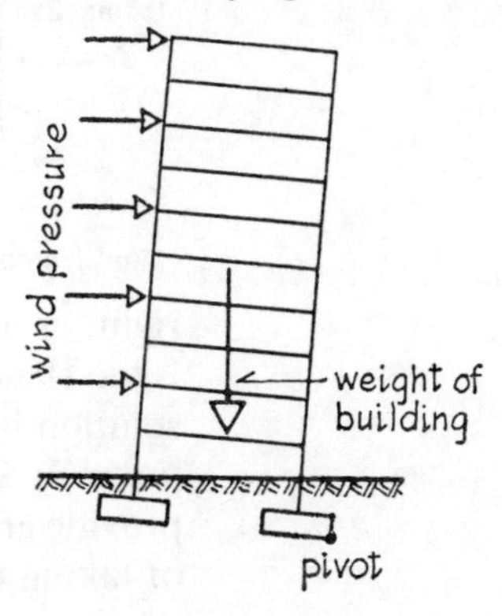

Fig. 11.5

ground level to about 1000 N/m^2 near the top of the building. Codes of practice and building by-laws give appropriate rules for calculating design wind pressures which take into account all the factors mentioned above.

One effect of wind pressure is a tendency to cause overturning of the whole building (Fig. 11.5). The moment exerted by the wind must be more than counterbalanced by the moment due to the weight of the building. Normally, the weight of the structure is sufficient to prevent its being overturned, but if calculations should

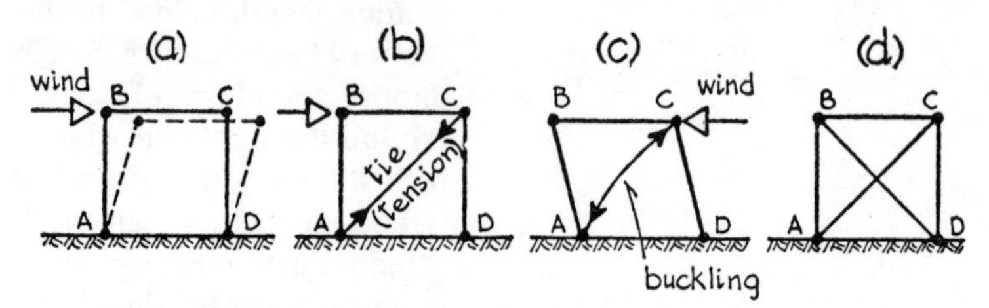

Fig. 11.6

indicate that overturning is a possibility, the building must be tied down to heavy foundations.

Assuming that the building as a whole is stable, provision must still be made to resist the 'racking' effect of the wind. Consider a structure with hinged joints, as in Fig. 11.6. The slightest horizontal force will cause failure as shown by dotted lines at (*a*). If, however, a member is connected to *A* and *C*, as at (*b*), it will be put into tension when the wind blows from the left and will prevent the racking effect shown at (*a*). When the wind blows from the

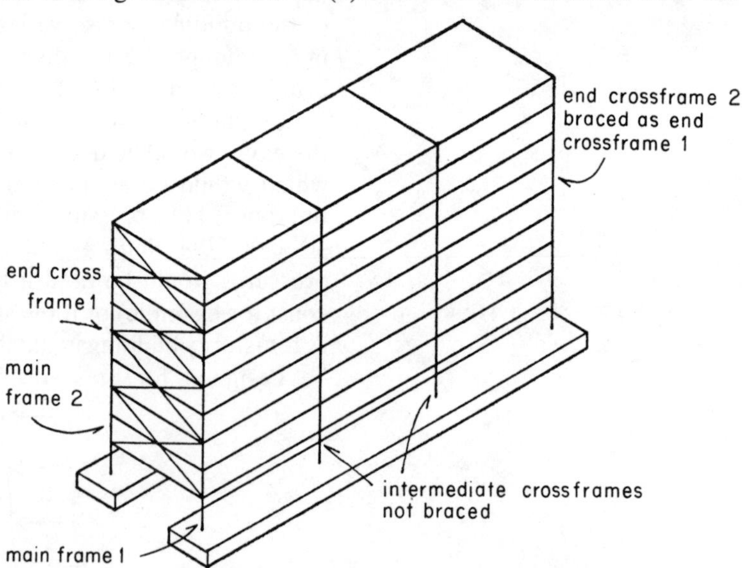

Fig. 11.7

right, as at (*c*), member *AC* will be put into compression, and if it is too slender to prevent buckling, failure will occur as shown. One solution is to make *AC* of sufficient size so that it can resist compression as well as tension, but it is usually more economical to provide another tie of small cross-section (flat bar or angle) capable of taking tension only. The finished structure is shown at (*d*). Tie

AC comes into action when the wind blows from the left, and tie BD when the wind blows from the right.

In a steel-framed building, the crossframes which interconnect the mainframes to form a three dimensional building are not usually stiff enough to resist the rocking action of wind. In a building as shown in Fig. 11.7, it is necessary to brace the end crossframes. Wind bracing of this type is only possible where it does not interfere with window openings, etc.; and when cross-bracing is impractical, the joints between beams and columns in the crossframes must be made rigid enough to resist the wind action. Examples of joints capable of resisting wind action are shown in Fig. 11.8.

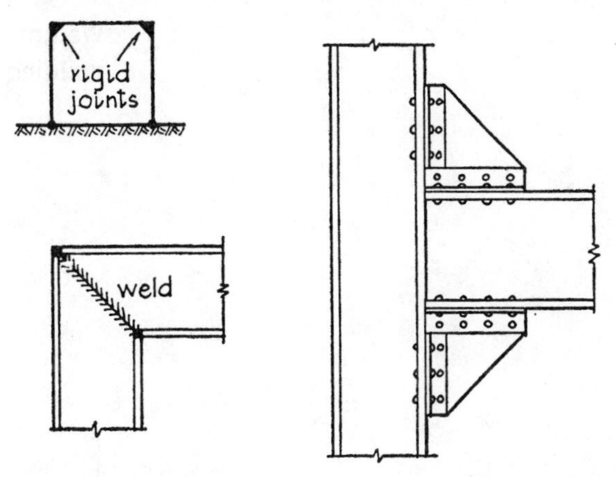

Fig. 11.8

In monolithic reinforced-concrete construction, the joints are rigid and cross-bracing is not required. The beams, columns, etc., must, however, be calculated to resist any additional bending moments and shear forces caused by wind over and above those caused by dead and imposed loads.

Earthquake loads

In many areas of the world, buildings must also be designed to resist earthquake forces. This is particularly true for Eastern Europe and for all countries surrounding the Pacific Ocean. Earthquakes result from sudden geologic movements within the earth's crust and as a consequence the ground shakes and buildings and other man-made structures sway and vibrate. If the ground motion is severe the vibrations within the building can be sufficient to cause damage and sometimes the complete collapse of the structure. Fortunately most earthquakes are small and a good proportion of those large ones that do occur are located under the ocean or in remote areas away from large centres of population. Nevertheless, every year for the last five years, there has been at least one destructive earthquake somewhere in the world. Seismologists agree that no country is totally immune from earthquake attack and since earthquake prediction is unreliable all structures should be designed to have at least some resistance to earthquake forces.

The accurate analysis of a building for earthquake loads is very complicated since not only is the load changing with time (dynamic) but no one knows for certain the magnitude and nature of the earthquake to be expected at any given site during the life of a structure. Furthermore, the building does not behave elastically because various joints and members yield (their material stresses exceed their elastic limit) due to excessive movements induced in the building even by moderate earthquakes.

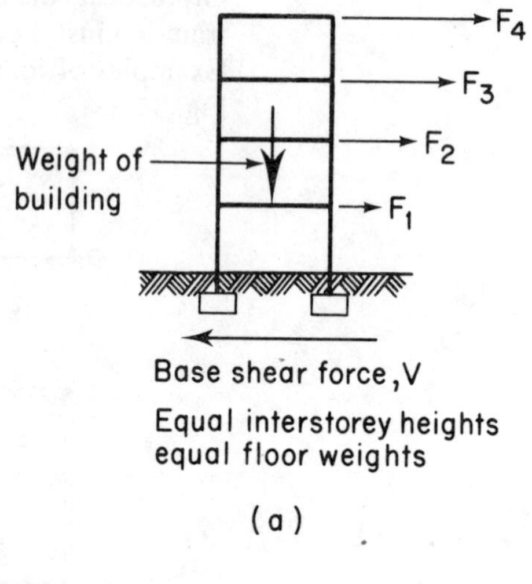

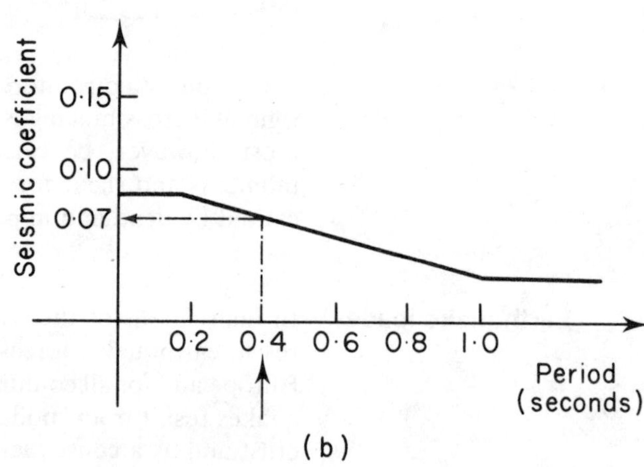

Fig. 11.9

Despite the complex nature of the problem a simplified approach is possible which is a first-order approximation only, but permitted by most codes for low-risk structures. In this method, equivalent static loads are calculated for each floor in the building so as to represent earthquake effects. The structure is then analysed in a conventional manner, just as for wind loading. The estimation of these equivalent forces is as follows. Consider the four-storey building shown in Fig. 11.9 (*a*). The lateral forces are related to the base shear, V, by an approximate code formula which for this uniform building would give $F_1 = V/10$, $F_2 = 2V/10$, $F_3 = 3V/10$ and $F_4 = 4V/10$.

The base shear is determined from the product of the total weight of the building and a seismic coefficient read from Fig. 11.9 (*b*). This coefficient is seen to depend on the fundamental period of vibration of the building, which for this simple frame would be about 0·4 s. Hence, the coefficient is about 0·07 and the base shear is then 0·07 × the total weight. Suppose all the floors had the same weight as the example given on page 138, then the total weight would be 4 × 394 kN or about 1600 kN. Therefore the base shear is 0·07 × 1600 = 112 kN and hence F_1 = 11·2 kN, F_2 = 22·4 kN, F_3 = 33·6 kN and F_4 = 44·8 kN.

By contrast the wind loading at each floor level would be about 9·4 kN, 9·4 kN, 11·3 kN and 6·6 kN respectively, assuming a basic wind speed of 40 m/s, normal exposure, an inter-storey height of 3·5 m and a frame spacing of 4·5 m.

Structure importance, structural type and ground conditions have been ignored in this simple example, but these factors would all be taken into account in an actual design, and the seismic coefficient modified accordingly.

It is worth noting that the equivalent forces just calculated are about one-quarter to one-eighth of the actual forces a building of this size might have to withstand under a moderate earthquake. Nevertheless the building would be designed for these lower forces but the building would be assumed to remain completely elastic. Under actual earthquake conditions therefore, the elastic limit would certainly be exceeded and the material would then yield but in doing so it would absorb energy (just like a car shock absorber) and dampen the violent motions of the building. The risk of major structural damage and possible collapse is thereby minimized. Of course, the joints and members have to be designed so as to allow yielding without fracture which is a difficult problem in structural detailing and one that requires all the skill and judgment the structural engineer has to offer.

12 The Foundations

One indisputable fact is this: whatever the shape of a building or structure and whatever the nature and number of its supports, the whole weight of the building must come down to, and be supported by, the ground. It is therefore essential that, for any proposed structure, sufficient knowledge be obtained of the nature of the supporting soil and its load-bearing capacity.

The purpose of a foundation is to convey the weight of a building to the soil in such a manner—

(1) That excessive settlement will not occur.

(2) That differential settlement of various sections of the building will not occur, thus causing cracks in the structure.

(3) That the soil will not fail under its load, thus causing collapse of the building.

Compared with structural materials, such as steel and timber, soil is difficult to investigate scientifically. It can vary considerably in its properties on one building site both in horizontal and vertical directions.

Until the 20th century, foundations were constructed mainly on the basis of experience. For important structures, deep trial pits were dug so that the soil could be examined for some distance below the surface, and sometimes loading tests on small areas at the bottom of the pits were made to estimate the safe load-bearing capacity of the soil.

Basic soil mechanics

Some time about 1920 there began a more scientific approach to the behaviour of soils, and a new science was born which is now known as *soil mechanics.* One of the earliest names, and a most important one, connected with this new science is that of Dr Karl Terzaghi, who made an extensive study of the properties of soils. Since 1920 much research has been carried out in many countries; many tests (site and laboratory) have been devised, and there is now a considerable literature on the subject of soil mechanics. Modern site investigations for important structures are carried out by specialist firms who have trained personnel; equipment for drilling and boring and for extracting samples of soil; and facilities for making site and laboratory tests.

All the material forming the crust of the earth likely to be affected by the presence of structures is divided by engineers into two major groups, *rocks* and *soils*. The term 'rock' is reserved for hard, rigid, and strongly cemented material, whilst 'soil' is applied to the comparatively soft and loose materials.

Soils (with the exception of organic deposits such as peat) are broadly divided into two groups, *cohesive* and *non-cohesive*. Silts and clays are cohesive, whilst granular materials such as sands and gravels are non-cohesive.

Owing to the weight of a building or other structure, there is bound to be a certain amount of settlement. It was stated in Chapter 1 that force cannot be applied to any material without

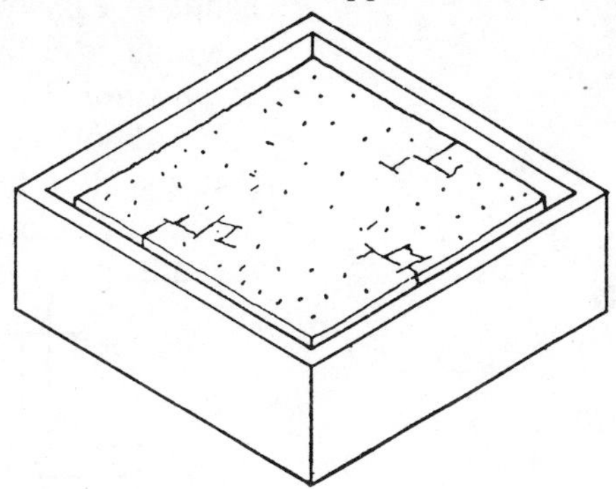

Fig. 12.1

causing deformation. The total settlement of a building takes place in two stages—

(1) Immediate settlement, as the building is being erected, due to elastic and plastic deformation of the soil (*see* 'Young's Modulus,' Chapter 1).

(2) Consolidation settlement caused by the squeezing-out of water contained in the pores of the soil and thus compressing the soil to a smaller volume.

From the settlement point of view, non-cohesive soils such as sands and gravels are not very troublesome since they are only moderately compressible and are very *permeable* (a permeable material is one which allows water to pass through it easily). Consolidation settlement due to squeezing-out of pore water therefore occurs very quickly during and soon after the erection of the structure, and is comparatively small. Special attention, however, must be given to loose sands, which can show appreciable settlement if subjected to vibration.

Cohesive soils such as clay have very low permeability, which means that squeezing of water from the pores due to consolidation settlement of the building is a slow process. Furthermore, the compressibility of most clays and silts is appreciable and there is a considerable volume reduction under pressure. The final settlement of a structure founded on clay may therefore not occur until some years after erection and must be allowed for in design calculations.

Clay is also a troublesome material when encountered in shallow foundations (for houses and other small buildings). Cohesive soils dry out in the summer and shrink; surface cracks can occur which may be up to a metre deep in severe cases. Many houses have cracked (particularly during hot dry summers) because their foundations were not taken deep enough.

The box (250 mm × 250 mm × 75 mm) shown in Fig. 12.1 was filled to capacity with wet London clay, which shrank as it dried to final dimensions of 225 mm × 225 mm × 68 mm, giving a volume reduction of 27 per cent.

Bulb of pressure

It is easy to appreciate that the foundation to a wall or column exerts a vertical pressure on the soil immediately beneath it, but it is not so easy to determine the depth of earth which is affected. Scientific investigations have brought forth the concept of the *bulb of pressure.* It is considered that the soil under a foundation is affected both horizontally and vertically as shown in Fig. 12.2, but that usually the effects of the imposed weight cease to be important below a depth equal to about 1½*b*, where *b* is the breadth of the foundation.

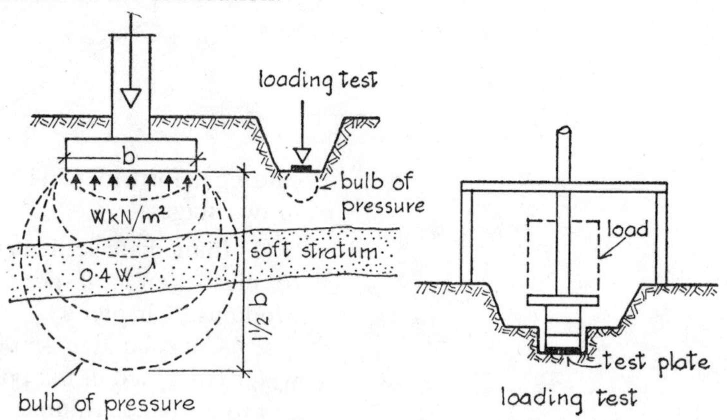

Fig. 12.2

It follows that boreholes for obtaining samples of soil for examination and testing should be taken at least to a depth equal to 1½ times the breadth of the foundation. Figure 12.2 also shows how a loading test could give inaccurate values for the bearing capacity of a soil. A loading test, carried out on a plate about 500 mm square affects the soil to a small depth (about 750 mm), and the possible soft strata shown in Fig. 12.2 would not be influenced by the test load. For the actual foundation, however, although immediately under the foundation the soil may be capable of supporting W kN/m², failure could result owing to the inability of the soft strata to support say 0·4 W kN/m². Loading tests are useful to determine the carrying capacity of a soil when investigations have shown that there are thick uniform deposits of granular soils, or soft rocks, or even stony clays. Where the soil consists of soft compressible silts or clays, the bearing capacity may be obtained from laboratory tests on samples.

When foundations are very close together, as in Fig. 12.3, their individual bulbs of pressure may overlap and the resultant combined bulb extends to a distance of $1\frac{1}{2}B$. Investigating boreholes should therefore be taken down to at least this distance.

An important property of a soil is its shear strength, and tests (both site and laboratory) have been devised to obtain this

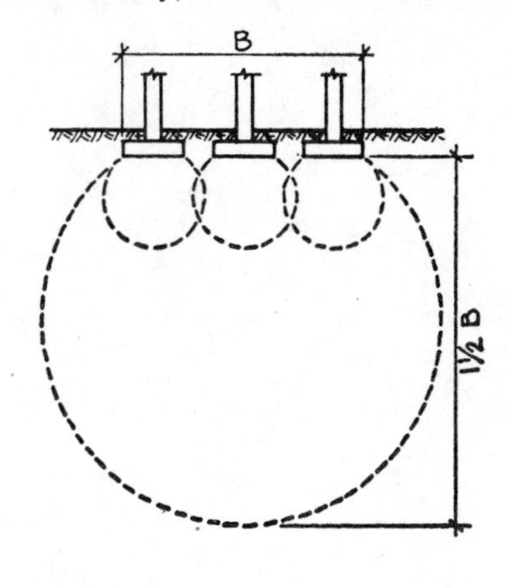

Fig. 12.3

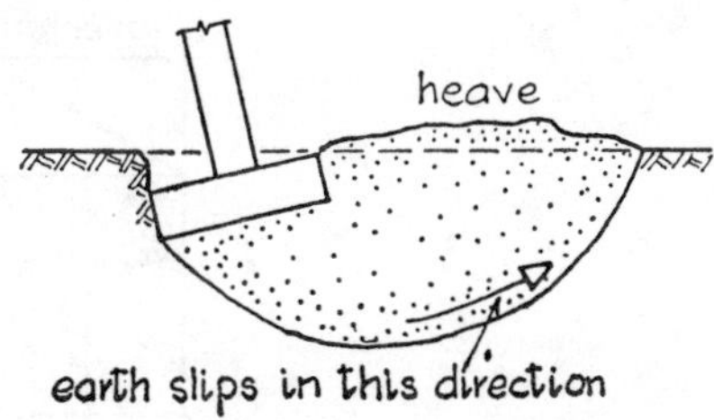

Fig. 12.4

strength, for both cohesive and non-cohesive soils. From a knowledge of the shear strength it is possible to calculate the bearing capacity of a soil. Figure 12.4 indicates a possible shear failure. The loaded foundation forces a block of soil to slide along a circular arc, with consequent heaving-up of the ground.

Safe bearing pressure

Another indisputable fact is that the minimum area of soil required to support the weight of a building cannot be varied by varying the type of foundation (except when piles are used). For example, a safe bearing pressure for many types of soil is 200 kN/m^2. If the weight of a building is 10 MN, this must be spread over a minimum area of 50 m^2 of soil. Similarly 100 m^2 is required if the weight is 20 MN, and so on. As will be demonstrated later, the *thickness* or depth of a foundation can differ according to the type adopted.

Concrete wall foundations

The load imposed on the soil due to the weight of a house is usually small, and an unreinforced (plain) concrete *footing* as shown in Fig. 12.5 is normally sufficient. The depth of the footing should be made at least equal to the overhang. If the wall carries heavy loads, thus requiring a greater area of support from the soil, which in turn requires a greater overhang of foundation beyond the edge of the wall and consequently a greater depth of plain concrete, it may be more economical to use a thinner footing reinforced with steel as shown in Fig. 12.6.

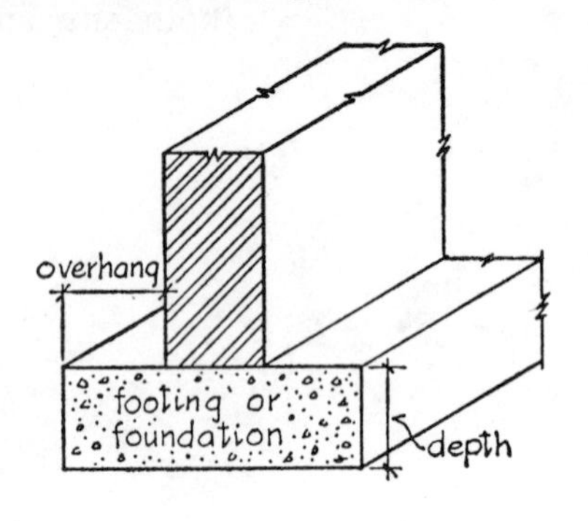

Fig. 12.5

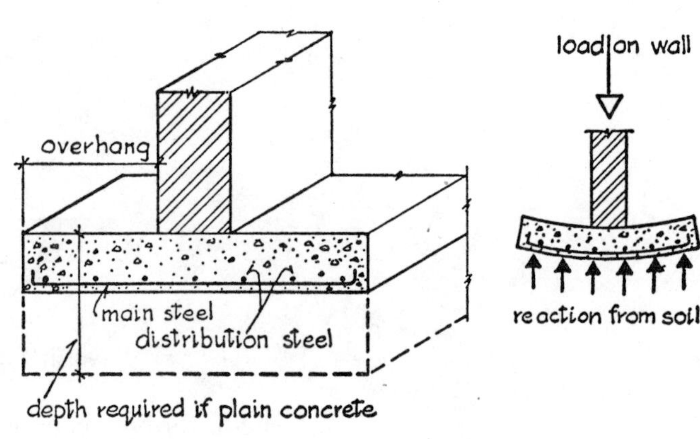

Fig. 12.6

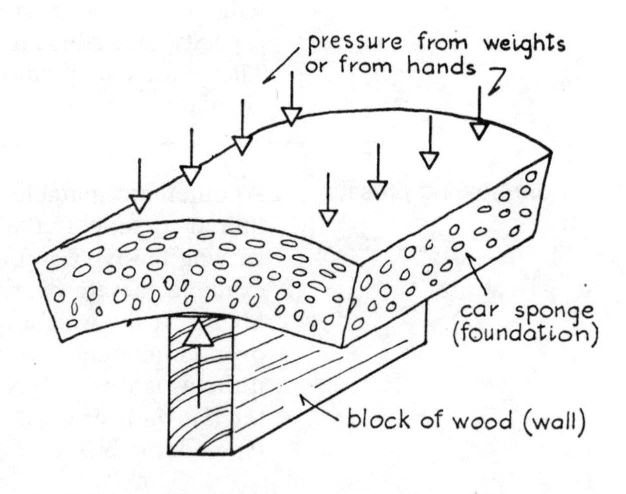

Fig. 12.7

The foundation is designed as a reinforced-concrete slab (double cantilever) which bends as shown owing to the downward load and the resulting upward reaction from the soil. The bending action can be visualized (in an upside-down fashion) by pressing a car sponge over a block of wood (Fig. 12.7).

Concrete column foundations

When practicable, the simplest and most economical solution is to provide each column in a building with its own base (Fig. 12.8). To avoid eccentricity of loading and consequently uneven pressure on the soil, the column should be located centrally on the base. If, for example, a reinforced-concrete column is 300 mm square and the total pressure on the soil due to the column load and the weight of the foundation is 500 kN, the area required at 200 kN/m² is 2·5 m².

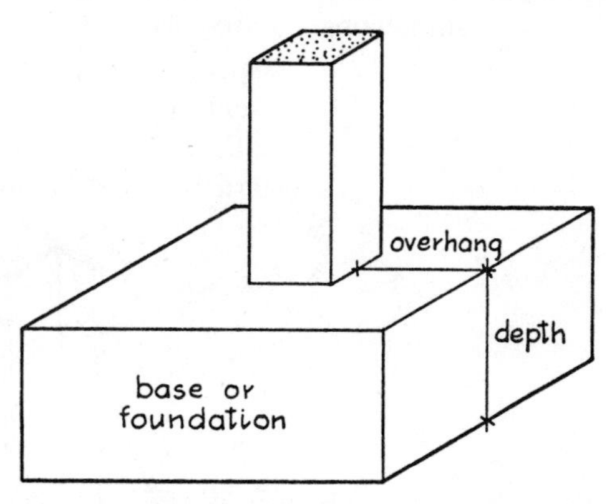

Fig. 12.8

The foundation will therefore be 1580 mm square, the overhang will be 640 mm, and the minimum depth of a plain concrete foundation will be 640 mm. For heavier loads, a reinforced-concrete base will probably be used, since it requires less depth of concrete and less excavation (at the expense of using a certain amount of steel).

Bending will occur as shown in Fig. 12.9, so steel reinforcement

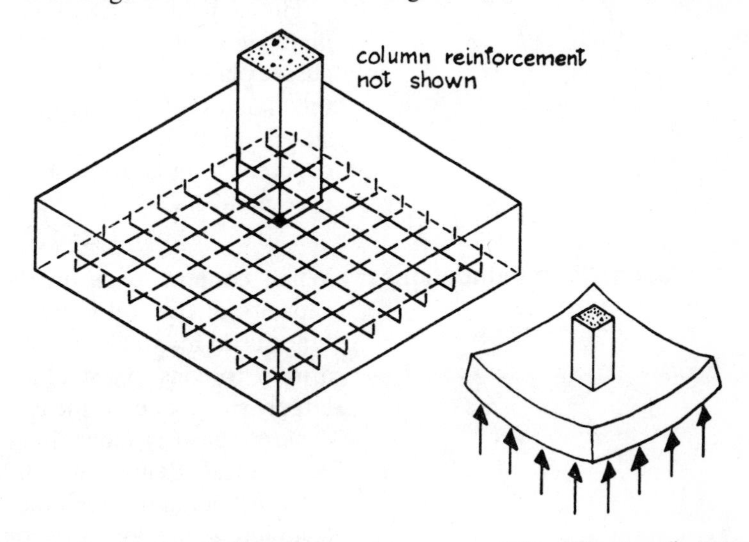

Fig. 12.9

is required in both directions as in two-way slabs (Chapter 6). It is not usual to supply shear reinforcement in foundations of this type, so the base has to be thick enough to enable the concrete to resist safely the shear (diagonal tension) stresses. A 450 mm square reinforced-concrete column (and its base) imposing a load of 2 MN on soil capable of safely supporting 200 kN/m² would require a foundation 3160 mm (10 m²). The thickness of a reinforced-concrete base as shown in Fig. 12.9 would be about 750 mm or even less. A plain concrete base would require to be at least 1355 mm thick. (Overhang of foundation beyond edge of column = ½(3160 mm — 450 mm) = 1355 mm.)

Steel base-plates for steel stanchions

A reinforced-concrete foundation as described above is also suitable for a steel stanchion, but a stanchion base is required to spread the load over a sufficient area of concrete. The cross-sectional area of a stanchion is comparatively small, and if it were allowed to rest directly on the foundation without a *spreading pad,* crushing of the concrete would result.

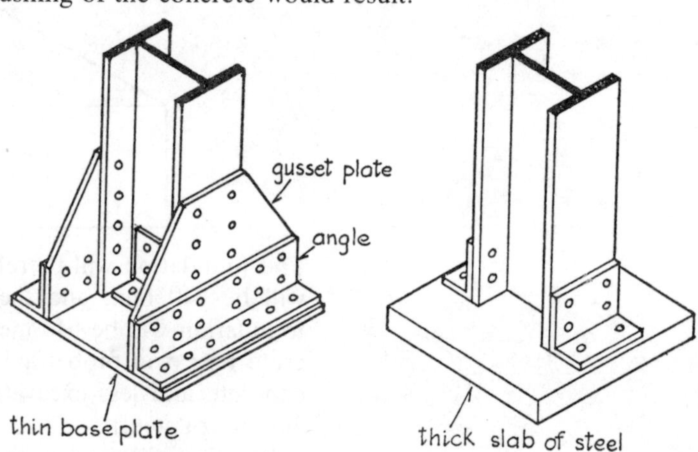

Fig. 12.10

The stanchion base may consist either of a comparatively thin plate of steel (10 mm to 20 mm thick) stiffened by angles and gusset plates, or a thick slab of steel connected only by angles to the flanges of the stanchion [Fig. 12.10]. For example, if the stanchion load is 2 MN and the permissible pressure on the concrete foundation is 4 MN/m², the minimum area of the steel base plate is 2/4 = 0.5 m².

Steel grillage foundations

When column loads are heavy and/or the permissible bearing capacity of the soil is small, steel grillage foundations are sometimes used for steel columns (Fig. 12.11). Since about 1940 (when a need to conserve steel arose) this type of foundation has been largely superseded (at least for moderate loads) by the reinforced-concrete base as shown in Fig. 12.9, which frequently is also more economical. Beams in both the top and bottom tiers are designed to resist bending moments and shear forces, and the grillage is completely enveloped in concrete.

Strip foundations

Two, three or more columns closely spaced may sometimes be given a common foundation consisting of a reinforced-concrete slab (strip) as shown in Fig. 12.12, or several steel beams may be encased in concrete.

If possible the centre of gravity of the foundation should be made to coincide with the centre-of-gravity line of the loads; therefore AG should be made equal to GB. This will mean that the pressure on the soil can be assumed to be uniform. The total load on the soil is $W_1 + W_2 + W_3 + W_4 +$ the weight of the foundation. This total weight divided by the permissible bearing pressure on the soil gives the minimum required area $ABCD$. Figure 12.12

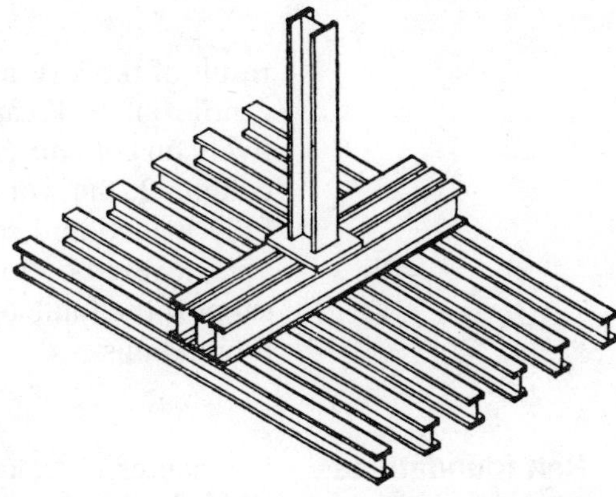

Fig. 12.11

indicates in an exaggerated manner the way in which the foundation tends to bend, and the thickness of the slab is made sufficient to resist the maximum bending moment. Steel reinforcement would be provided in the bottom of the slab at some places (when tension occurs at the bottom) and in the top of the slab at other places.

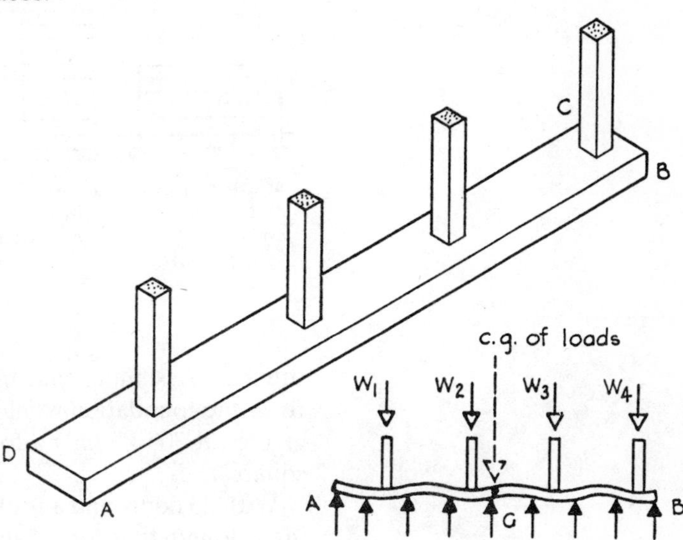

Fig. 12.12

Cantilever foundations

Sometimes, owing to the nearness of the building boundary line or the foundations of existing buildings, it may be impossible to place the foundation of an individual column directly under it. One solution is to combine the foundations of two columns, as shown in Fig. 12.13, by means of a foundation beam. Calculations must be made to ensure that there is not likely to be uplift of column 1 as a

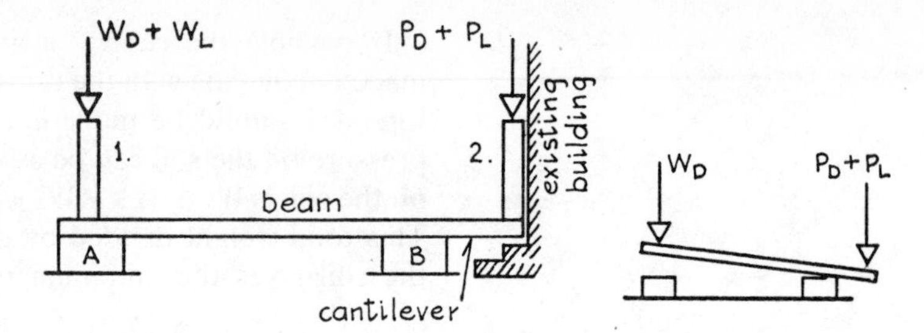

Fig. 12.13

result of the leverage effect of the column loads on the end of the cantilever. If W_D and W_L are the total dead and live loads respectively on column 1, and P_D and P_L are the dead and live loads on column 2, the worst case for lifting tendency is when column 2 is fully loaded and column 1 has only its dead load. The total soil area covered by foundation blocks *A* and *B* must be sufficient to support the total loads from both stanchions plus the weight of the foundations.

Raft foundations

In Chapter 11, calculations showed that the load on column *C* due to 10 floors and a roof was nearly 4·2 MN. At 200 kN/m² on the soil this column would require a foundation of 21 m² (about 4·6 m

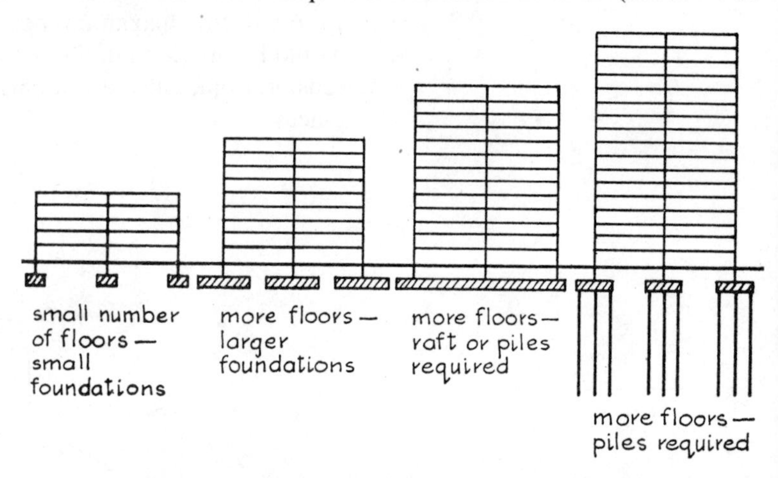

Fig. 12.14

square). Assuming that the building consisted of 5 floors and a roof, the foundation would be required to support about 2·3 MN, giving (at 200 kN/m²) a foundation area of 11·5 m² (about 3·4 m square).

With 20 floors and a roof, the total load is about 7·9 MN, requiring a foundation for column *C* of 39·5 m² (about 6·3 m square). Therefore as the loads get bigger the required foundations get bigger and the limits of the foundations of different columns get closer together until the foundations occupy practically the whole of the available ground space (*see* Fig. 12.14). In such cases the foundation may be made of one solid slab (raft) of reinforced concrete with or without beams (although it may probably be more practical in many designs to adopt piling).

By using beams between columns as in Fig. 12.15 (*b*), the slab itself can be thinner than in Fig. 12.15 (*a*). Although column *C* of Fig. 11.4 has been taken as an example in discussing raft foundations, it is improbable that a raft would be used for such a heavily loaded multi-storey building. A better example of the use of a raft foundation would be that of a smaller building on soil having a very low bearing capacity. Even with lightly loaded columns, the area of spread required might mean that individual foundations to columns would practically touch one another. A thin raft could be used, since the upthrust from the soil causing bending in the raft slab would be small.

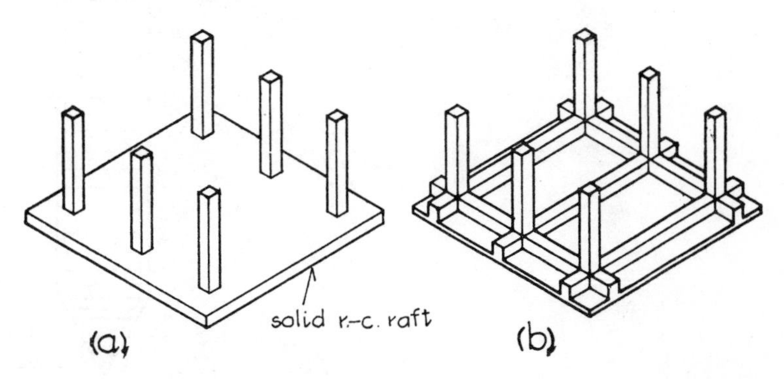

Fig. 12.15

In the building at Millbank, London, the foundation consists of a 3·36 m thick reinforced-concrete raft supported on 163 'bored' piles of 915 mm diameter at an average depth of 27·45 m below ground level. The total design load for the foundations was 488 MN.

Pile foundations

For heavily loaded columns in multi-storey buildings, piles would probably be used. Even if it were desired to use a raft (without piles), this would be impossible when the total weight of the building and its imposed loads divided by the permissible pressure of the soil gave a required area of spread greater than the area of the building site.

Most piles are of reinforced concrete, either precast or cast *in situ* in previously bored holes, but timber piles (which date from before Roman times) are still used in some countries where timber is plentiful and cheap.

Precast piles

Precast piles (Fig. 12.16), which are usually square or octagonal in cross-section, are driven into the soil by repeated blows from a falling weight or from a steam hammer. The piles are driven in until a certain number of blows produce only a small further penetration which has been predetermined by calculations in accordance with the loads the piles will be called upon to support.

A simple illustration is that of knocking a timber stake into soil (Fig. 12.16). The deeper the stake is driven into the ground, the greater is the frictional resistance and the harder it is to drive the stake downwards. When a pile is driven a sufficient distance, the resultant load it has to carry is usually supported partly by frictional force on the sides of the pile and partly by the bearing resistance of the soil under the foot of the pile.

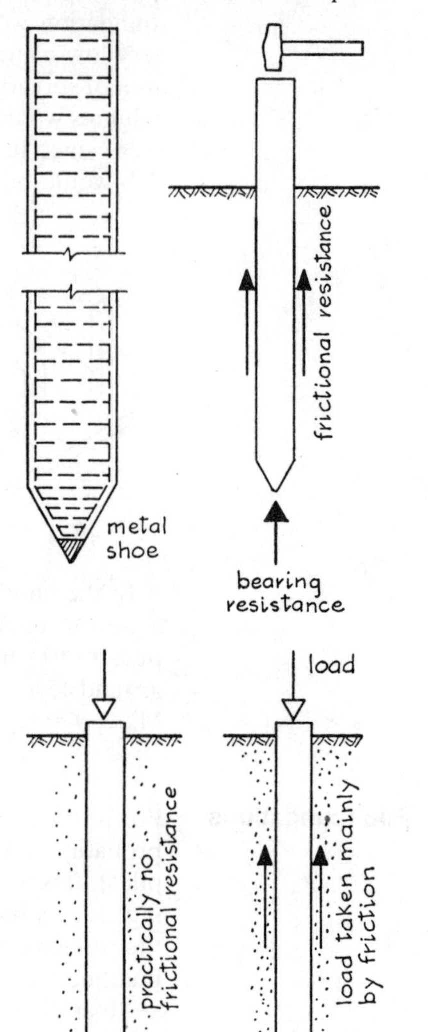

Fig. 12.16

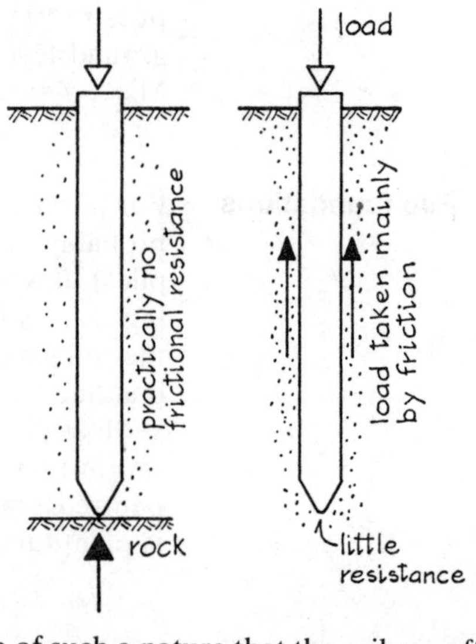

Fig. 12.17

If the strata are of such a nature that the soil can offer negligible frictional resistance, the pile must be driven until it meets a hard stratum such as rock capable of supporting the full load (Fig. 12.17). Alternatively the soil may be of such a nature that most of the resistance to the downward action of the weight of the building is provided by frictional resistance.

In supporting their loads, piles act as columns, except that the soil provides lateral restraint.

Another method, as used by the Franki Compressed Pile Co., consists in driving a steel tube to the required depth and filling it with *in-situ* concrete (Fig. 12.18). First, the bottom end of the tube is sealed by a temporary gravel plug. The tube is then driven to the required depth by blows on this plug from a long heavy cylindrical

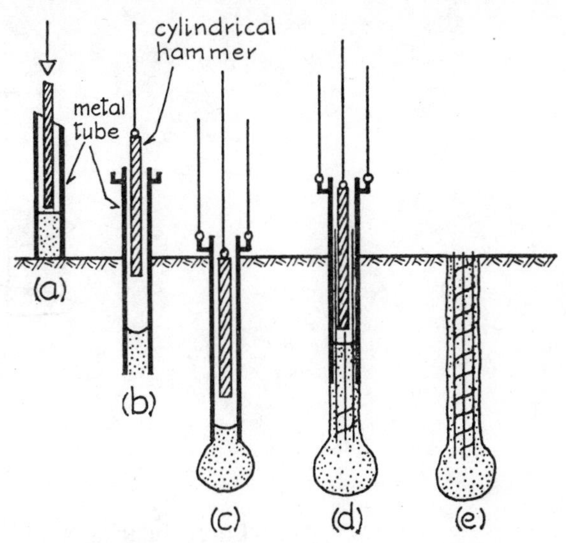

(a) consolidating gravel to form a solid plug.
(b) driving the tube through the soil.
(c) forming the base.
(d) gradual removal of tube as pile is formed.
(e) completed pile.

Fig. 12.18

hammer. Concrete is then placed in the tube and rammed so that a bulbous foot is produced (thus increasing bearing resistance). As the rammed concrete approaches the ground surface, the tube is gradually withdrawn.

Bored piles

Driven piles displace and compress the soil. When piles are bored *in situ*, the soil is removed by special coring tools to form cylindrical holes equal to the depths of the prepared piles. Steel reinforcement is then inserted and concrete rammed in to form the pile, which frequently has a bulbous base due to heavy ramming of the first batch of concrete. The equipment used for installing bored piles is lighter in weight than that required for driven piles. In addition bored piles are useful where headroom is restricted so that it is impossible to have long lengths of precast piles projecting above ground level previous to being driven.

Cylinder piles

Cylinder foundations 2 m or more in diameter have been employed for many years, the holes being made by excavating soil by hand or by mechanically operated grabs. During the last few years new methods have developed for forming these large-diameter holes. In one system, first used in the United States, augers (drills), as shown in Fig. 12.19, varying from 1 m to 2·3 m in diameter, are used for boring holes which may be as much as 25 m deep or more. The bottoms of the holes are usually enlarged by special belling-out tools to a diameter of 5 m in the case of the 2·3 m diameter cylinders, and 2·3 m for the 1 m diameter cylinders.

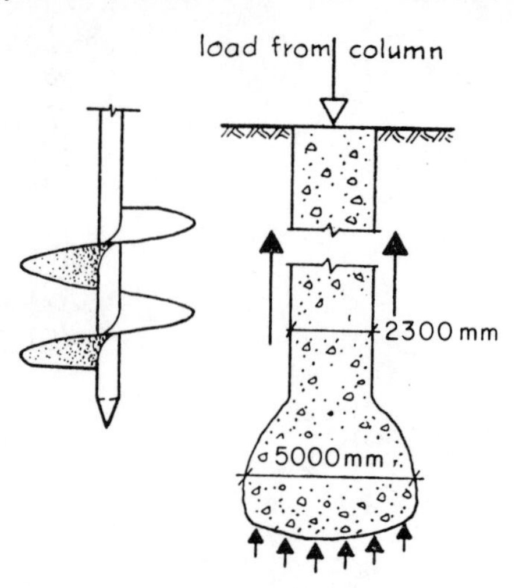

Fig. 12.19

Large bearing areas are thus obtained; one of these piles in the London clay was designed to support a load of 20 MN. One advantage of a large-diameter cylinder pile is that a heavily loaded column can often be supported directly by one pile, thus saving the cost of a pile cap, which is required when the column has to be supported by a number of small-diameter piles.

Pile caps

When small-diameter driven or bored piles are used, the number of piles required to support one column can be obtained by dividing the column load by the safe load for one pile. The load from the column must be transferred to the piles by means of a foundation called a *pile cap*. Examples are given in Fig. 12.20. Reinforcement (which is not shown in the sketches) is required in the cap to resist both bending and shear stresses.

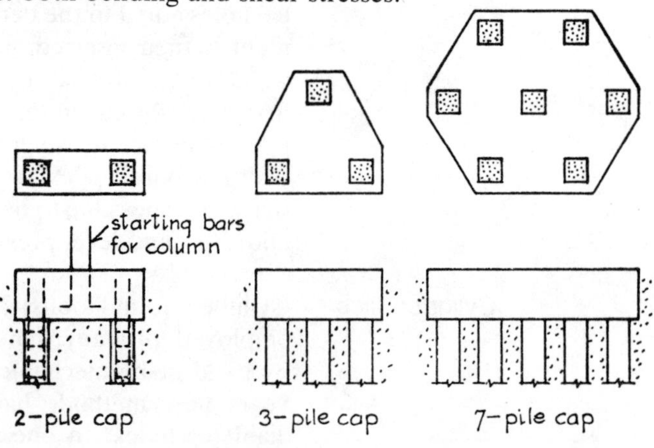

Fig. 12.20

13 The Modern Arch and the Portal Frame

The masonry arch built up of *voussoirs* (bricks or blocks of stone) has to be made of such a thickness that only compressive stresses are developed. The thin mortar joints may be capable of transmitting very small tensile stresses, but it will be appreciated that no reliance can be placed on this doubtful supposition. Masonry arches, therefore, must be of considerable thickness, the thickness varying according to the spans of the arches and the loads they have to support. Palladio adopted for his bridges a thickness at the crown of 1/15 of the span of the arch.

Steel, reinforced concrete and timber are capable of taking tensile stresses and can be constructed without joints or, in the case of steel and timber arches, with joints capable of transmitting tension. Larger spans with thinner and more graceful outlines are possible with these materials than are possible with masonry. Depending on the ratio of live load to dead load, the thickness of a reinforced-concrete arch may be about 1/60 to 1/90 of the span, that of a timber arch about 1/60 to 1/120 of the span, and that of a steel arch about 1/90 to 1/180 of the span.

Arches

Arches capable of taking bending moments and therefore tensile stresses are classified as *three-pin, two-pin* and *completely rigid* (no pin). The pin joints are constructed as explained in Chapter 10.

The three-pin arch

The 3-pin arch (Fig. 13.1) has one hinge at the crown and one at each support. Its advantages are—

(1) Calculations are easier than for the other types.

(2) No bending moments are caused at the abutments and the crown because hinges cannot resist moments.

(3) Differential settlements of the supports (the abutments) do not appreciably affect stresses, since the pins or hinges enable the arch to take up the slightly different shape consequent upon settlement.

(4) The pin joints enable the arch to adjust itself to expansions and contractions due to changes in temperature.

A disadvantage is that bending moments away from the pins are larger than in the 2-pin and completely rigid arches.

As with the masonry arch, there are horizontal thrusts at the supports, and the shallower the arch the greater are the thrusts. The abutments can be relieved of these horizontal thrusts by

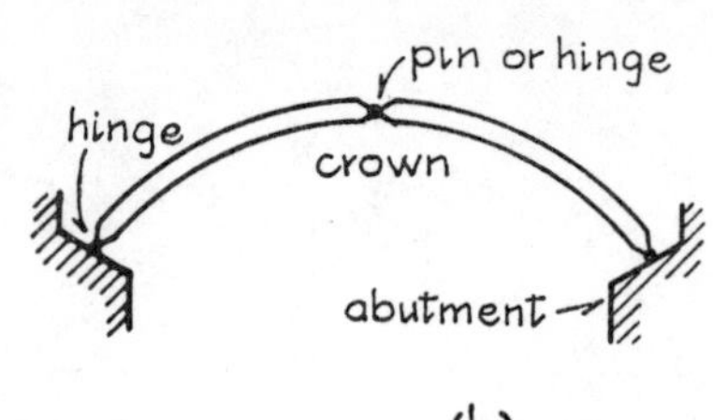

Fig. 13.1

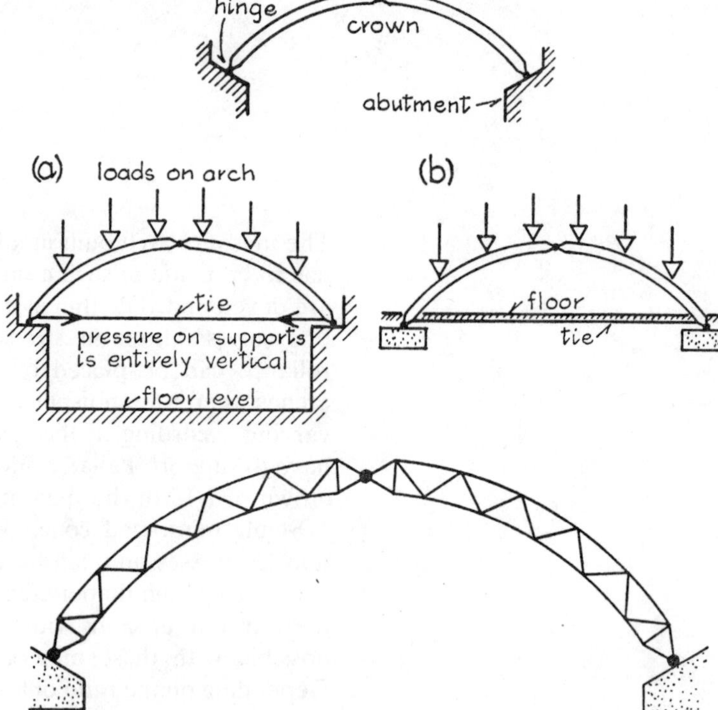

Fig. 13.2

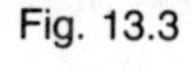

Fig. 13.3

inserting a tie member, but this might be aesthetically unacceptable in the example of Fig. 13.2 (*a*). Where the arch springs from ground level as at (*b*), the tie member is concealed and does not interfere with the appearance of, or the free space under, the arch. Arches need not be solid; Fig. 13.3 shows a triangulated steel arch.

The two-pin arch

This type of arch [Fig. 13.4 (*a*)], which is frequently used, has pin joints at the abutments only. There is a more even distribution of bending moments than in the 3-pin arch, and this can lead to

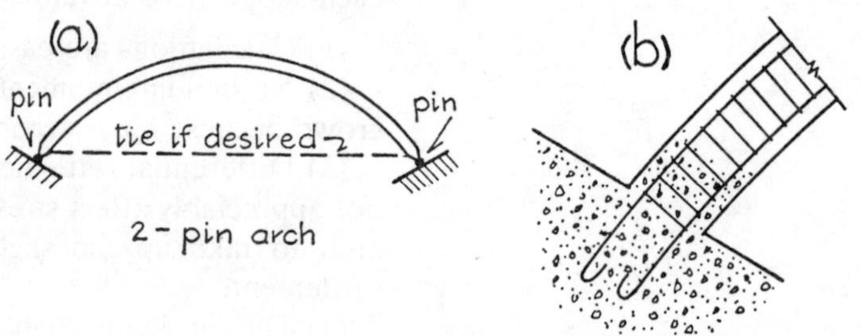

Fig. 13.4

greater economy of material. Undue settlement, or differential settlement of the abutments, can have more serious effects than in the 3-pin arch.

The rigid arch

In this type there are no pin joints, the arch being monolithic with its foundations [Fig. 13.4 (*b*)] or joined to the supports by rigid connexions when the arch is of steel. A disadvantage of the rigid arch is that the foundations are subjected to moments. (A more detailed discussion of the differences between types of arch follows shortly under the heading 'Portal Frames.')

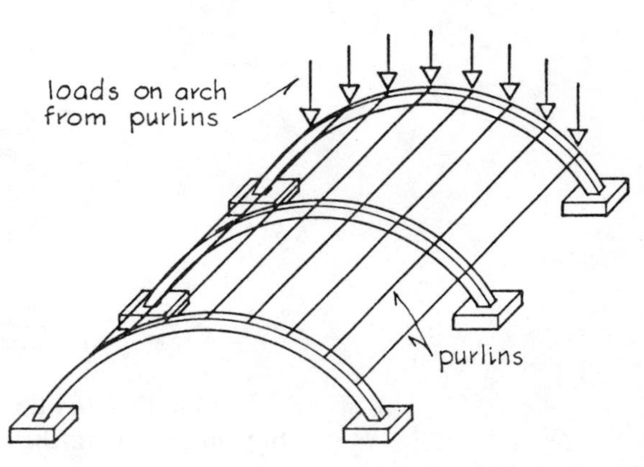

Fig. 13.5

One way of enclosing space by using arches is shown in Fig. 13.5. The purlins act as beams spanning between the supporting arches, and each purlin is designed to support its share of the load due to the roof covering and wind pressure. Each arch is then designed for the loads transmitted to it by the purlins.

Portal frames

Portal frames are, in effect, arches, but whereas true arches are curved, portal frames may consist of straight members. Also, like arches, they can be 3-pin, 2-pin or completely rigid, and they exert horizontal thrusts on their foundations. Figure 13.6 shows some usual shapes for portal frames.

Consider the three structures shown in Fig. 13.7. Although they have certain points of similarity, the design procedures would be different, particularly with respect to the portal frame (*c*). The

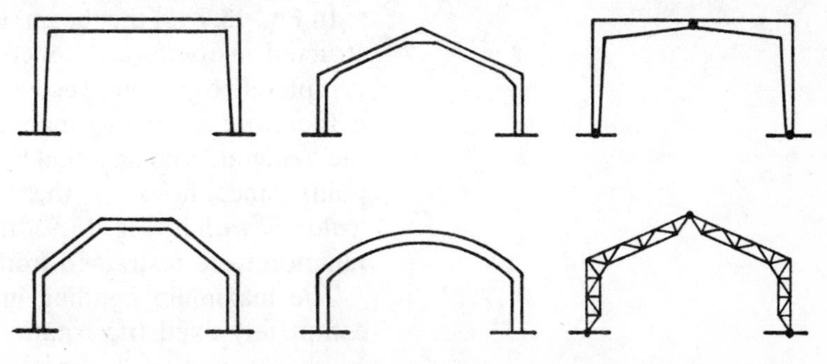

Fig. 13.6

beam of Fig. 13.7 (*a*) is freely supported at its ends, and the manner of bending is shown exaggerated in Fig. 13.8 (*a*). The brick pier would probably be designed for an axial load, although it would be wise to take into account the possible eccentricity due to the beam bearing on the edge of the pier.

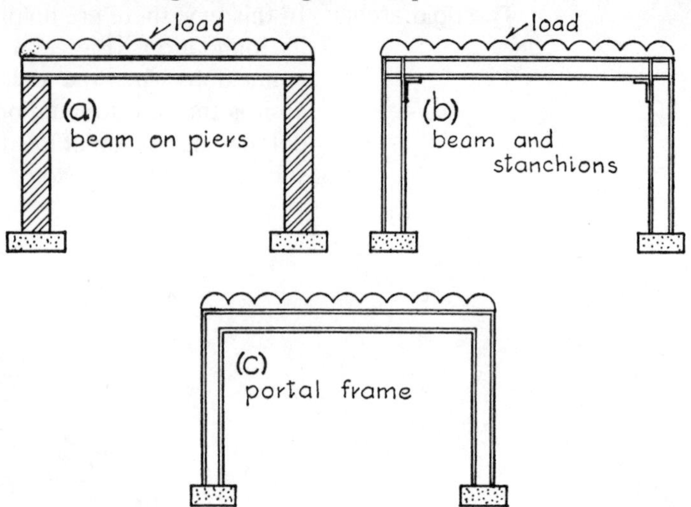

Fig. 13.7

In Fig. 13.7 (*b*) the beam appears to be fixed to the stanchion, but, in fact, with the type of connecting angle normally used, the joint is virtually pinned and cannot resist bending moments. The horizontal leg of the connecting angle would yield as shown in Fig. 13.8 (*b*), so that the beam would be designed on the assumption that it was freely supported at its ends. Eccentricity of the load would be taken into account when designing the stanchion, which has to resist a moment of $\frac{1}{2}W \times e$ in addition to the direct load of $\frac{1}{2}W$.

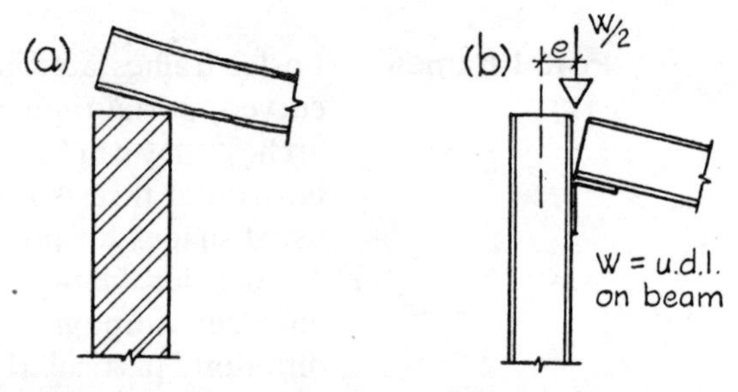

Fig. 13.8

In Fig. 13.7 (*c*) the 'beam' is monolithic with the 'column' if constructed in reinforced concrete, or joined to the column with a completely rigid joint such as a weld if constructed in steel. There is therefore a bending moment in the 'beam' at its junction with the 'column,' and an equal bending moment in the 'column' at this point. Since, however, the 'beam' in bending pulls the top of the 'column' with it (Fig. 13.9), the bending moment is less than if the junction were restrained from rotating.

The maximum bending moments in beams freely supported, completely fixed (no rotation), and partly fixed (some rotation) are compared in Fig. 13.10, assuming a uniformly distributed load on the beam and equal spans in all cases. The maximum bending moment for the beam of Fig. 13.10 (*a*) is $\frac{1}{8}WL$, where W is the total uniformly distributed load and L is the span. Let this bending moment be denoted by Ms. There are no bending moments at the supports. At (*b*) the maximum bending moment is at each of the

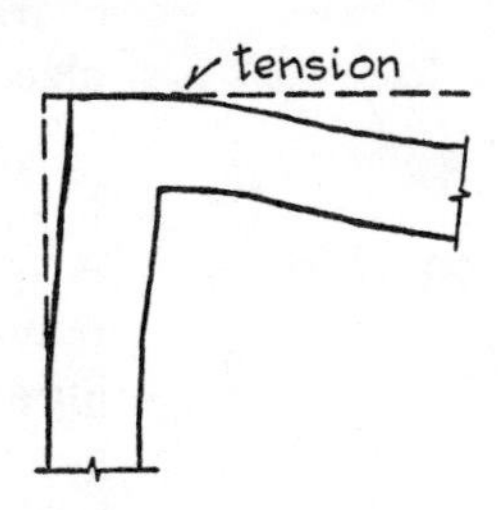

Fig. 13.9

fixed ends and is $\frac{2}{3}Ms$. The bending moment at mid-span is only $\frac{1}{3}Ms$. A smaller beam than that at (*a*) can therefore be used for (*b*) when loading and spans are identical.

The bending moments when the ends are partly fixed, as shown in Fig. 13.10 (*c*), depend on the amount of movement or rotation

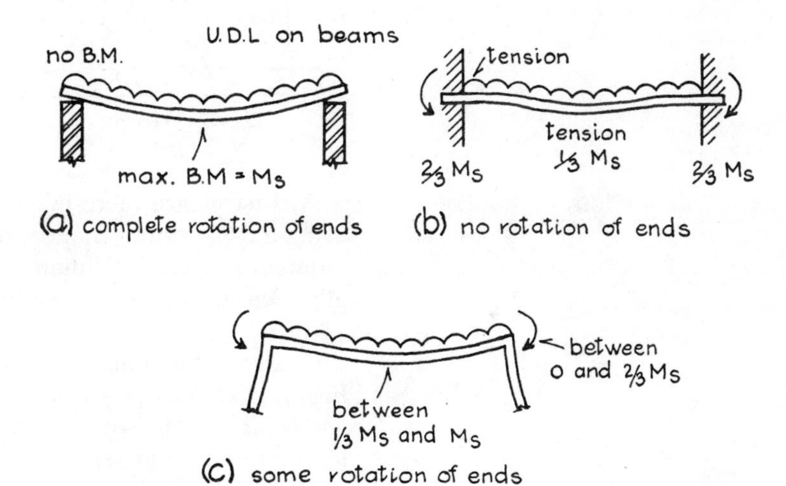

Fig. 13.10

at the end of the 'beam' (except when the portal frame has 3 pins). If the 'beam' is short and big (and therefore very stiff), and the 'column' is long and slender so that it can offer very little resistance to being 'rotated,' the ends of the 'beam' will behave practically as if they were freely supported. The bending moments at the junctions (or knees) will be very small, and that at mid-span will be nearly

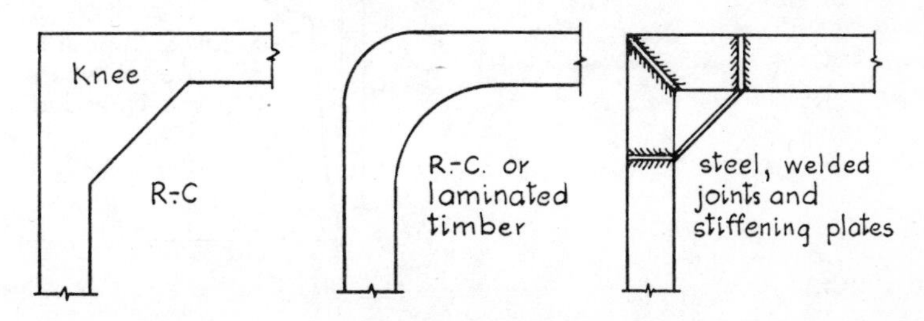

Fig. 13.11

equal to *Ms*, the bending moment for a freely supported beam. On the other hand, if the 'beam' is long and slender and the column very stiff, the resistance to rotation of the joint will be great, and the bending moment at the knee will approximate to ⅔*Ms*.

In practical structures, the bending moment at the knee is invariably greater than that at mid-span, and this gives rise to the characteristic form shown in Fig. 13.11, where more material is placed at the knee than elsewhere.

The three-pin portal frame

A frame of this type when carrying a uniformly distributed load bend is shown (exaggerated) in Fig. 13.12 (*a*), the 'beam' behaving as a double cantilever (unlike the beam in Fig. 13.10). There is no bending moment at mid-span; the maximum bending occurs at the knee, and is equal (for the loading shown) to that which occurs at mid-span of a freely supported beam.

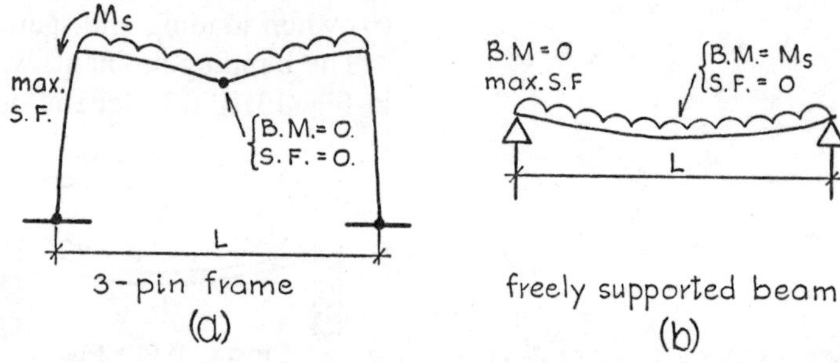

Fig. 13.12

At first glance there might not appear to be any advantage in using a 3-pin portal frame, but since both the maximum bending moment and the maximum shear force are at the knee, aesthetically pleasing structures can be erected of the type shown in Fig. 13.13.

Compare the 3-pin frame with the simply supported beams of Figs 13.13 (*b*) and (*c*). Since the maximum bending moment in (*b*) and (*c*) is at mid-span, the shape shown in (*b*) is a bad one since the least amount of material is where the bending moment is a maximum. At (*c*) the shape is better because the material gets less towards the supports as the bending moment gets less. The maximum shear force occurs, however, at the ends of the beam, and since concrete is weak in shear (diagonal tension), such a beam is not a good structural solution. It may be possible in steel because

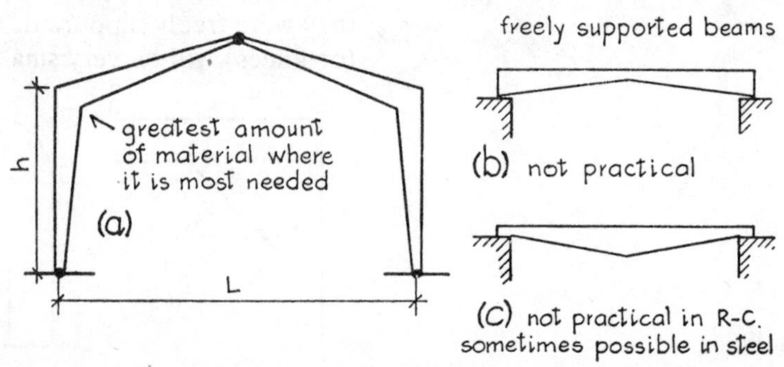

Fig. 13.13

Steel portal frames (*Fleming Bros (Structural Engineers) Ltd*)

of the high shear strength of the material, and crane girders in factories are sometimes of this shape.

In all portal frames, because of the displacement or rotation of the ends of the members, thrusts are caused at the supports or abutments. As with true arches, the shallower the frame, i.e. the greater the ratio L/h, the greater is the horizontal thrust. A tie can be inserted between the supports in the manner indicated in Fig. 13.2.

The two-pin portal frame

The 2-pin portal frame bends as indicated in Fig. 13.14 (*b*), and the bending moments are more evenly distributed than in the 3-pin frame, thus leading to greater economy of material. Since there is a considerable bending moment at mid-span, the 'refined' shape of

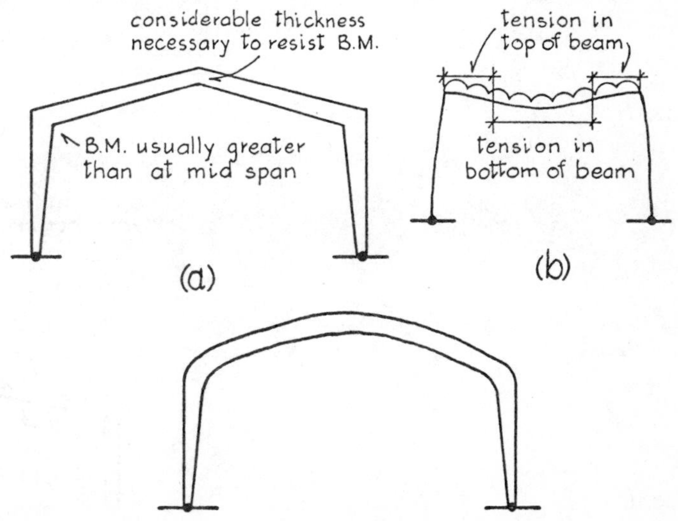

Fig. 13.14

Fig. 13.15

Fig. 13.13 is not possible. The 'beam' may be horizontal, or inclined as in Fig. 13.14 (*a*), or may be of the arch shape shown in Fig. 13.15.

Rigid (no-pin) portal frame

Because of the rigid connexion with the foundations the 'rigid' portal frame bends as shown in Fig. 13.16 (*b*), and a sufficient thickness must be supplied at the feet of the 'columns' to resist the bending moments.

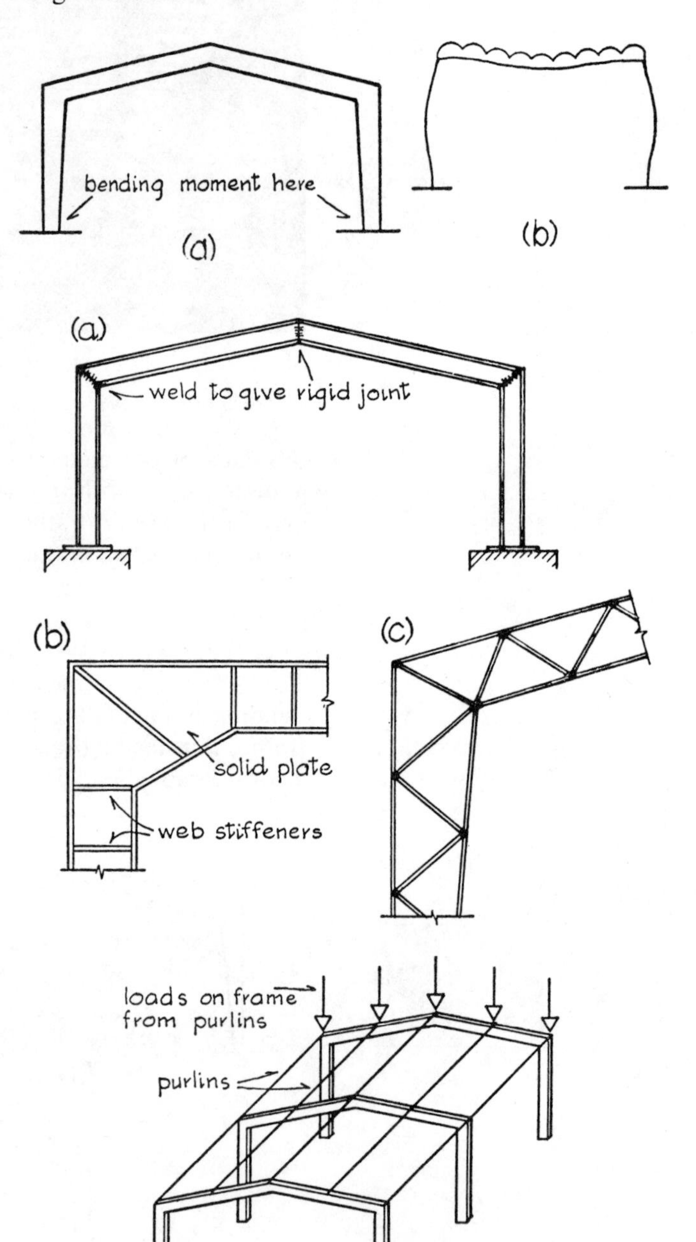

Fig. 13.16

Fig. 13.17

Fig. 13.18

For small spans, a steel frame may be built up of lengths of the same joist section as shown in Fig. 13.17 (*a*), but for large spans, solid plate or triangulated forms may be used.

Because of the cleaner lines, architects often prefer portal frames to triangulated roof trusses supported on walls or columns. Usually the roof covering is supported by purlins which transfer the loads to the portal frames as shown in Fig. 13.18. In reinforced-concrete construction the roof spanning between portals sometimes consists of a reinforced-concrete *shell* (*see* Chapter 14).

Timber portal frames (*Timber Development Association*)

14 Folded Slabs, Shells, Hyperbolic Paraboloids

It has been stated in previous chapters that the shape of individual structural members such as beams and columns is an important factor in their design. Until fairly recently, however, little thought was given to relating shape to the design of a structure as a whole. Starting about 1920, and at first with slow progress, new and very important constructional systems have been developed. These new constructions can all be included under the general heading of *space structures*, but since during recent years the word 'space' has become associated so much with man's exploration of outer space, the term *three-dimensional* (3-D) *structures* might be more appropriate.

Most of the structures and members of structures described in previous chapters are designed on the *one-plane*, i.e. *two-dimensional*, principle. Referring to Figs 8.8 and 13.18, this means that the members (called *purlins*) which support the roof covering are designed first, then each of the roof trusses (or arches, portal frames, etc.) is treated as a structure in one plane and designed to support its proportion of the roof loads transmitted to it by the purlins. The purlins spanning from truss to truss are not assumed to contribute in any way to the strength of the roof trusses. The design of a three-dimensional or space structure is based on the behaviour of the structure considered as one complete unit and not as being built up of separate independently designed components.

Nature, too, has given inspiration to modern designers, particularly with regard to curved structures. For example, an eggshell because of its curved shape is very strong considering the thinness of the shell and its brittle nature. Similarly, because of curves and corrugations, shells of sea animals have a high strength-to-weight ratio.

For purposes of discussion, space structures will be classified in this book as:

The folded slab or plate.
The barrel vault or shell (also the shell dome).
The hyperbolic paraboloid.
Braced domes, including geodesic domes.

The single-layer and double-layer grid.
Suspended cable structures.

The folded slab

A very simple experiment can be performed with a sheet of A4 paper. In Fig. 14.1 (*a*) the sheet will bend a great amount and its self-weight may even be sufficient to cause it to collapse into the space below. If the sheet is folded as at (*b*), it becomes very much more rigid and is capable of supporting its own weight and perhaps a little superimposed load. The sheet, however, tends to flatten under the superimposed load due to 'arch action,' but if this flattening tendency is prevented by glueing both ends to sheets of

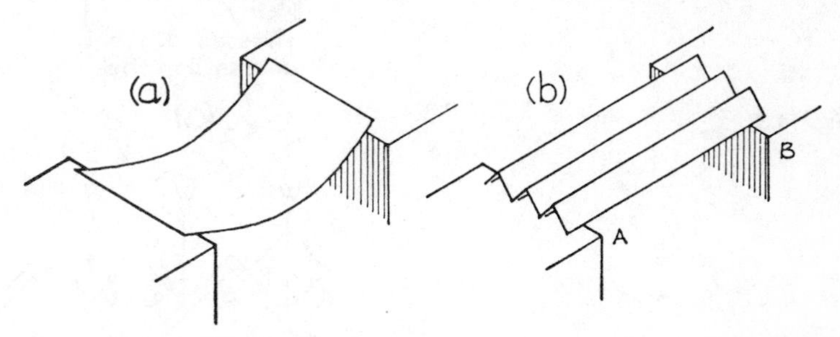

Fig. 14.1

paper or cardboard (diaphragms) as shown in Fig. 14.2 (*a*), the 'roof' becomes capable of supporting still more load. The structure acts as a beam spanning between *A* and *B*.

This principle (without the end diaphragms) has, of course, been used for many years in the common corrugated galvanized steel sheet [Fig. 14.2 (*b*)]. Only recently, however, has it been realized that a large-span roof can take advantage of this shape and be designed as a 3-D structure.

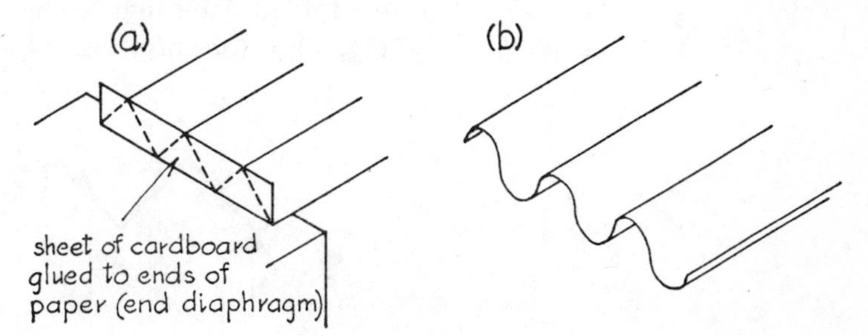

Fig. 14.2

Stressed skins

The folded slab is an example of *stressed skin* construction, the principle of which is as follows. A horizontal slab as shown in Fig. 14.3 (*a*) has very little stiffness and therefore little resistance to bending when subjected to vertical loads due to its own weight and any superimposed loads. The same slab loaded as at (*b*) has great stiffness, the stresses being taken by the 'skin.' This could be demonstrated by loading a small piece of plywood or hardboard in the manner shown in the diagram.

Figure 14.4 represents the cross-section of a folded-slab roof. W is the vertical load due to the weight of the roof and superimposed load such as snow. The force W can be resolved into two components (*see* Chapter 3), a component P acting parallel to the roof slope and a component R acting at right angles to the slab as at (a).

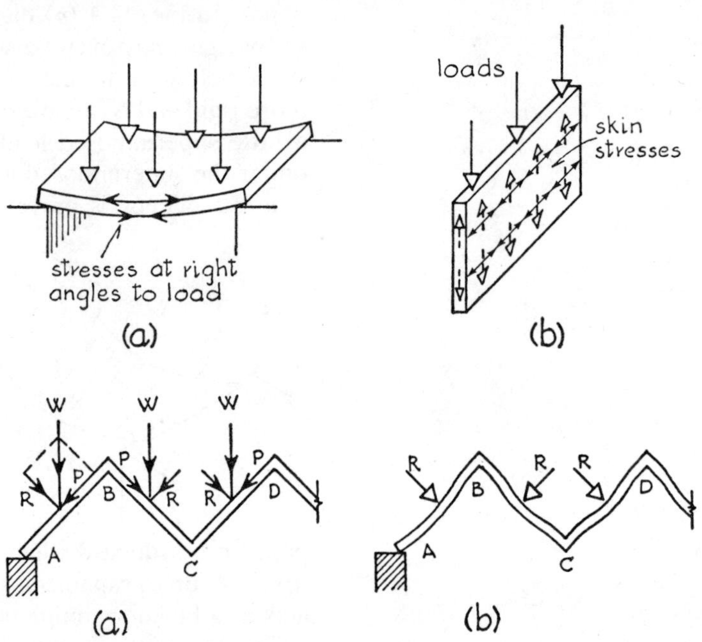

Fig. 14.3

Fig. 14.4

Force P is resisted by skin stresses as in Fig. 14.3, and only force R causes bending of the slab due to spanning between A and B (and between B and C, etc.). It is to be noted that each fold (such as B, C or D) acts as a support or beam for two adjacent slabs, so that the folded slab can be compared to a continuous beam as shown in Fig. 14.5. (*See also* Fig. 14.4 (b) for manner of bending.)

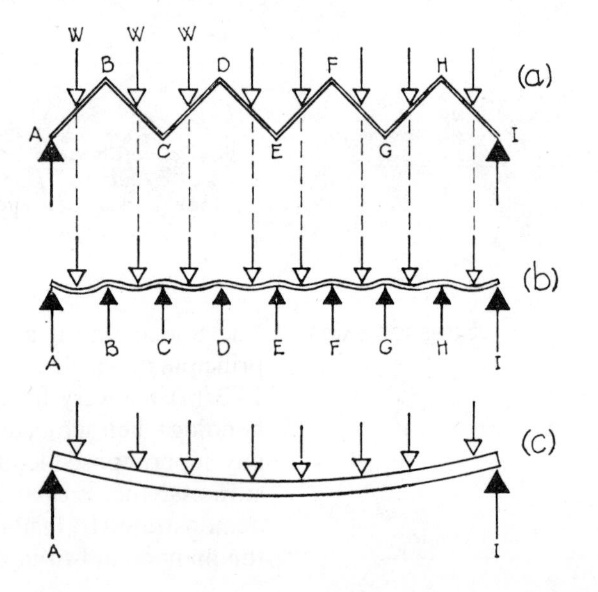

Fig. 14.5

Compare this method of construction with a horizontal slab spanning from A to I as shown in Fig. 14.5 (c). A very much thicker slab would be required than at (a) because of the slab bending as shown. Another important difference between the constructions of Figs 14.5 (a) and (c) is that in (c) the slab is designed

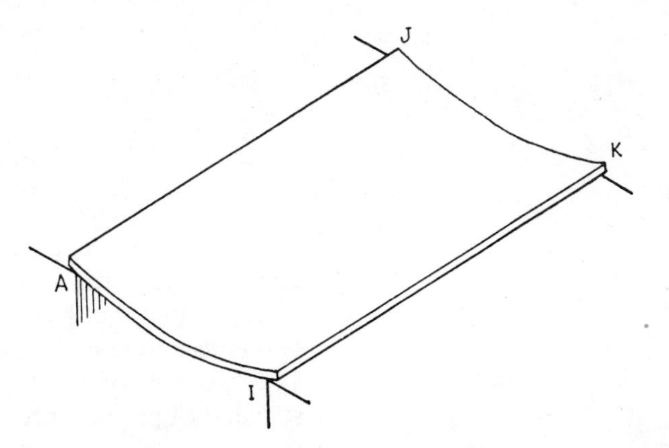

Fig. 14.6

as spanning from A to I and would be supported continuously along the sides as shown in Fig. 14.6, the main supports running from A to J and from I to K.

With a folded slab, the whole of the roof acts as a beam spanning from the front (AI) to the back (JK) as in Fig. 14.7. The edge beam and its supports a, b, c, along the sides are necessary only to enable the first 'local slab' to span from IK to HL.

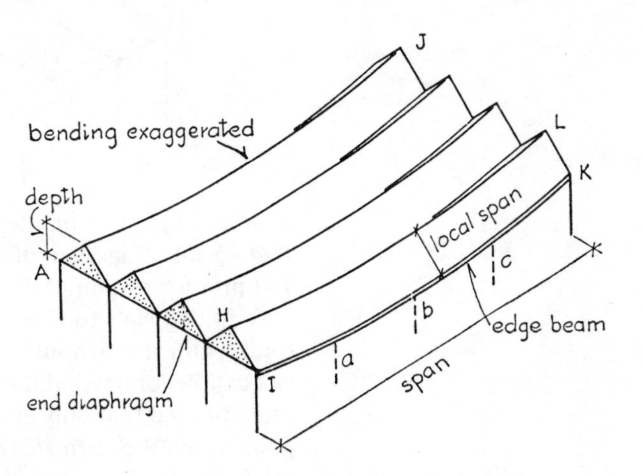

Fig. 14.7

The structural analysis of such folded slabs or stressed-skin construction is complex, but put simply, and to sum up, the roof can be considered as a huge beam spanning from front (AI) to back (JK), the strength of the beam being derived from its hollow shape and its depth. The slabs forming the roof, since they have only a small span (called 'local span' in Fig. 14.7) can be quite thin, and their functions are to resist, by acting as small-span beams, the load components acting at right angles to the roof (components R

in Fig. 14.8), and to resist by skin stresses the components P acting parallel to the roof.

The load components P act similarly to the thrust of an arch and tend to cause the roof to flatten; this is prevented by tying the two ends (AI and JK) by means of diaphragms which resist the thrusts

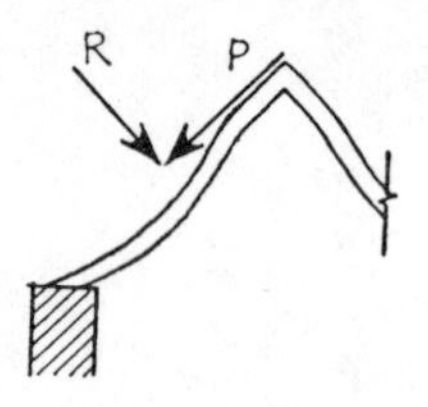

Fig. 14.8

from the whole length IK. When the roof consists of reinforced-concrete slabs the diaphragms are also of solid reinforced concrete, so that the thrust from the whole of the roof is taken at the two ends and none by the edge beams and columns *a, b, c,* of Fig. 14.7. A roof of the type shown in Fig. 14.7 was recently constructed of precast, prestressed slabs 50 mm thick. The span (I to

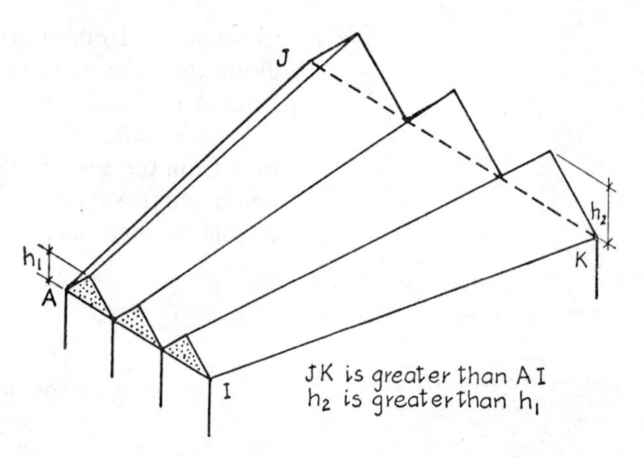

Fig. 14.9

K) was 20 m, and the depth of roof was 1220 mm. It is estimated that 65 mm thickness of slab is sufficient for a span of 30 m, and 100 mm for a span of 40 m.

A folded-slab roof need not be of uniform depth and uniform total width throughout its span (Fig. 14.9), and if desired the roof can cantilever beyond the end diaphragm (Fig. 14.10). It should be remembered that only at each end [and perhaps at an intermediate point when the span IK (Fig. 14.9) is very large] is the roof closed

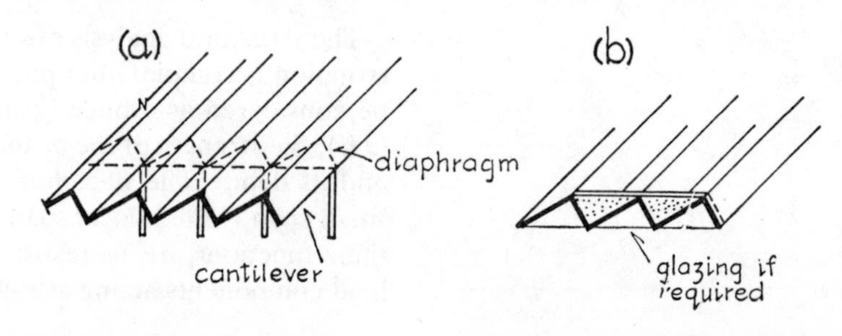

Fig. 14.10

by diaphragms. The remainder of the roof is completely hollow and unobstructed by ties or other members. Even at the ends, the diaphragms, if desired, can be at the top as in Fig. 14.10 (*b*), thus enabling glass to be used. Other folded-slab shapes in addition to that of Fig. 14.7 are possible, a few types being shown in Fig. 14.11.

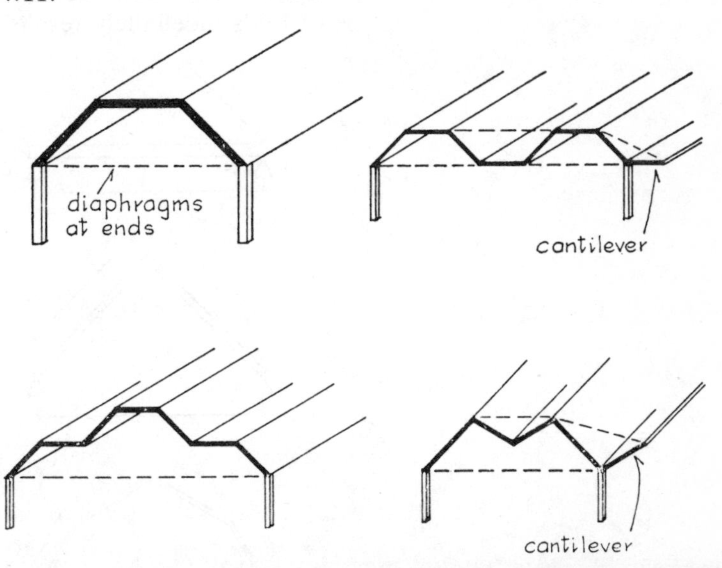

Fig. 14.11

Folded-slab roof at Canterbury, Kent (*Cement & Concrete Association*)

The modern barrel vault or shell roof

Assume that a roof is required to cover a certain width of opening l and length L. If the width l were considerable, a very thick slab would be required for Fig. 14.12 (*a*). By designing as a folded-slab roof as at (*b*), the span for the design of the slab (the local span) is reduced to l_1 and a thinner slab would suffice. The shape at (*c*) reduces the local span to l_2 giving a still thinner slab, and at (*d*) the local span is reduced still further (l_3). The greater the number of folds, the less is the local bending, so it is an advantage to reduce the local span to a length which is capable of being bridged by the thinnest practicable slab. (It must be remembered also that a certain minimum thickness is required to enable the roof to act as a hollow beam spanning from back to front.) Increasing the number of folds indefinitely results in the curved roof of Fig. 14.12 (*e*),

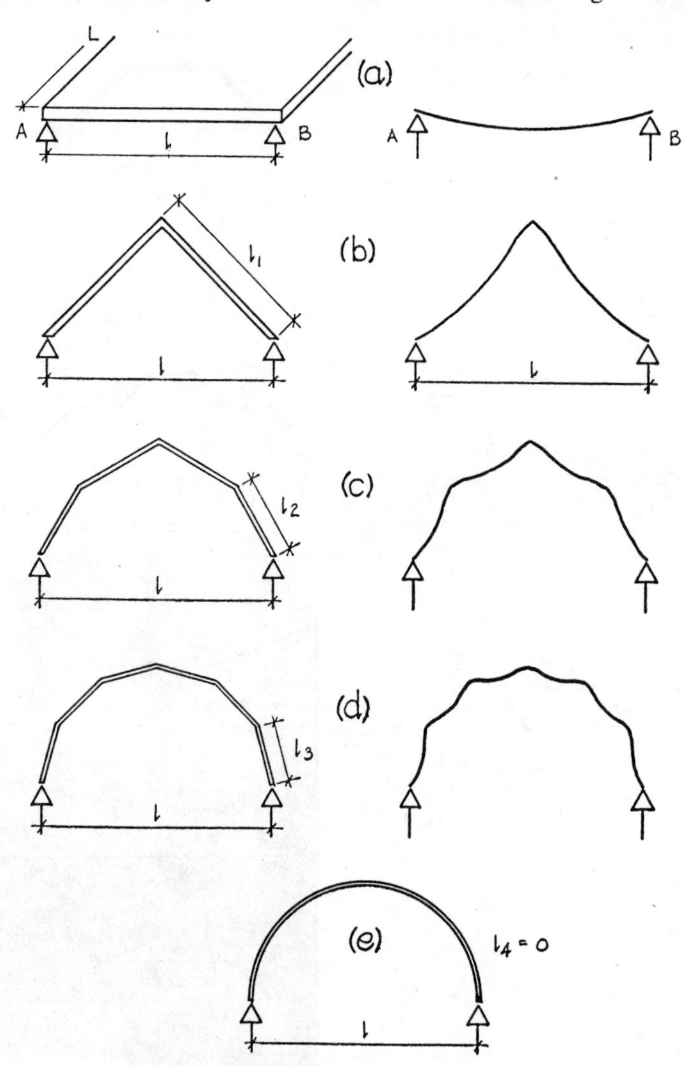

Fig. 14.12

where the local span has been reduced to zero. This type of roof is known as a *barrel vault* or a cylindrical *shell roof* (compare with an eggshell), and when constructed in reinforced concrete these roofs are usually about 50 to 75 mm thick. Pieces of paper can be folded to give the shapes of Fig. 14.12 and to illustrate the advantages of many folds.

It was mentioned in Chapter 4 that the masonry barrel vault imposes thrust along its whole length, behaving in effect like a series of arches (Fig. 14.13) and requiring large abutments. The shell roof could be called the modern barrel vault, and like the folded slabs already dealt with, requires end diaphragms. These diaphragms resist the thrust from the whole roof, which then behaves as a beam spanning from front to back.

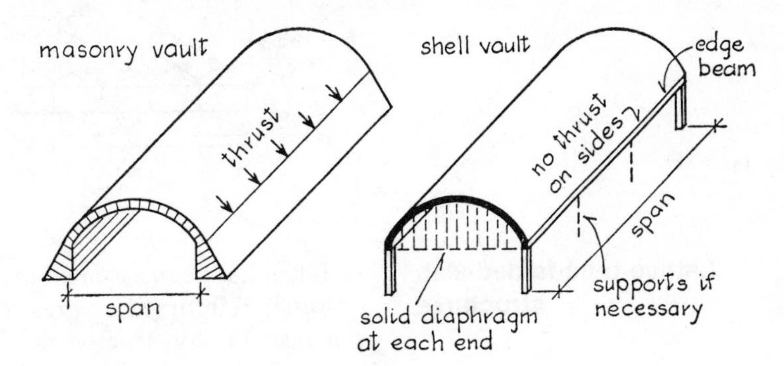

Fig. 14.13

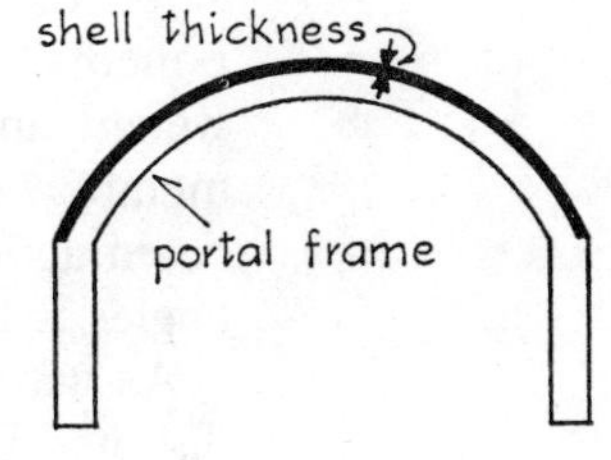

Fig. 14.14

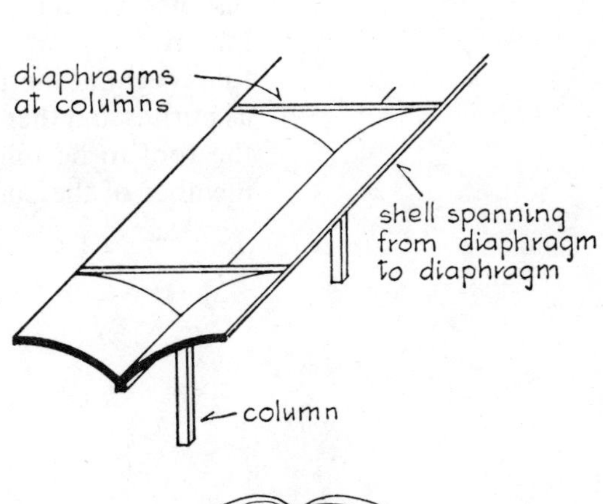

Fig. 14.15

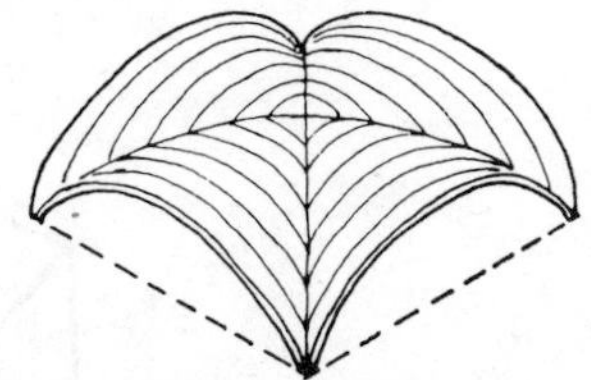

Fig. 14.16

Instead of a solid diaphragm, a rigid portal frame (Fig. 14.14) can be used to resist the thrust from the roof.

There are various possible adaptations of the shell roof, one of which, the butterfly roof (Fig. 14.15), is a structure suitable for railway platforms. A shell roof formed by two intersecting barrel vaults is shown in Fig. 14.16.

As an example of the large areas without any internal columns which can be covered by shell roofs (and other folded-slab structures), Fig. 14.17 shows a roof consisting of 10 barrel shells each 12 m wide and having a span of 30 m.

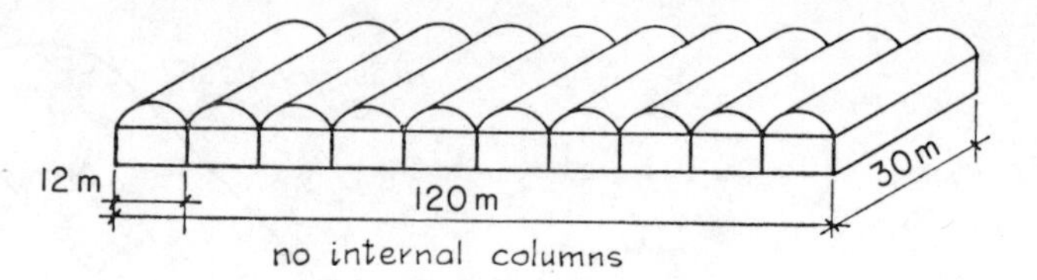

Fig. 14.17

Lattice-type folded-slab structures

So far, it has been assumed that the slabs or shells have been solid, of either reinforced or prestressed concrete. These types of roof can also be constructed in steel and the same basic principles apply. Referring to the simple structure shown in Fig. 14.18, each slope of the roof acts as a lattice girder in resisting the components, acting parallel to the roof, of the forces due to self-weight and the weight of the covering. In addition to acting as members of the lattice truss, the main struts are designed for the beam action due to the components (*R*, Fig. 14.4) acting at right angles to the roof ('local bending').

As with slab roofs, the smaller the local spans, the better. The distance between two purlins is usually dictated by the roofing material, e.g. about 1375 mm for asbestos-cement sheets. If the folds occur at these points, the chords of the lattice girders also act as purlins and there is no bending in the main struts. Note that, for the roof to be fully effective, each fold must be supported by a member of the end diaphragm truss (Fig. 14.19).

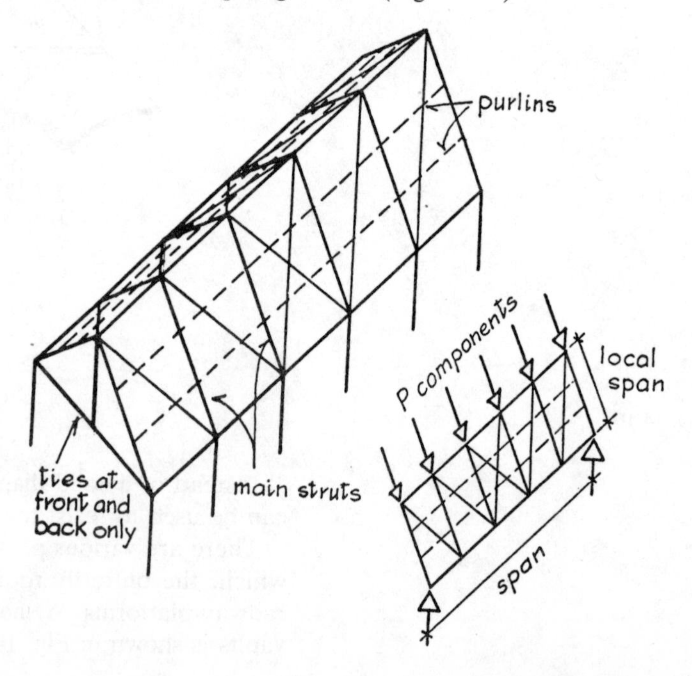

Fig. 14.18

All the shapes shown in Fig. 14.11 are possible by using steel lattice construction, an example being given in Fig. 14.20. If it is decided to dispense with columns *d* and *e* in order to obtain a large opening, as for example in aircraft hangars, a lattice girder spanning from *A* to *I* can act as the end diaphragm (Fig. 14.21).

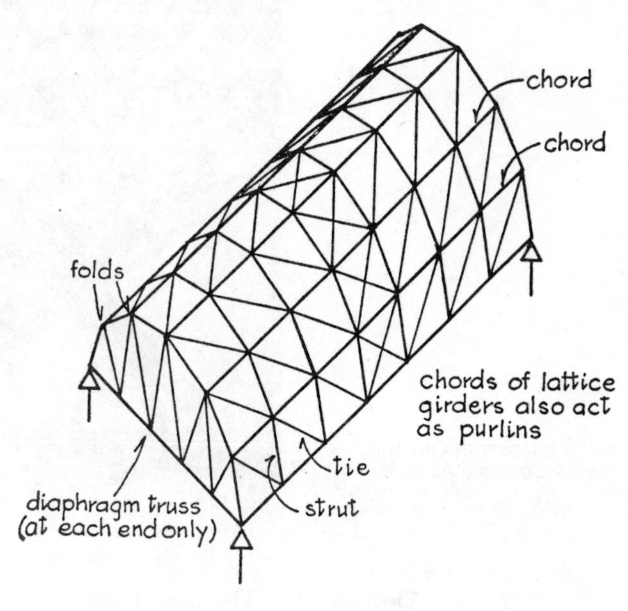

Fig. 14.19

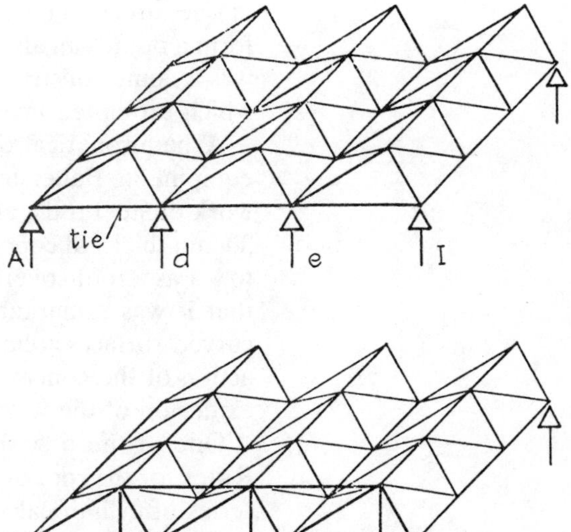

A
tie
d
e
I

Fig. 14.20

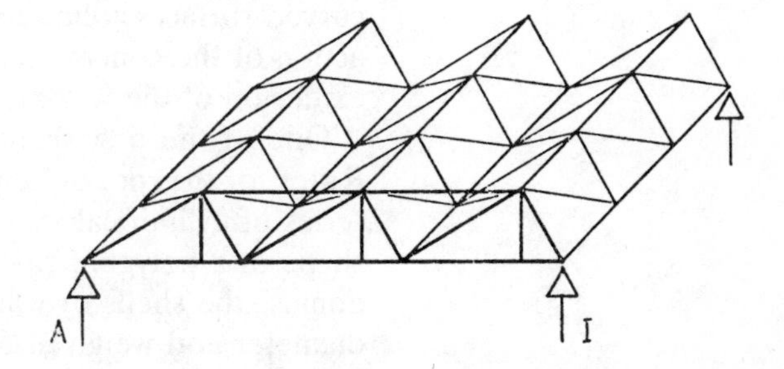

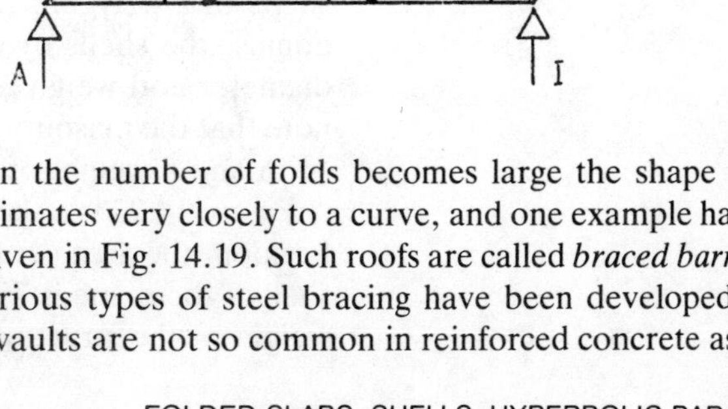

Fig. 14.21

When the number of folds becomes large the shape of a roof approximates very closely to a curve, and one example has already been given in Fig. 14.19. Such roofs are called *braced barrel vaults*, and various types of steel bracing have been developed. Braced barrel vaults are not so common in reinforced concrete as in steel.

Nervi's aircraft hangers at Orbetello, Italy (*Cement & Concrete Association*)

Domes

The barrel vaults discussed so far are examples of shells curved in one direction only. Shell roofs which curve in two directions to form a dome can also take advantage of skin stresses, and in fact it was a dome constructed in 1922 by a German, Walter Bauersfeld, which prompted investigations into shell structures.

A hemispherical dome was required for the testing of optical equipment. Bauersfeld constructed the dome with a lattice network of steel rods, and covered the steel with a thin layer (about 30 mm thick) of concrete. The purpose of the concrete was merely to act as a roof covering, but tests on the completed dome showed that it was remarkably strong. Experiments and calculations on curved surfaces followed, which took into account the combined action of the concrete and the steel reinforcement, and led to an extension of the work to barrel-vault roofs.

One of the first shell domes was designed by Dischinger and Ritter for the roof of Leipzig Market Hall. This dome consists of a series of cylindrical shells which intersect to cover an area in the shape of a polygon, ribs being provided at the intersections. The domes, the shells of which are 89 mm thick, are each 75·6 m in diameter and weigh 21·5 MN. For comparison it is interesting to note that the masonry dome of St Peter's, Rome, has a diameter of 40 m and a weight of 100 MN.

Figure 14.22 is a sketch of a covered tennis court at Wimbledon, London, the area covered being 1415 m^2. The reinforced-concrete shell of the dome is 76 mm thick with a span of 53·4 m. The whole weight of the structure is supported by four corner columns.

A similar construction was used for a factory at Brynmawr, South Wales, where the main production area was covered by nine domes each 25·3 m by 19·4 m, and 76 mm thick. The domes are supported by rigid end frames with columns at the four corners of each dome. A sketch impression of part of the roof is given in Fig. 14.23.

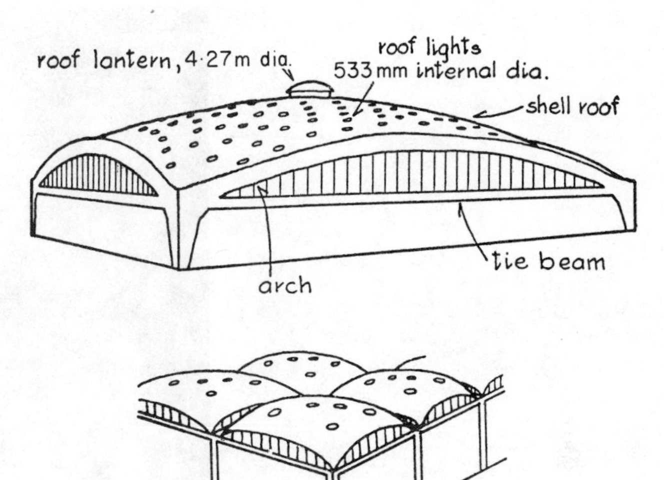

Fig. 14.22

Fig. 14.23

Two large dome structures have recently been completed in the United States of America. The King County Stadium in Seattle is covered with a scalloped concrete spherical dome that spans 210 m and the Louisiana Superdome in New Orleans is a ribbed steel spherical dome that is 207 m in diameter. The Superdome might be better described as a braced dome which is discussed further in Chapter 15.

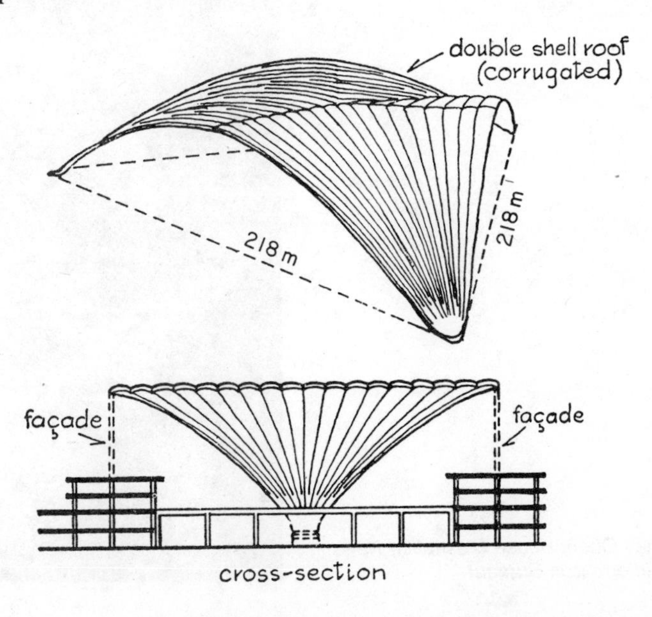

Fig. 14.24

A domed structure may also spring from ground level as in Fig. 14.24, and the shell itself can consist of corrugations to give extra strength. This is the principle of the Exhibition Hall in Paris, which covers about 2·06 ha. The building is an equilateral triangle with a length of side of 218 m. (To convey some idea of the immense size, it may be mentioned that the maximum length of a football pitch is

Sydney Opera House (*Australian News and Information Bureau*)

Sydney Opera House (*Australian News and Information Bureau*)

119 m.) The shell is in three fan-shaped sections to form a groined vault (without ribs). Other types of domes are discussed in Chapter 15.

The roof structures enclosing the Sydney Opera House and Concert Hall appear to be a complex series of shells but in fact they are a set of spherical triangles, each one standing on a vertex and leaning against its neighbour. The shells themselves consist of a series of concrete ribs radiating from the podium and reaching 55 m above the floor at their highest point. The greatest span is 57 m and the overall length of the larger hall, measured from tip to tip of the end shells, is 121 m. The shell radius is 75 m and varies in thickness with distance from the rib centre-line and height above the podium.

Sydney Opera House (*Australian News and Information Bureau*)

The hyperbolic paraboloid

The early structures in reinforced concrete (*see* Chapter 11) followed a traditional pattern, usually consisting of flat horizontal slabs and beams supported by vertical columns. Concrete, however, is a free form material and can be made to take the shape of practically any mould or formwork into which or on which it is poured. The shell roofs already described illustrate a departure from traditional post and lintel construction, and the hyperbolic paraboloid, which makes efficient use of materials by relying on shape rather than on weight, is one of the newest forms of shell roof construction. Its use is not confined to reinforced concrete; hyperbolic-paraboloid roofs have been constructed in timber and even in plastics.

The principle of this type of shell roof is illustrated in Fig. 14.25. $ABCD$ is a square on plan; F is vertically above C; and at the other diagonally opposite corner, E is vertically above A; so that the original square has been 'warped' into the doubly curved surface *EBFD*. Now divide all the sides into an equal number of parts and join by straight lines from *BF* to *ED*, and from *EB* to *DF*. Although these sides are joined by straight lines, the surface formed is curved in two directions. The curved line *BD* has been put in the sketch merely to illustrate that the surface curves up from the low corner *B*

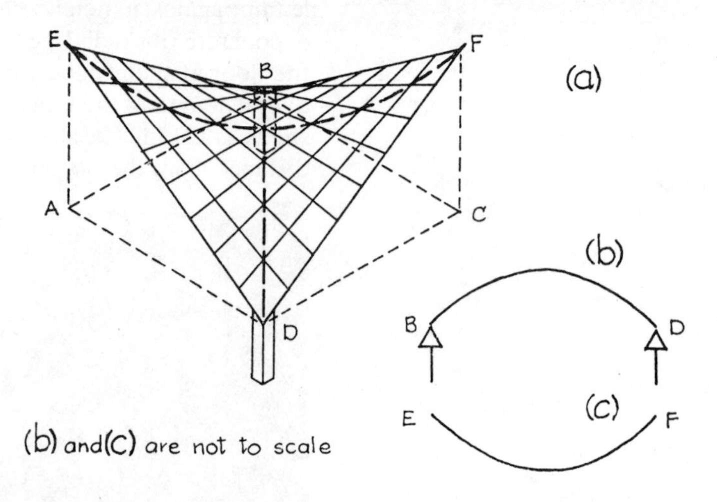

Fig. 14.25

and then down to the other low corner *D*, forming a parabola (concave looking upwards). The cross-section cutting the two corners *B* and *D* is shown in Fig. 14.25 (*b*). On the other hand, the surface curves down from the high corner *E* and then up to the other high corner *F*, forming a parabola (concave downwards), the cross-section being shown at (*c*). All cross-sections parallel to the

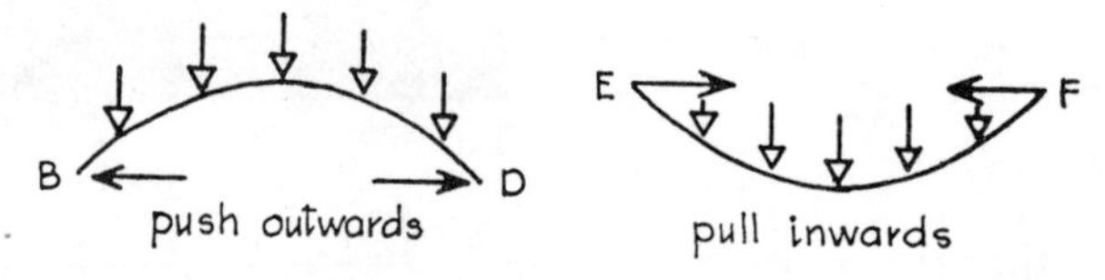

Fig. 14.26

diagonals *BD* and *EF* are parabolic. Although this doubly curved surface appears to require complicated *formwork* (timber panels or other material on which to place the wet concrete forming the roof) the formwork can actually consist of straight timber joists spanning along the straight lines shown in Fig. 14.25 and covered by flexible plywood panels to form the complete curved surface to receive the concrete.

Usually, *edge beams* are provided along the four straight edges *BF, FD, DE* and *EB* to provide restraint for the shell. The weight of

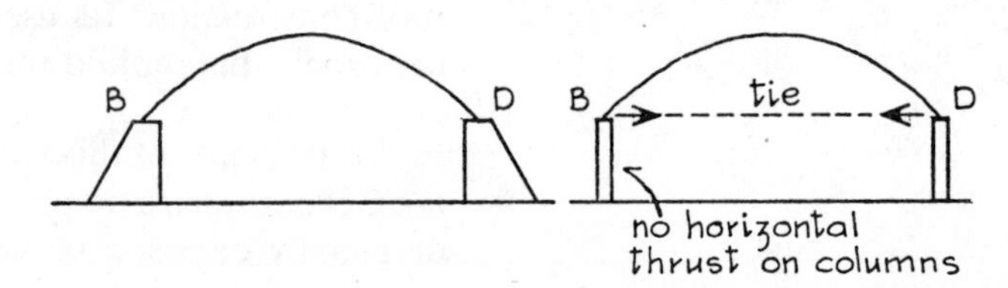

Fig. 14.27

the shell and any superimposed load tends to push the roof downwards, so that *B* and *D* tend to get further apart, and *E* and *F* tend to get closer together (Fig. 14.26). All diagonal cross-sections such as *BD* act as arches, and all diagonal cross-sections such as *EF* act as suspension cables. The shell is therefore in compression along the curved surface *BD* (and all lines parallel to this low diagonal) and in tension along *EF* (and all lines parallel to the high diagonal). The

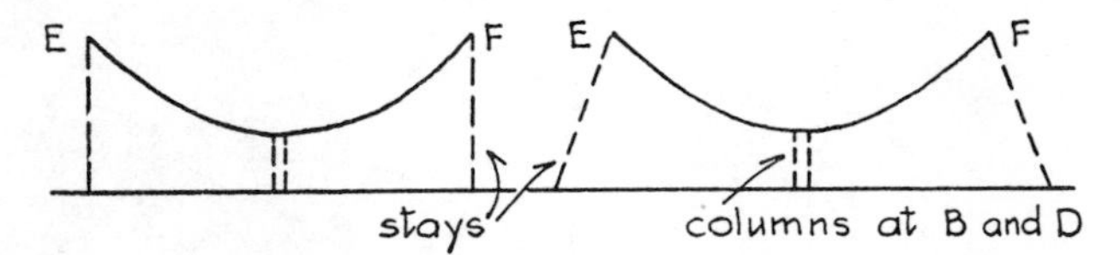

Fig. 14.28

load is shared between these 'arches' and 'suspension cables.' It is necessary to support the roof only at two points such as *B* and *D*, but because of the arch action there must be either buttresses at *B* and *D* or a tie joining these two points (Fig. 14.27).

With either buttresses or ties, any tendency of the 'suspension cable' *EF* to sag (Fig. 14.25) is resisted by the 'arch' because *EF* cannot sag without pushing out *B* and *D*. Buttresses and ties to the

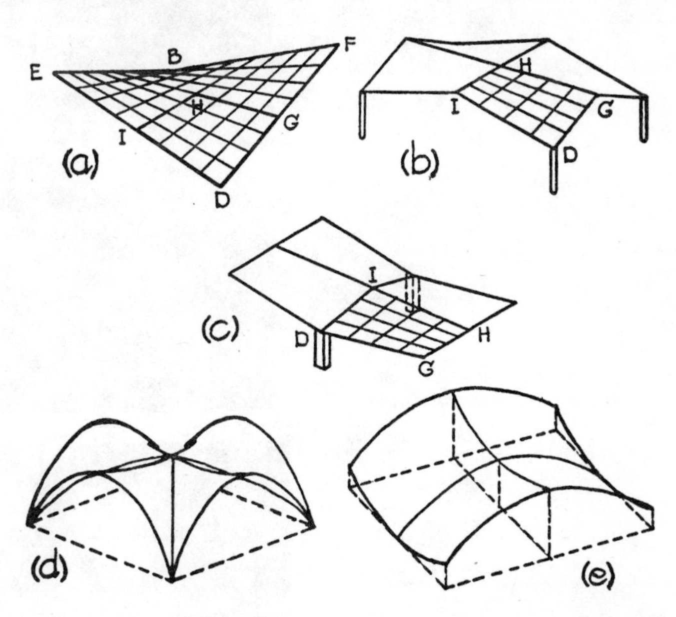

Fig. 14.29

'arch' can be dispensed with if points *E* and *F* are held in position as shown in Fig. 14.28. Stays as shown, firmly attached to supports, prevent *EF* from sagging. Points *B* and *D* of the 'arch' cannot move outwards if points *E* and *F* are held in position; therefore there are no thrusts at *B* and *D*. The stresses in the shell itself are small, and the roof can be very thin, whether it consists of timber or reinforced concrete. The roof of the Cosmic Ray Pavilion at the University of Mexico has a thickness of only 16 mm. Normally, however, the shell thickness is in the range 60 mm to 75 mm in order to give sufficient protective cover to the steel reinforcement.

Candela's Restaurant, Mexico (*Cement & Concrete Association*)

Elm Grove Church, Wisconsin, USA (*Cement & Concrete Association*)

School at Southwark, London (*Cement & Concrete Association*)

Many types of roof can be constructed by combining hyperbolic parabola units or parts of units in various ways. Figure 14.29 (*a*) shows the entire warped surface of one hyperbolic parabola; *DGHI* represents one quadrant. Four (or more) similar quadrants can be combined as at (*b*). (Edge beams and ties are not shown.) By combining the quadrants in a different way, a structure can be formed as at (*c*).

An example of a roof similar to that shown in (b) is the Ponce Coliseum in Puerto Rico. This 10,000 seat arena encloses an area of 6000 m^2 with four, 102 mm thick, hyperbolic paraboloid shells. Supported on four buttresses, the entire roof was post tensioned in place before the temporary falsework was removed. The edge girders span 42 m from the buttresses and stiffen what are most probably the longest cantilevered shells in the world.

Two further applications of hyperbolic paraboloid surfaces are the groined vault [Fig. 14.29 (*d*)] and the saddle roof (*e*).

15 Braced Domes and Geodesic Domes

A great deal of publicity has been given in the architectural press in recent years to *geodesic domes,* which were first developed by R. Buckminster Fuller, born in 1895 in the United States. These domes will be discussed in more detail later in the chapter. It should be remembered, however, that domes built up with timber bracing members were used in the Middle Ages, and steel-framed domes date back to about the middle of the 19th century. Schwedler designed a dome, erected in Berlin in 1863, which had a span of 40 m.

One of the simplest types of braced dome consists of a number of polygonal rings joined by meridional members; the quadrilaterals thus formed are divided into triangles by other bracing members. In order to explain the principle, Fig. 15.1 (*a*) shows a braced dome in which the polygonal rings have four sides only (forming a square). Usually the polygons have 6, 8 or more sides, an impression of an eight-sided dome being given at (*b*). There are

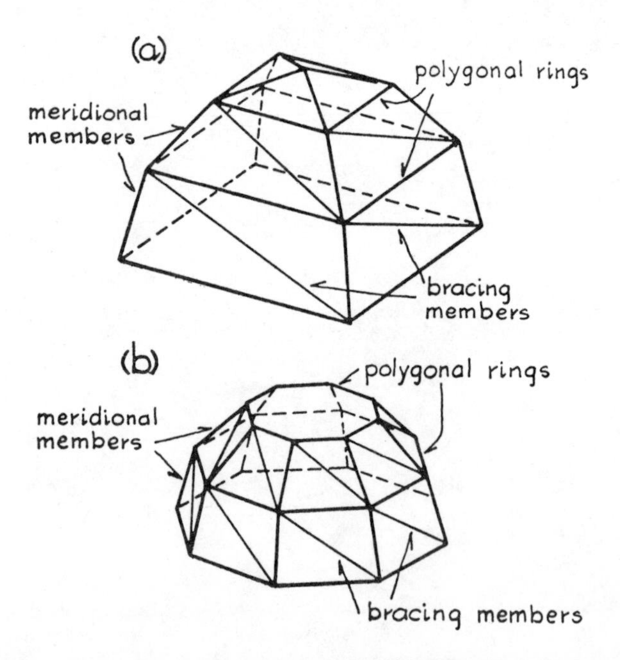

Fig. 15.1

many types of braced dome, and for detailed information reference may be made to a booklet published by Dr Z. S. Makowski of the University of Surrey, England.

Geodesic domes

'Geo' comes from the Greek and means 'earth'; hence 'geography,' 'geology,' etc. In connexion with domes, the word 'geodesic' can be taken simply as meaning 'spherical,' but a more detailed explanation of what constitutes a geodesic dome will be given later. Although some writers have pointed out that satisfactory braced domes of various types have been used for many years, there is no doubt that the work of Buckminster Fuller has aroused a new interest in dome construction in many architects and engineers. In view of this interest and of the personality of Mr Fuller, a few notes about the man and his ideas may not be out of place in this volume.

On the 5th June, 1958, the R.I.B.A. Discourse was given by Mr Fuller, and in introducing him, Sir Hugh Casson stated—

> 'He has been variously described in public as an engineer, a mathematician, a chemist, an inventor, a designer, a writer, a philosopher, a scientific idealist, a prophet, and a crank. . . . I believe that he prefers his own title of "comprehensive designer," a name to which after some 40 years of original thought and most lively invention in all fields, particularly spherical geometry, light-alloy fabrication and the chemistry of plastics, coupled of course with—to us as architects—his more famous experiments in transport and housing, he is surely as much entitled as any man alive.'

Many new terms have been introduced into technical literature by Mr Fuller, such as 'energetic synergetic geometry'; 'dymaxion structures'; 'tensegrity.' His work and philosophy is much too extensive and complicated to be dealt with adequately here, and the reader is referred to *The Dymaxion World of Buckminster Fuller,* by Robert W. Marks (Reinhold Publishing Corporation, New York).

In the lecture referred to above, Mr Fuller defined *synergy* as the 'integrated behaviours of nature and the behaviour of a whole system unpredicted by the behavour of its components or any sub-assembly of its components.' For example, the two gases oxygen and hydrogen when combined in certain proportions give water, and this could not be predicted by the behaviour of the individual gases. Similarly, when certain metals are combined the strength of the resulting alloy can be much greater than the sum of the separate strengths.

One meaning of 'energetic' is 'science of energy,' and Fuller's creed is that Nature has recognizable patterns of energy relationships which can be transformed into usable forms. Nature builds her structures so that internal forces act invariably in the direction of minimum effort, and thus 'maximum gain of advantage from minimum energy input' is obtained. What may be true for the regular patternings of force in minute structures such as crystals can also be true for man-devised structures such as bridges and buildings. For example, the triangle is the plane figure which has maximum rigidity accomplished with least effort, and symmetrical triangular systems provide the most economical energy flow. The

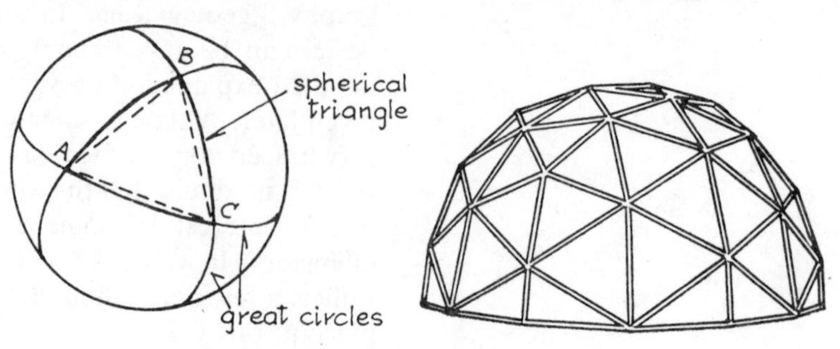

Fig. 15.2 Fig. 15.3

geodesic dome is normally a network of triangles, so that the maximum strength and rigidity are obtained with the minimum amount of material. Fuller's domes are local applications of a comprehensive system which he calls 'energetic geometry' and which deals with 'the most economical relationships of points in universe and their transformation tendencies.'

A spherical triangle (Fig. 15.2) is a portion of the curved surface of a sphere, formed by the intersection of three great circles.

Kaiser Aluminium dome at Honolulu (*Kaiser Aluminium Company*)

(A great circle has a diameter equal to the diameter of the sphere.) If the vertices *A*, *B*, *C* of the triangle are joined by straight lines (chords) a plane triangle is formed, and a network of such triangles can be constructed to form a geodesic dome as shown in Fig. 15.3. The greater the number of triangles, the greater resistance is offered to collapse of the structure. The first geodesic dome dates from about 1948, and in 1952 the Ford Motor Co. constructed under licence a 28·4 m aluminium and plastic dome over their Dearborn Rotunda building.

The Kaiser Aluminium Co. (United States) was one of the first firms to take out a licence under Buckminster Fuller's patent, and their first construction was an aluminium-skinned dome for a concert hall in Honolulu. The dome is 44.2 m in diameter and was erected in 22 hours. Within 24 hours of the landing from America of the dome components, a concert was given by the Hawaiian Symphony Orchestra to an audience of 1832.

In the geodesic dome shown in Fig. 15.3 the strength of the dome is given entirely by the triangular framework. Light-weight roof covering such as thin aluminium sheeting or some form of plastic would not be assumed to contribute towards the strength of the structure. The Kaiser Aluminium dome is a space or 3-D truss which makes use of the roof covering as a stressed skin as well as employing the advantages of the geodesic principle. According to the Kaiser Aluminium Co., the aluminium panels account for 'approximately 75 per cent of the structural integrity of the space

An interior at Fort Worth, Texas, USA
(*Kaiser Aluminium Company*)

truss.' This system uses diamond-shaped panels of aluminium sheet which incorporates struts along the edges forming integral parts of the diamonds.

Figure 15.4 (*b*) represents one diamond sheet. If this shape is cut from paper or cardboard and then folded along the diagonal OG, a 'bent' diamond results with points A, B as high corners and points G, O as low corners. Now, referring to Fig. 15.4 (*a*), which represents part of an actual dome, six of these panels are joined together at each low point, e.g. O, G, H, etc., by an aluminium gusset, and three panels are joined at each high point by an aluminium hub. Also, joining the high points are exterior struts such as AB, BC, CD, etc.

It is to be noted that geodesic domes, although extremely light in weight, are very stiff and rigid. Deflexion is not normally a problem as it is with conventional beams and trusses; therefore it is possible to use materials which have low elastic moduli. Aluminium is very suitable for geodesic domes, whereas it would not be practicable in most cases for beams. Plastics, which also have low elastic moduli (i.e. a small stress produces comparatively large alterations in dimensions), are suitable for geodesic domes because of the great rigidity and freedom from deflexion of the domes. In 1955, Fuller began producing for the U.S. Air Force 16·8 m diameter domes made of fibre-glass plastic. These were required in Canada and

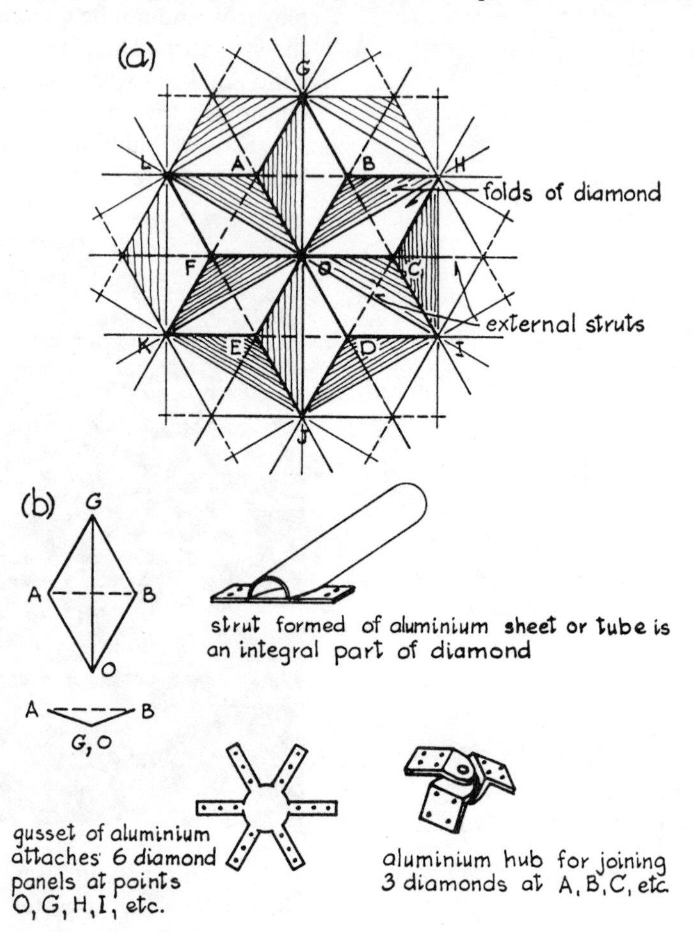

Fig. 15.4

Alaska for radar installations on the DEW (Distant Early Warning) Line, and metal construction was not permissible because it reflects radar beams. The erection time on the site for one dome was 14 hours, and the domes withstood a test load equivalent to wind velocities greater than 350 km/hr.

What was at that time (1958) the largest clear-span structure ever to be erected was a steel-skinned geodesic dome designed by Fuller's company for the Union Tank Car Co. at Baton Rouge, Louisiana. The total clear span of the dome is 117 m, with a rise at the centre of 39 m and a total weight of 12 MN. Compare again with the masonry dome of St Peter's, Rome (diameter, 40 m; weight, 100 MN).

Statements have been made that, if the struts of large geodesic domes are covered with a transparent plastic skin instead of an opaque material, the domes tend to become invisible. That is, the struts are so slender compared with the size of the dome that they are hardly noticeable. It has also been stated that the geodesic dome is not limited in size, and that diameters measured in miles are possible. The future may see houses and their gardens or even entire cities covered by transparent domes. (Further references to domes are made in Chapters 14 and 16.)

16 Grid Roofs and Floors

Rectangular grid system

In Chapter 6 it was explained that a slab can be designed to span in two directions, the two layers of steel sharing the load between them. This principle can be extended to beams which support the slab when the total area to be covered is too large for a beamless floor. For example, the arrangement shown in Fig. 16.1 would be suitable for an area (*ABCD*) of about 24 m × 24 m. Each panel of 6 m × 6 m could be covered with a two-way slab.

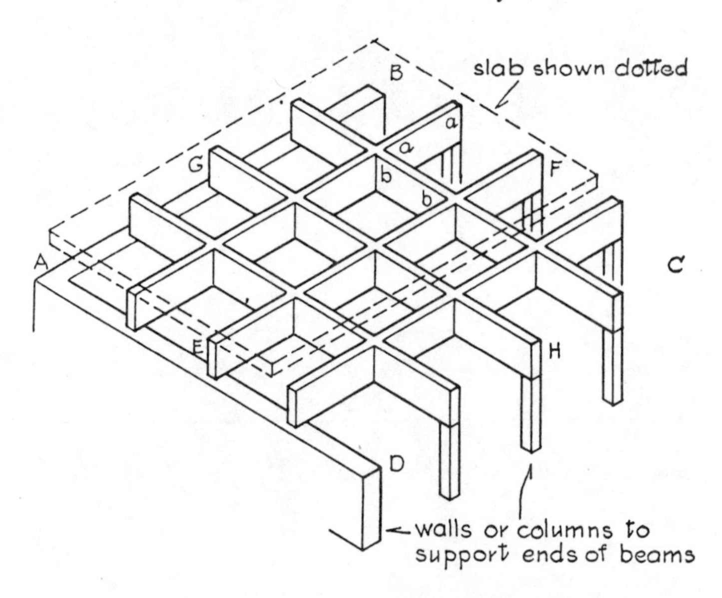

Fig. 16.1

The reinforced-concrete beams spanning in one direction are monolithic with the beams spanning in the other direction, and the load is shared between them. The slab is not shown in Fig. 16.1, but if it is cast at the same time as the beams each beam can be designed as a T-beam. This type of construction is suitable also for prestressed concrete by using precast units such as *aa* and *bb* and stretching the cables (two sets at right angles to each other) on the site. Like the two-way solid slabs described in Chapter 6, this system acts to best advantage when the area covered is square or nearly so. If the area is 'very oblong' the short-span beams will take a high proportion of the load and will not be helped much by the long-span beams. This can be demonstrated by considering that there are two intersecting beams only, crossing at mid-span.

In Fig. 16.2 (*a*), if the beams are of equal length and equal cross-section, each will support ½W, where W is the total point load at mid-span. At (*b*), if beam EF is twice the span of beam GH, ⁸⁄₉W will be carried by the short beam and ¹⁄₉W only by beam EF. This is because the deflexions of the two beams where they cross are equal, and, since it is more difficult to deflect a short beam to a certain value than it is to deflect a long beam by the same amount, the short beam must carry a greater proportion of the load. Furthermore, as was stated in Chapter 5, deflexions are proportional, not to the lengths of the spans, but to the cubes of the lengths. When the load is uniformly distributed and there are several beams, as in Fig. 16.1, the principle (with modifications) still applies.

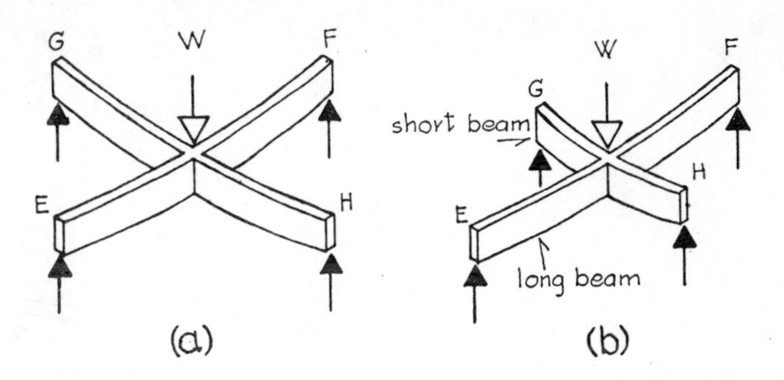

Fig. 16.2

Returning to Fig. 16.1, EF and GH are the most highly stressed beams, since they are at the middle of the floor and they bend similarly to the beams of Fig. 16.2.

Diagonal grids

The 'egg-crate' construction shown in Fig. 16.1 is not so efficient as the diagonal system of Fig. 16.3, particularly if it is required to support the grid at four points only, since the diagonal grid has greater torsional rigidity (resistance to twisting).

At first glance it might appear that this is a worse arrangement than that of Fig. 16.1, because the diagonal beams AC and BD are of greater span than beams EF and GH of Fig. 16.1 (assuming equal areas to be covered). Since, however, all the beams are of equal cross-section, the shorter beams are stiffer than the longer ones and help to carry them. In fact, beams JK and ML behave almost as rigid supports for beam AC, so that, instead of the beam bending as shown in Fig. 16.4 (*a*), it bends as at (*b*). The maximum moments in the long beams are therefore much less than for beams simply supported at the ends.

With beams in three directions, as shown in plan only at the top of Fig. 16.5, the grid becomes 3-way, and this system is stiffer than the 2-way diagonal grid. Figure 16.5 also shows (in plan) 3-way grids for triangular and hexagonal areas.

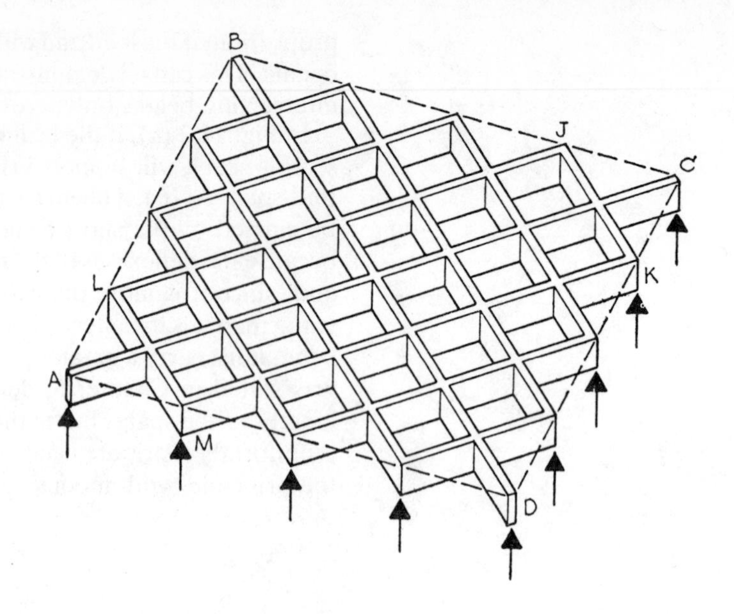

Fig. 16.3

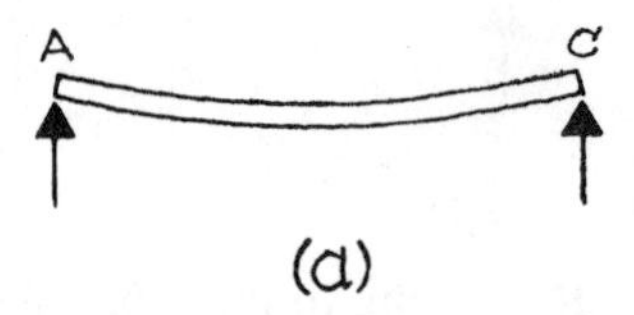

Fig. 16.4

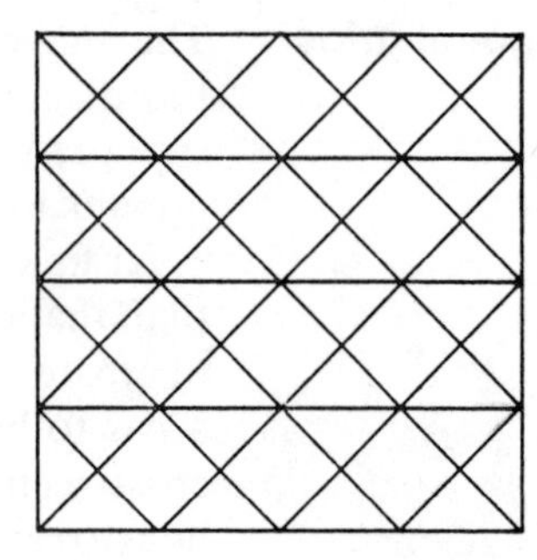

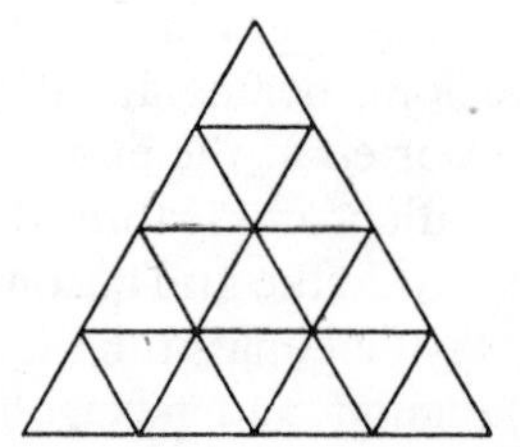

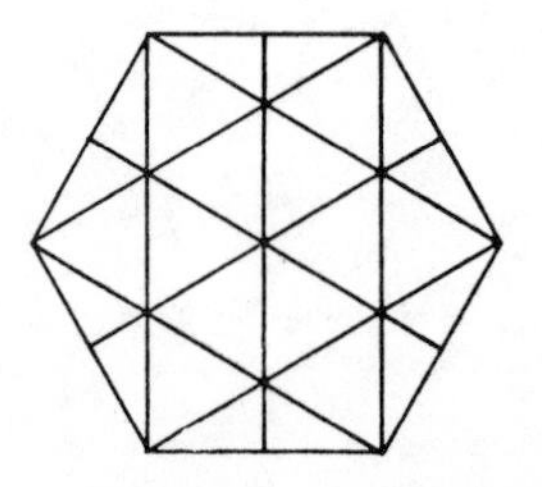

Fig. 16.5

Prestressed roof of restaurant, Festival Hall, London (*Cement & Concrete Association*)

Double-layer grids

Since about 1945, a new system of roof construction (usually metal) has developed which is known under various names such as *Three-dimensional frames, space frames, space decks, double-layer grids.* There are many types of double-layer grid, the simplest consisting of a two-way lattice construction as in Fig. 16.6. The example shown has eight intersecting lattice girders (there could be a much greater number), and the structure can be thought of as consisting of a top and bottom layer interconnected by vertical and sloping members.

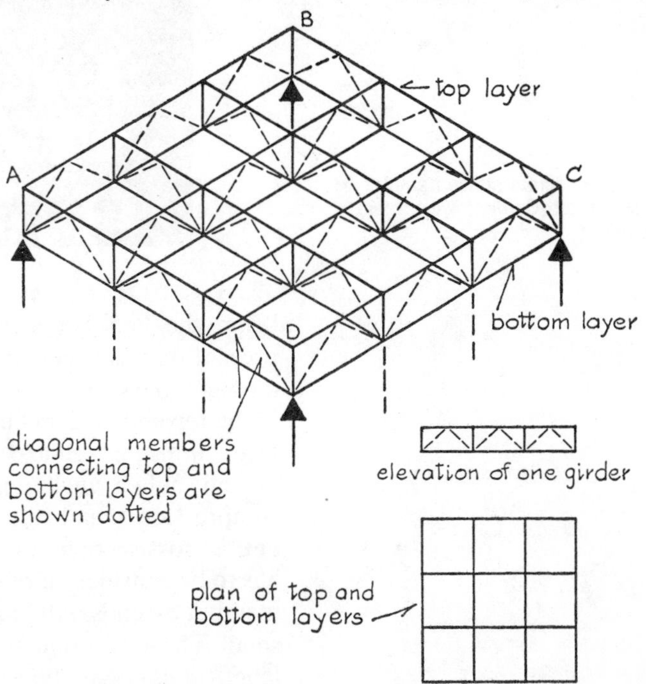

Fig. 16.6

Comparing it with the beam grid of Fig. 16.1, it will be seen that solid beams have been replaced by 'hollow' beams, i.e. by lattice girders, and referring to Chapter 8 it will be realized that the members of the girders are mainly subjected to axial forces, namely tension and compression. Thus more efficient use is made of the material, since the stress in any member is uniform.

For example, the solid beam of uniform cross-section (Fig. 16.7) does not make very good use of the material because it has to be made big enough to resist the maximum bending moment, which for symmetrical loading occurs at mid-span. Towards the ends of the

Reinforced-concrete space frame at Gatwick Airport (*Cement & Concrete Association*)

beam the bending moments get smaller and more material is present than is necessary. Even at the point of maximum bending moment, the stress over the cross-section is not uniform, being large in the top and bottom fibres and small near the neutral axis. Therefore, in a rectangular solid cross-section beam as used in a grid system, only a small amount of the material is fully stressed. The compression member of Fig. 16.7 is stressed uniformly over its entire cross-section, and all the material is fully utilized. Buckling has to be considered, of course, but in a grid system a compression member extends only from joint to joint and the buckling length is small. The tension member again makes full use of the material, and length does not matter since axial tension cannot cause buckling.

Referring to Fig. 16.8 (in which all the members are not shown), the top-layer members, like the top surface of a slab, are in compression (when supported as shown) and the bottom layer members are in tension. The maximum stresses are in the members which are near mid-span. The maximum stresses in the diagonals connecting the top and bottom layers occur in those members near the supports. In spite of these varying stresses, grid roofs are often built

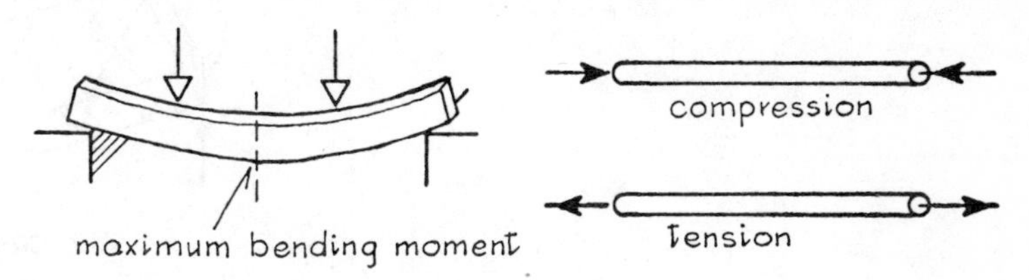

Fig. 16.7

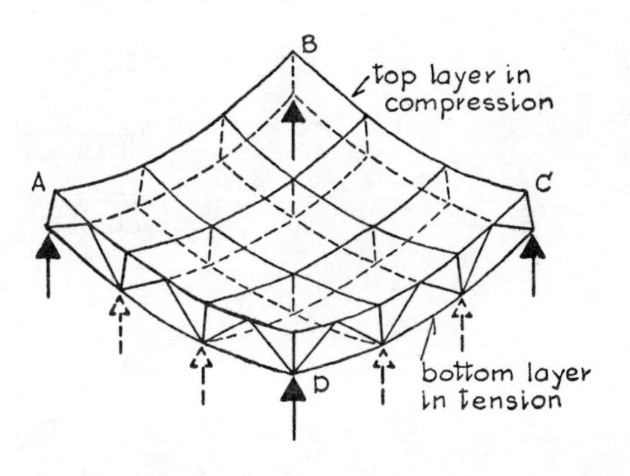

Fig. 16.8

up of a large number of short members of identical overall cross-section, but when hollow tubular members are used, allowance is sometimes made by varying the thickness of the tubes. Although a grid roof may be supported at four points as in Fig. 16.8, a stiffer structure results with extra columns (shown dotted).

Three-way lattice grids

A stiffer and stronger structure than a 2-way grid is obtained by using three intersecting rows of lattice girders as shown in Fig. 16.9. Because of the large number of intersecting members, these grids are difficult to depict pictorially, so at (*a*) only the lattice construction of six girders is shown, all the other girders, which are similarly triangulated, being represented by dotted lines in the top layer. *JK* is a common member of lattice girders, *EF, GH* and *BD*. Similarly, *HL* is a common member of lattice girders *AD, HG* and *EH*. All vertical members (such as *HL* and *JK*) are of equal length, and all sloping members are equal. One of the primary aims in designing double-layer grids is the use of as few different types of member as is possible.

Although double-layer grids have been constructed in reinforced concrete and also in timber, most grids are of steel, or aluminium alloys. Because of the stiffness of these grids, deflexions are small,

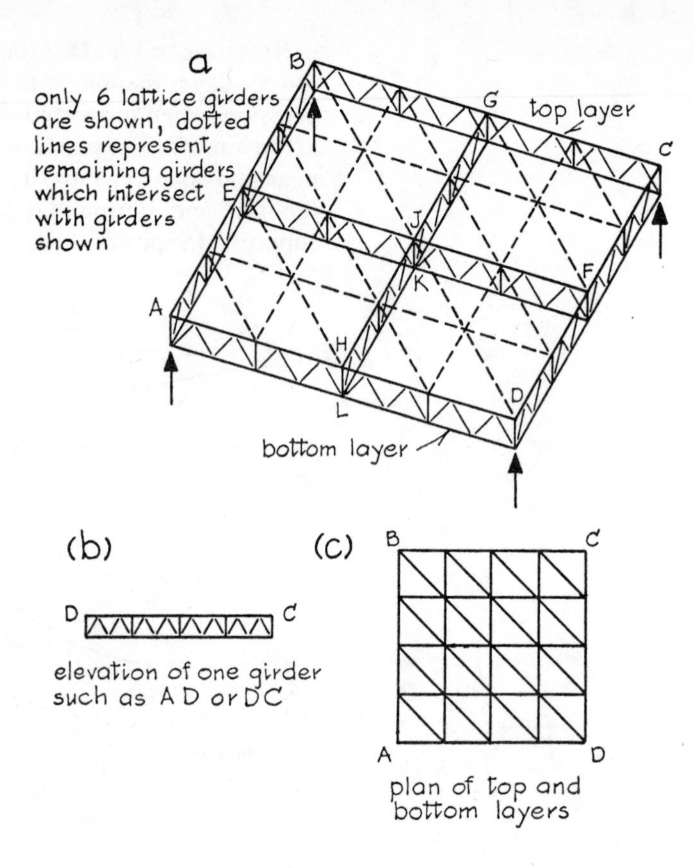

Fig. 16.9

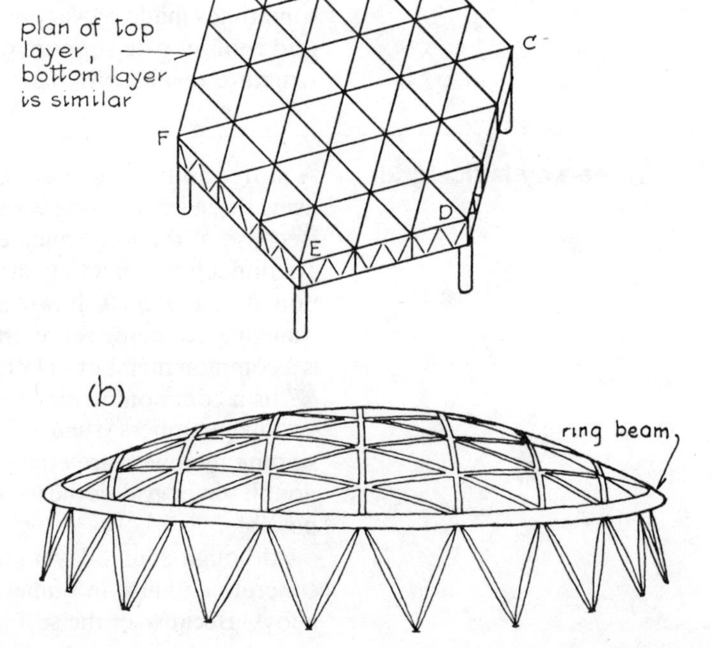

Fig. 16.10

and aluminium alloys, in spite of their low elastic moduli compared with that of steel, are quite suitable. Figure 16.10 (*a*) shows a 3-way hexagonal grid; the members of three lattice girders only are shown (*FE, ED* and *DC*), but all other lines represent girders built up in a similar manner.

Double-layer grids need not be flat; they can be curved to form domes and to give in fact more rigid structures than flat grids [Fig. 16.10 (*b*)]. Like all domes, however, thrust is exerted outwards on the supports and must be resisted by adequate buttresses. Alternatively, a ring beam can be provided to absorb the thrust and thus to give only vertical loads on the columns. This principle of 3-way lattice construction was used for the Dome of Discovery at the Festival of Britain in 1951, the entire roof construction being in aluminium alloy. The diameter of the dome, which at that time was the largest in the world, was 111·3 m, and its weight was only 3 MN. The main ribs were triangular in cross-section and were latticed.

Space grids

The double-layer grids so far discussed and illustrated consist of intersecting lattice girders. Even stronger and stiffer constructions are the space grids the basic principle of which is as follows.

If six members are joined as shown in Fig. 16.11 (*a*), a very stiff frame results. Further members can be added to form a triangular 'tube' as at (*b*), and by interconnecting other tubes a square or other shaped roof can be obtained. Such a roof can be thought of as built up of interconnecting pyramids having triangular bases as in Fig.

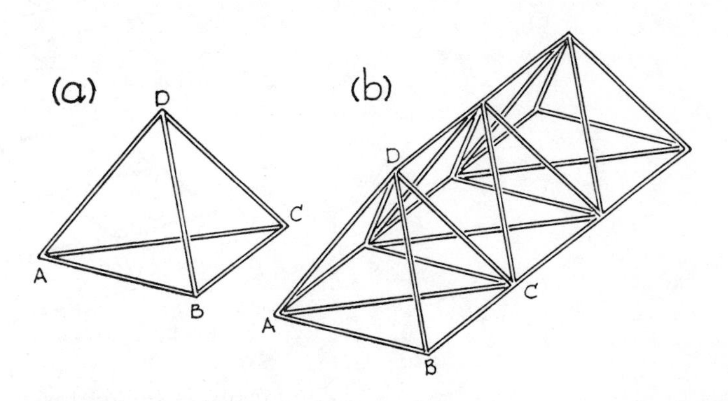

Fig. 16.11

16.11, or (as will be demonstrated later) square or hexagonal bases. Because of their great stiffness and rigidity, space grids are very suitable for covering large areas in factories, exhibition buildings, swimming pools, etc. Grids measuring 100 m × 100 m without any internal columns have been constructed.

A few bays of one of the simplest types of grid are shown in Fig. 16.12. Both the top and bottom layers consist of square bays. Each bay of the top layer (shown by thick full lines) can be considered as the base of an inverted pyramid having its apex in the bottom layer. All the apexes a_1, a_2, a_3, etc., are joined by members to form a square grid [shown by thin full lines in Fig. 16.12 (*a*)]. Sloping members (shown dotted) connect the top layer with the apexes of the pyramids in the bottom layer.

Figure 16.12 (*b*) is an 'exploded' view showing the bottom layer separated some distance away from the top layer and showing one pyramid only (apex e_1).

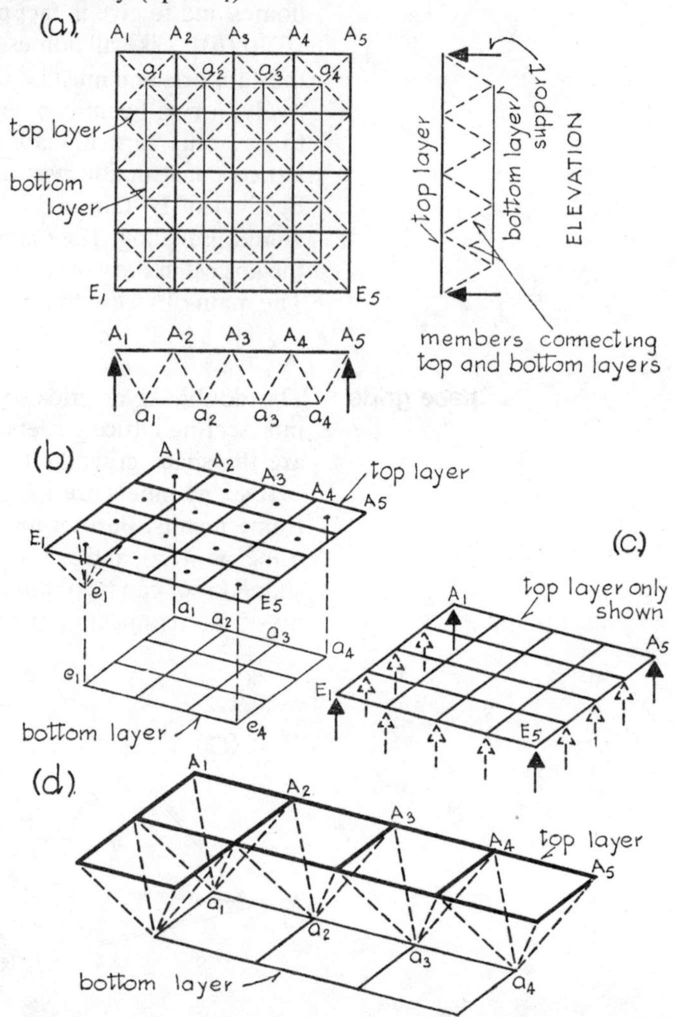

Fig. 16.12

Figure 16.12 (*d*) shows the four pyramids in one row. In this grid system, all members are of equal length, which makes for ease of fabrication. The top layer would be, of course, closed with a suitable roofing material, and the grid may be supported at all external points as in Fig. 16.12 (*c*), which arrangement gives smaller stresses than if supports are at the corners A_1, A_5, E_1, E_5 only.

It should be remembered when speaking of these external supports that an actual grid roof would have many more bays than the four given for illustration purposes in Fig. 16.12. Alternatively, the supports could be under the bottom layer. This grid is not completely triangulated, and it can be made even stiffer and stronger by introducing horizontal diagonal members in both the top and bottom layers as shown in Fig. 16.13.

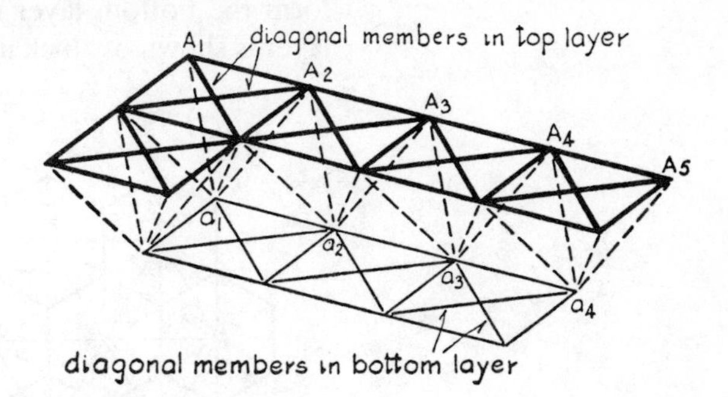

Fig. 16.13

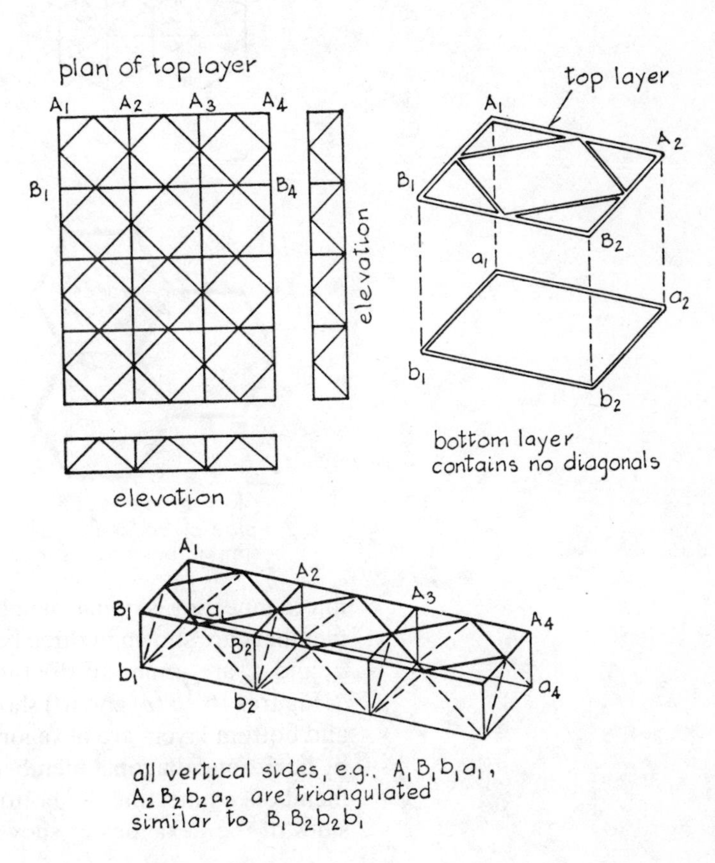

Fig. 16.14

Another type of grid is shown in Fig. 16.14; this can be thought of as a 2-way lattice grid with the top-layer compression members stiffened by horizontal diagonals. The plan of the bottom layer is similar to that of the top, except that there are no horizontal diagonals.

During recent years double-layer grids having hexagons in one or both layers have been used for roofs. Some types are shown in Fig. 16.15.

Figure 16.15 (*a*) shows a few bays of a roof consisting of intersecting hexagonal pyramids. From each corner of the hexagonal bases in the top layer, members slope downwards to meet at apexes a_1, a_2, etc. All the apexes are connected together by other members to form the bottom layer made up of equilateral triangles. The top layer is shown by thick lines, the bottom layer by thin lines, and the

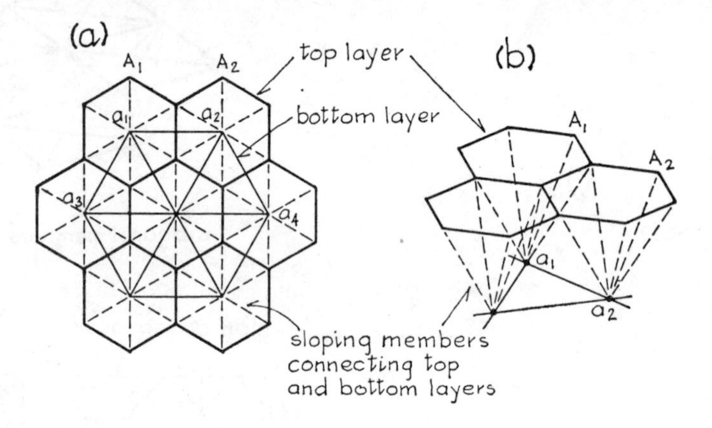

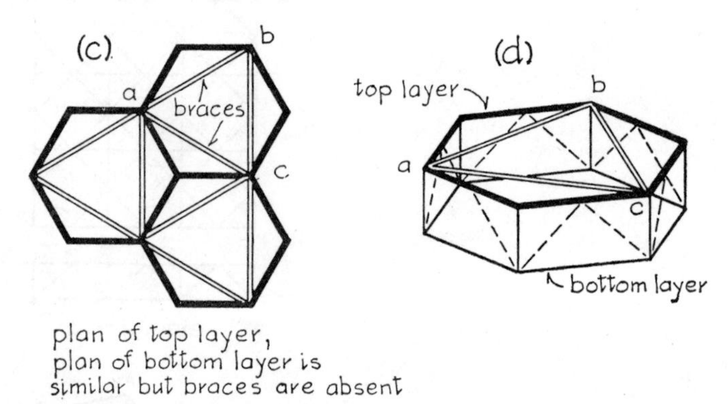

Fig. 16.15

interconnecting sloping members by dotted lines. These sloping members are shown in three bays only, but all other apexes such as a_3 and a_4 are joined to the top layer in a similar manner.

Figures 16.15 (*c*) and (*d*) show another grid in which both the top and bottom layers are hexagons, only the top layer being stiffened by horizontal diagonal members such as *ab, bc* and *ca.* The sloping members connecting the bottom to the top layer occur only in the sides of the hexagons as shown at (*d*).

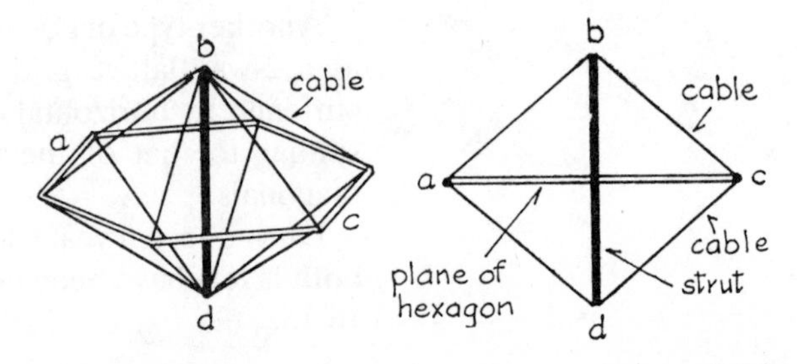

Fig. 16.16

A Canadian, Jeffrey Lindsay, has built domes of the Buckminster Fuller type, using units as shown in Fig. 16.16: *b* is above the plane of the hexagon and *d* is below the plane.

The grid systems already described use bars (or tubes or angle sections) of aluminium or steel interconnected at the *nodes*. Different methods of connecting bars at joints have been evolved and various patents have been taken out.

Stressed-skin grid roofs

Referring back to Fig. 16.12, the roof is built up of a number of bars or tubes to form interconnecting pyramids, the apexes a_1, a_2, etc. [Fig. 16.17 (*a*)] being joined by other members to form a grid in the bottom layer. The sloping members connecting the top and bottom layers act as struts (compression members) and must be strong enough to resist buckling.

A new system called Pyradex Roof Construction has been developed which makes use of the stressed-skin principle. Instead of bars, aluminium sheets, welded or otherwise, joined at the edges a_1B, a_1C, etc., are used as shown in Fig. 16.17 (*b*). The whole of

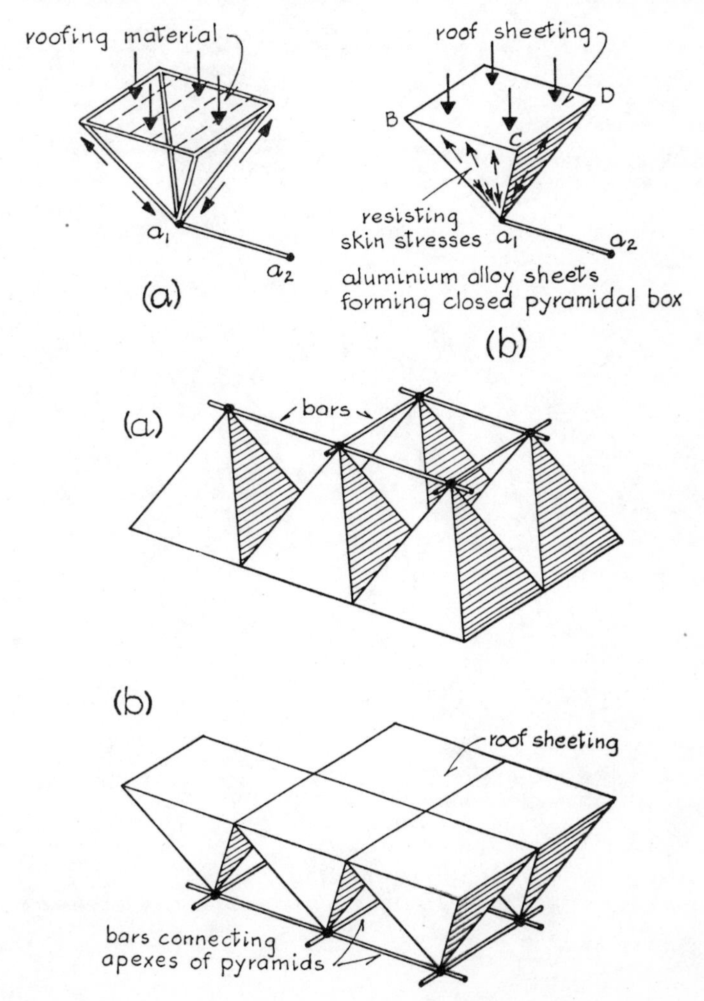

Fig. 16.17

Fig. 16.18

each sheet such as a_1BC or a_1CD acts as a strut, and although the sheets are thin, buckling cannot occur because each triangular sheet is restrained by its neighbours. This type of construction is very strong and rigid and is suitable for plywood and plastics as well as steel and aluminium. A number of such pyramids connected together with their apexes joined by bars can be utilized to form a three-dimensional roof. With pyramids having triangular or square bases, the points may be placed up or down.

In the type shown in Fig. 16.18 (*a*), the sheets forming the sides of the pyramids also act as the roofing material. The bottoms of the pyramids can be left open. At (*b*) the top sheeting is an integral part of the structure and helps to give greater stiffness. The bars joining the apexes of the pyramids can be arranged so as to form either a two-way or a three-way grid. These roofs possess acoustic advantages, since the hollow pyramids act as sound baffles and would reduce noise in buildings such as restaurants, railway stations, and swimming pools. Pentagonal and hexagonal pyramids can also be used.

17 Cable-Suspended Structures

Steel-rope cables have been used for suspension bridges for about 80 years, and ropes of other materials have been used for many hundreds of years. Only recently, however, has this principle of suspension been applied in the supporting of roofs and other members of buildings.

A rope, of course, is only capable of taking tension, and as mentioned previously, tension members make very efficient use of their material since the cross-section is uniformly stressed. Furthermore, length does not matter because there are no problems of buckling to

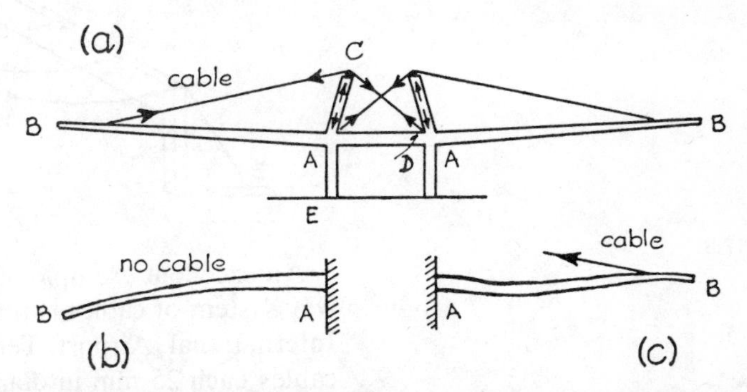

Fig. 17.1

consider. Modern steel cables, which are built up of separate wires, use cold-drawn steel, which has a breaking stress of about 1850 N/mm^2. (Compare with ordinary mild steel, which has a breaking stress of about 450 N/mm^2.)

One manner in which steel cables have been used is illustrated in Fig. 17.1, which is the principle of an aircraft hangar at Kansas City. The cantilever AB would bend as shown at (*b*) if it were unsupported at any point except A. The cable supports the cantilever so that the bending is as shown at (*c*), and a much smaller beam is required. The cable, of course, exerts a pull on the supporting member at C, but by anchoring it at D, with a suitable inclination of the cable the reaction in the supporting member AC can be rendered almost entirely axial. Similarly, by skilful arrangement of the members, the load on the column AE can be such that there is no, or very little, bending moment.

As explained previously, an arch exerts thrust [Fig. 17.2 (*a*)]. A suspension cable (which can be thought of as an inverted arch) exerts a pull [Fig. 17.2 (*b*)], and this must be resisted by adequate anchorages. In Fig. 17.2 (*c*) the pull of the cable causes bending of the column, but by anchoring the cable as at (*d*), it can be arranged that the column is under axial thrust only.

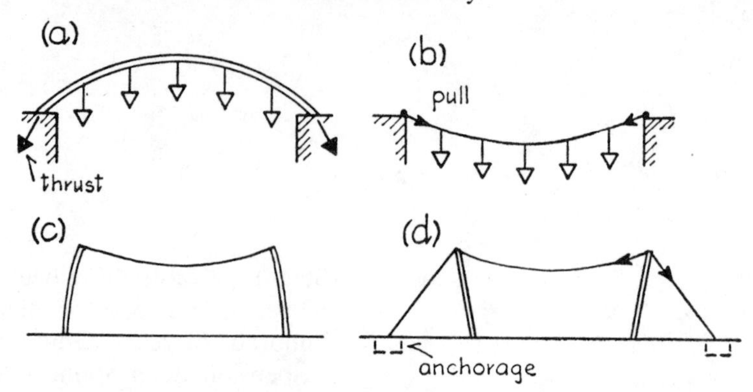

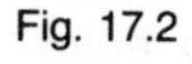
Fig. 17.2

Figure 17.1 illustrates cables used as ties in relieving the stresses in roof-supporting members, and enabling large spans to be achieved. In recent years, structures have been erected in which the cables form part of the actual roof structure.

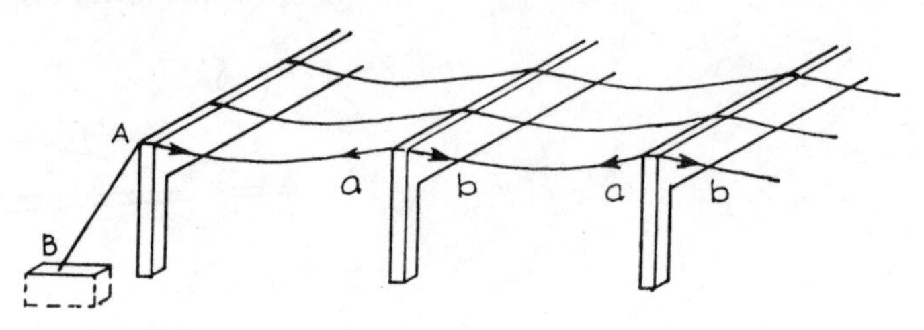

Fig. 17.3

An excellent example of a rectangular roof supported by a one-way system of cables is that spanning the great hall of the Dulles International Airport Terminal in Washington, D.C. Pairs of cables each 25 mm in diameter span 49 m across the hall at 3 m centres, and carry cast-in-place concrete segments which eventually form the roof covering. The cables are anchored in edge beams which are in turn supported by columns around the perimeter of the building. Bending moments in the colums are reduced by leaning the columns outwards at the roof line.

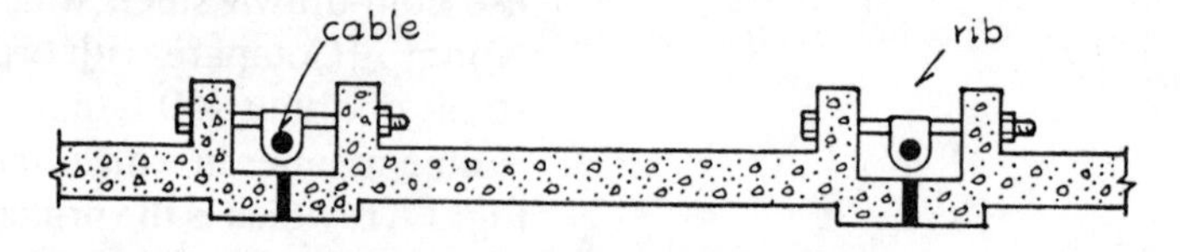

Fig. 17.4

Figure 17.3 is a simplified illustration of the use of cables in a roof curving in one direction only. The horizontal effects of the tension (*a* and *b*) in the cables cancel each other out, so as to give only vertical loads on the intermediate supporting beams and columns.

At an end column either a stay *AB* has to be provided or the beam and column must be designed to resist the cable tension.

One method of forming the roof covering of a cable structure is to use precast-concrete slabs as shown in Fig. 17.4. These units are attached to the cables, and sandbags or other loadings are placed on the roof equal to double its normal working load. This load causes the cables to stretch, and whilst in this state the ribs are filled with concrete. After the concrete has hardened, the sandbags are removed from the roof, causing the cables to 'spring back' and to put the concrete into compression.

One of the great problems in suspended-cable roofs is the prevention of *flutter*. Wind forces tend to cause the roof to vibrate, and if the forces coincide with the natural period of vibration of the roof dangerous movements can occur. (Flutter has been the cause of the collapse of more than one suspension bridge.) An analogy is the simple garden swing. If even a small force is applied in a rhythmic manner as the swing moves to and fro, large movements can soon be built up. The same action occurs when carrying a pail of water or a

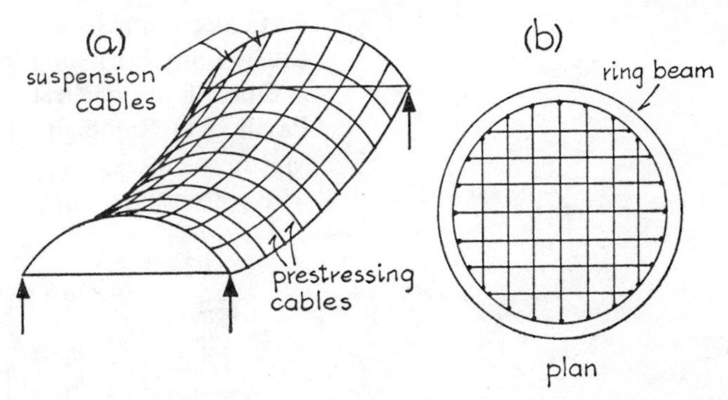

Fig. 17.5

tray of wine-glasses. If there is a certain rhythmic movement of the person carrying the vessels, the liquid is set into motion and will spill over.

Designers adopt different methods in the construction of their roofs to minimize the likelihood of flutter. In doubly curved roofs one set of cables is usually prestressed, i.e. the cables are tightened before the roof receives the load [Fig. 17.5 (*a*)]. The end supports must, of course, be capable of supporting the tension from the cables. In circular structures the pull from the cables is often resisted by a ring beam. If the cables are prestressed (i.e. pre-tightened) this roof can be compared roughly with a tennis racket [Fig. 17.5 (*b*)].

The roof of the Arizona Stat Fair Coliseum is a cable-suspended circular roof which, when viewed in plan, looks like Fig. 17.5(b). However, the ring beam is warped vertically so as to form a saddle-shaped roof that is 116 m in diameter. The distorted shape induces double curvature into the roof similar to that shown in Fig. 17.5(a). When the cables are prestressed against one another the entire network is better able to withstand wind vibrations. The ring beam must of course be designed to resist the precompression due to the cable forces particularly during construction when unevenly distributed cable loads could induce large bending moments in the ring making it more susceptible to a buckling failure.

The American Pavilion at the Brussels Exhibition of 1958 used the principle illustrated in Fig. 17.6 (*a*). This can be compared with a bicycle wheel as shown at (*b*), where the forces due to the tightened spokes tend to tear the hub apart and to cause the rim to buckle

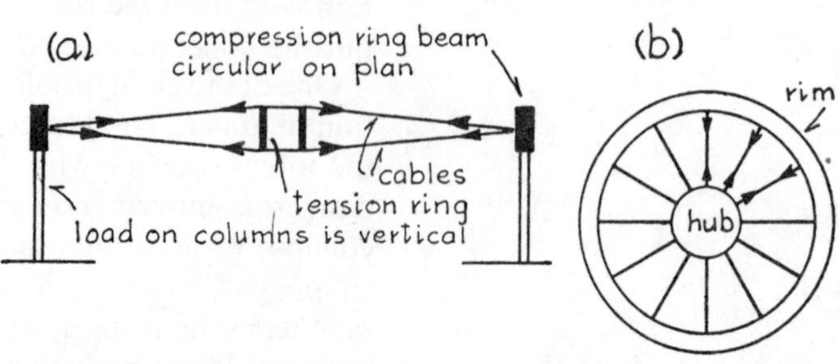

Fig. 17.6

inwards. The wheel can, however, be supported horizontally on vertical columns under the rim and can then be loaded without causing any horizontal pull on the supports.

One of the earliest examples of a large cable structure is the Pavilion at Rayleigh, North Carolina, designed by the architects Matthew Nowicki and W. H. Dietrick [Fig. 17.7 (*a*)]. The steel-wire cables are supported by two inclined reinforced-concrete

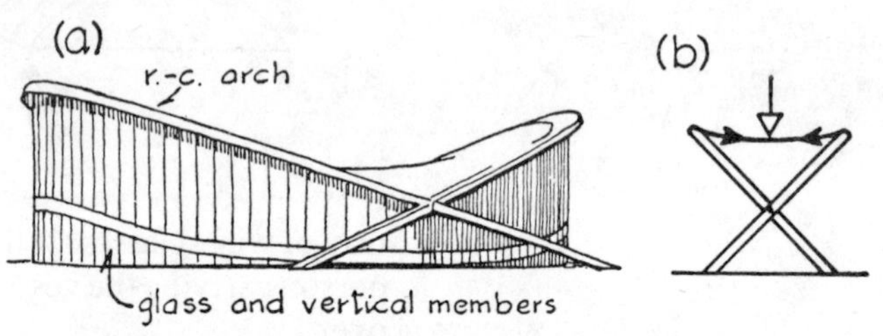

Fig. 17.7

arches rising to a height of 27·5 m. To stop flutter, secondary cables running transverse to the main cables are prestressed to hold down the latter. The roofing material is of corrugated steel sheets, insulated on top. The principle can be roughly compared to the manner in which a canvas stool supports its load [Fig. 17.7 (*b*)].

A new system of roof construction comprises two main cables, an upper and lower, braced together by diagonal stays or ties in order to make the truss rigid and to eliminate distortion under the application of live and wind loads. These diagonals are secured to the main cables by specially designed clamps. They are

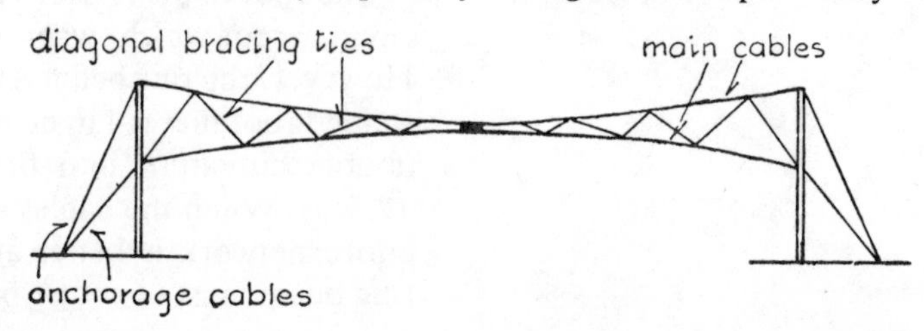

Fig. 17.8

pretensioned to a calculated stress by simple adjustment of the stay length. By this means a positive tension is maintained both in the cables and in all the diagonal stays for all conditions of loading to which the roof may be subjected. Such loads, apart from self-weight, may be from wind, snow or other source, either uniformly or unsymmetrically distributed [Fig 17.8].

A variety of coverings can be used. One that has been employed with great economy and success is a light trough type of decking fixed to the cables by U-bolts surmounted by insulation board and mineralized roofing felt. Ceilings can be attached to or slung from the lower cables.

A spectacular roof system, similar to that just described, was erected to cover 75,000 m² of stadia at the 1972 Olympic Games in Munich. The continuous tent-like roof was suspended from towers and guyed masts and consisted of a network of cables covered with

Olympic stadia, Munich 1972

translucent Plexiglas. Stiffness against wind and snow loads was again achieved through double curvature and by stressing the net against heavy edge cables.

By far the most common application of cable roofs is in single storied buildings like sporting arenas and exhibition halls. However, the compound arch-suspension system shown diagrammatically in Fig. 17.9 has great potential for multistoried buildings. *ABC* is an arch and *ADC* is a cable of inverse curvature to the arch. The weights suspended from the arch cause an outward thrust which is balanced by an inward pull due to the weights suspended from the cable. It can therefore be arranged that the loads on the columns are entirely vertical (except for pressures due to wind). In an actual building, some floors can be suspended from the arch by vertical cables, and other floors suspended from the main cable of the suspenarch. At ground level the whole of the area served by the building would be completely free of internal columns or cables. The supporting columns of the suspenarches can be of hollow box-construction to contain services, lifts, stairs, etc., for the entire building. The cables, of course, would have to be suitably protected from the effects of fire by a casing of concrete or other material.

The Federal Reserve Bank in Minneapolis is at least one example of this system in practice. The cable spans 83 m and supports 11 floors from towers 61 m high. The first floor is suspended 9 m above an open, column-free plaza which extends between the two towers.

At present the building does not include the arch and the horizontal component of cable tension is taken by a 9 m deep truss spanning between the towers. At some future date a 6-storey addition will be made to the top of the building which will be suspended from the arch which spans the same distance as the cable. Part of the compression in the truss will then be relieved by the outward thrust of the arch.

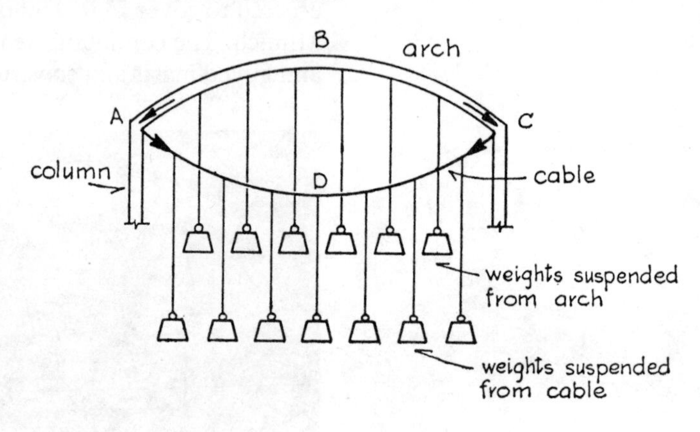

Fig. 17.9

18 Dams and Retaining Walls

Dams Dams built across valleys in order to store water or across rivers to raise the water levels have been used from very early times. The earliest dams were earth constructions, and this type is still used today, although concrete dams are probably more common. Figure 18.1 is the cross-section of a dam 46·7 m high built in the United

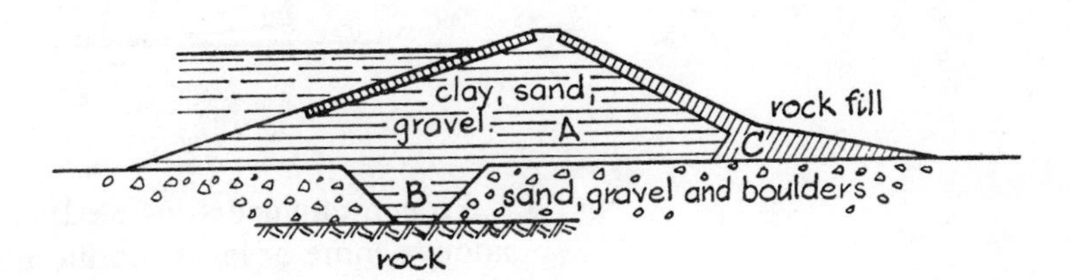

Fig. 18.1

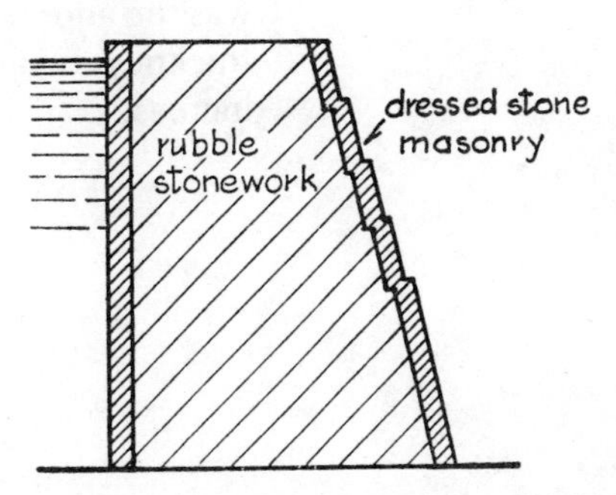

Fig. 18.2

States in 1937. Clay, sand and gravel (A) were placed in layers about 150 mm thick and rolled. B represents a cut-off trench, the purpose of which is to prevent seepage under the dam, and C represents rock fill.

A gravity dam resists the water pressure by virtue of its size and weight, and (ignoring earth dams) may be either solid concrete or a hollow wall with buttresses. Many dams are now built in an arch form, and the water pressure is resisted mainly or partly because of the curved shape. Before the advent of Portland-cement concrete, dams were built of masonry; Fig. 18.2 is the cross-section of a dam built in Spain in the 16th century.

As mentioned in Chapter 1, water exerts pressure equally in all directions, so that on a vertical wall the force is horizontal. The total force on a metre length of dam 45 m high is about 10 MN, so that on a 50 m length the total force is about 500 MN. This pressure tends to push the dam forward and also to overturn it about its toe (Fig. 18.3). The resultant force *R*, obtained by combining the water force *P* with the weight of the wall *W* should fall within the middle third of the base if tensile stresses are to be avoided. (Further discussion on stability of walls and dams will be found later in the chapter under the heading 'Retaining Walls.')

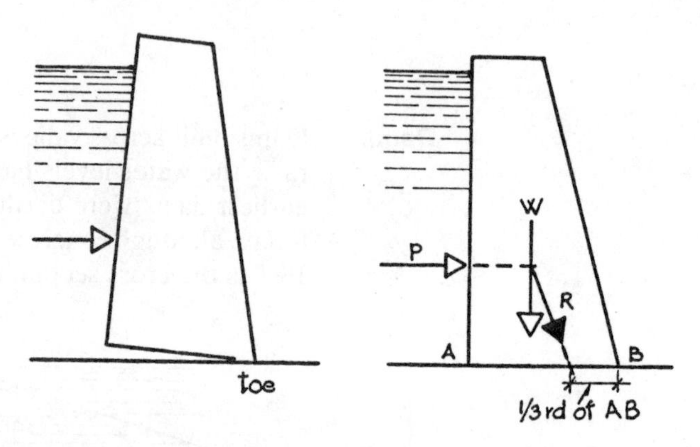

Fig. 18.3

French engineers in the 19th century were among the first to adopt a more or less scientific approach to the design of dams, and the result of their work and of research by others such as Rankine was the adoption of dams of the shape shown in Fig. 18.4, where the thickness is increased towards the base as the lateral water pressure increases.

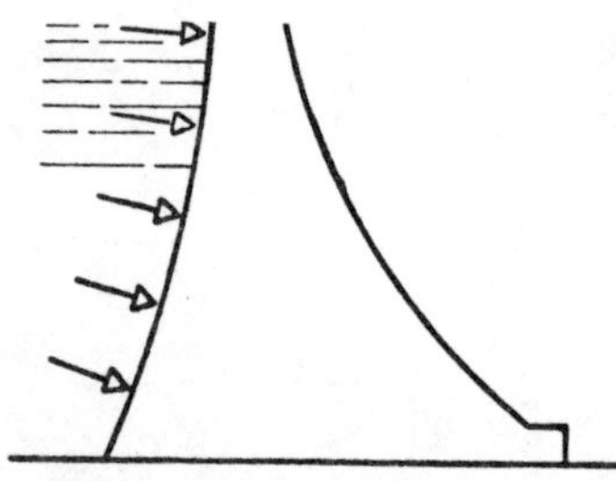

Fig. 18.4

Uplift forces do not appear to have been allowed for in early designs and some failures resulted. Uplift is due to water penetrating the masonry joints or even the pores of concrete, and although the actual amount of water may be small, the forces exerted are large, particularly near the base of the dam, and are taken into account in modern designs.

With the introduction of reinforced concrete, the buttress type of dam began to be constructed, two forms of which are shown in Fig. 18.5. The sketches merely illustrate the principle of construction and do not represent the correct number of struts, etc. Figure 18.5 (*a*) shows an inclined arch dam with buttresses joined by struts. At (*b*) a flat slab is used instead of arches. Fully reinforced-concrete dams of the type shown in this diagram were constructed during the early days of reinforced concrete, but the present tendency is towards the use of bigger unreinforced buttresses.

The reason for sloping one or both faces of a wall can be seen by reference to Fig. 18.6, where, for ease of explanation, simplified

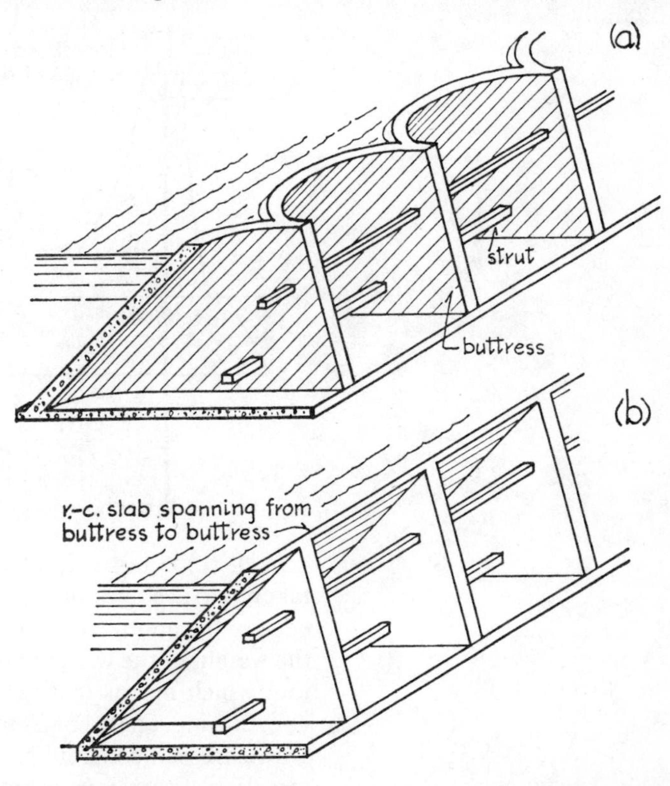

Fig. 18.5

shapes and solid walls have been assumed. All the walls shown have a height of 4 m and a base width of 3 m. Taking the density of concrete as 22·6 kN/m³, the weight of each wall is 135·6 kN for a metre length, and the resultant water pressures per metre run of wall are shown in the diagrams. The centre of gravity of wall (*a*) is 2 m from the toe of the wall, that of wall (*c*) is 1·5 m from the toe, and that of wall (*c*) is 1 m. In wall (*a*), by combining the water pressure with the weight of the wall (by the parallelogram of forces) the

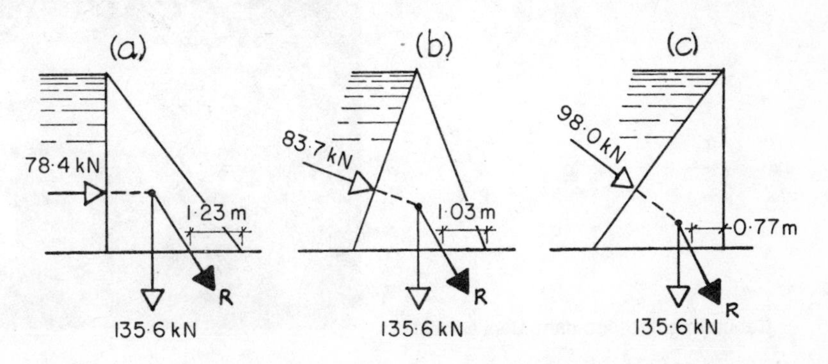

Fig. 18.6

resultant force R cuts the base at 1·23 m from the toe and is therefore inside the middle third of the base (no tensile stresses anywere). Wall (b) has its centre of gravity nearer the toe than wall (a), but this is partly compensated for by the fact that the water pressure is inclined; the resultant force cuts the base at 1·03 m from the toe (within the middle third).

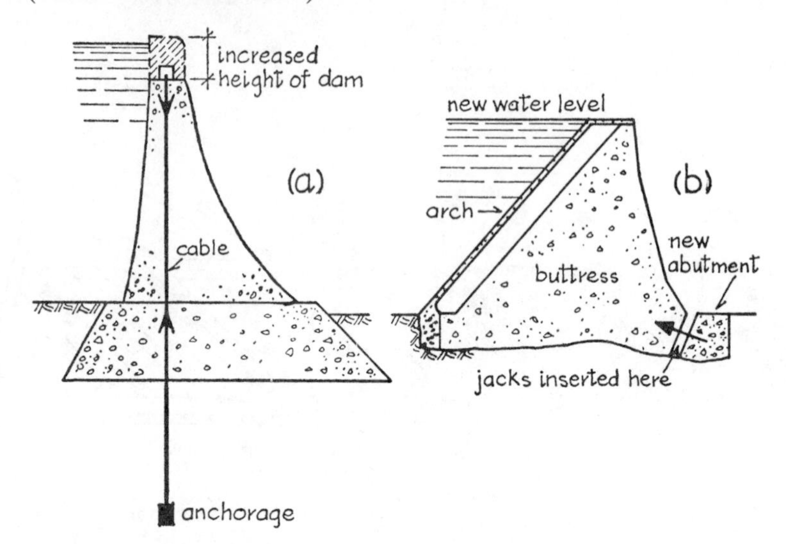

Fig. 18.7

Wall (c) is not a good solution. Even before water pressure is taken into account, the resultant weight of the wall is at the extreme edge of the middle third. When the water pressure is combined with the weight of the wall, the resultant cuts the base at 0·77 m from the toe, which means that tensile stresses are caused and to prevent overturning, an inclined buttress would be required as in Fig. 18.5.

Prestressed concrete has not been used extensively in dam construction. Freyssinet improved the stability of an existing dam (the

Concrete buttressed dam, Beni Bahdel

Cheurfas Dam in Algeria) to enable it to be increased in height from 30 m to 33 m [Fig. 18.7 (*a*)]. Holes were drilled at 4-metre centres right through the dam into sandstone below foundation level. Each cable, of a total number of 37 throughout the length of the dam, consisted of 630 steel wires stressed to 780 N/mm², giving a total prestressing force for each cable of about 10 MN.

Another example of prestressing, but by jacks and not cables, is the Beni Bahdel Dam [Fig. 18.7 (*b*)]. Whilst the dam was under construction it was decided to make it 61 m high instead of the originally planned 54·3 m. The arches of the dam were supported by buttresses which were estimated to be strong enough for the additional load due to the increased height of water. To counter-

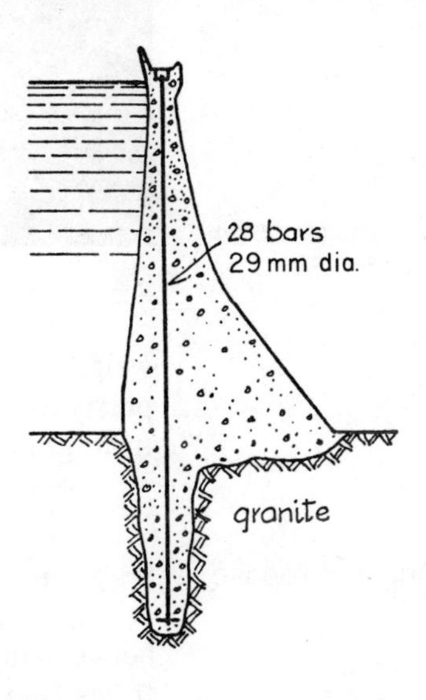

Fig. 18.8

act the extra pressure on the foundations, new abutments were constructed and flat jacks inserted to prestress the buttresses. After jacking, the joints were filled with concrete and the jacks were removed.

The Allt-na-Lairige Dam in Scotland, built in 1956, is shown in simplified cross-section in Fig. 18.8, twenty-eight bars each 29 mm in diameter being used in each prestressing duct. The object of prestressing was to put considerable compressive stresses into the concrete to counteract the tendency of the lateral water pressure to cause tension. It is estimated that this dam, because of prestressing, used about 60 per cent of the concrete which would have been required for an ordinary gravity dam.

Retaining walls

The design of large dams is usually the concern of specialist engineers; the average engineer or architect is more often concerned with walls of comparatively small height for retaining earth. The pressure exerted by earth cannot be determined with the same exactitude as pressure exerted by a fluid, but empirical formulae have been derived based on the angle of repose of the earth and its

Benmore dam, New Zealand (*The High Commissioner for New Zealand*)

weight. In addition, special formulae have been derived to estimate pressures due to clay behind a wall, since clay behaves differently from granular materials such as gravels and sands.

Angle of repose

If a granular material like sand is tipped on the ground, it will pile up in a heap as illustrated in Fig. 18.9 (*a*). The greater the friction between the grains of earth, the larger will be the angle of repose A. It has been stated that the pressure exerted by water is always at right angles to the surface with which it is in contact. Due to friction of the earth on the back of the wall the direction of the pushing force on a retaining wall will be inclined at an angle B, which if more definite information is not available is usually taken as equal to about one-half or two-thirds of the angle of repose. From Fig. 18.9 (*b*) it will be seen that the pressure behind the wall is due to the weight of a wedge of earth which, if the wall were removed, would slip to allow the earth to take up its natural slope. Consequently the greater the angle of repose of a material, the less is the pressure exerted (Fig. 18.10).

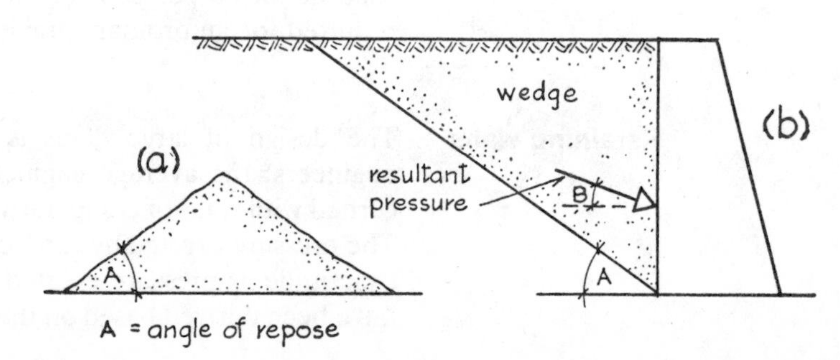

Fig. 18.9

Most codes for earth-retaining structures give approximate values for the angles of repose. For example, loose sand has an angle of repose of about 30° to 35°, whilst the values for rock filling are about 35° to 45°.

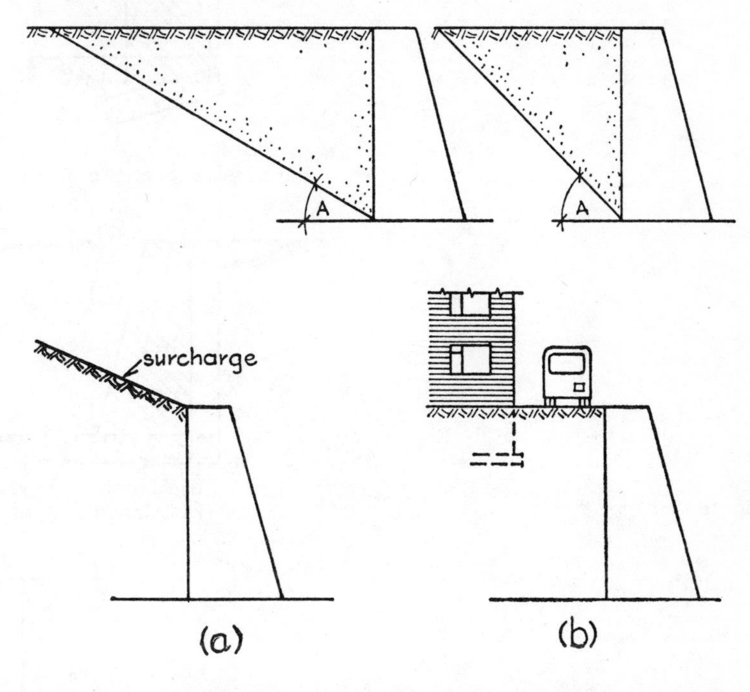

Fig. 18.10

Fig. 18.11

The walls shown in Fig. 18.10 are retaining a 'level fill' of earth. Increased pressures must be allowed for when there is a surcharge [Fig. 18.11 (*a*)] or when there are buildings or traffic-carrying roads near the top of the wall [Fig. 18.11 (*b*)].

For the present, pressure conditions of the type already illustrated will be discussed. Retaining walls in the basements of buildings will be dealt with later.

Gravity walls

Most walls are either of solid concrete (or, less frequently, of masonry) or of reinforced concrete. The former type, known as *gravity walls*, must be thick and heavy as in Fig. 18.11, so that only compressive stresses (or at the worst only very small tensile stresses) are developed in the concrete. Reinforced-concrete walls are more slender because of their ability to resist tension. The design of a gravity wall must ensure:

(1) That the pressure on the earth at the toe of the wall is not greater than the earth can safely resist. Excessive pressure will lead to the settlement of the toe, thus increasing the tendency to overturning [Fig. 18.12 (*a*)].

(2) That the resultant thrust on the soil preferably falls inside the middle third of the base, thus ensuring that there are only compressive stresses [Fig. 18.12 (*b*)].

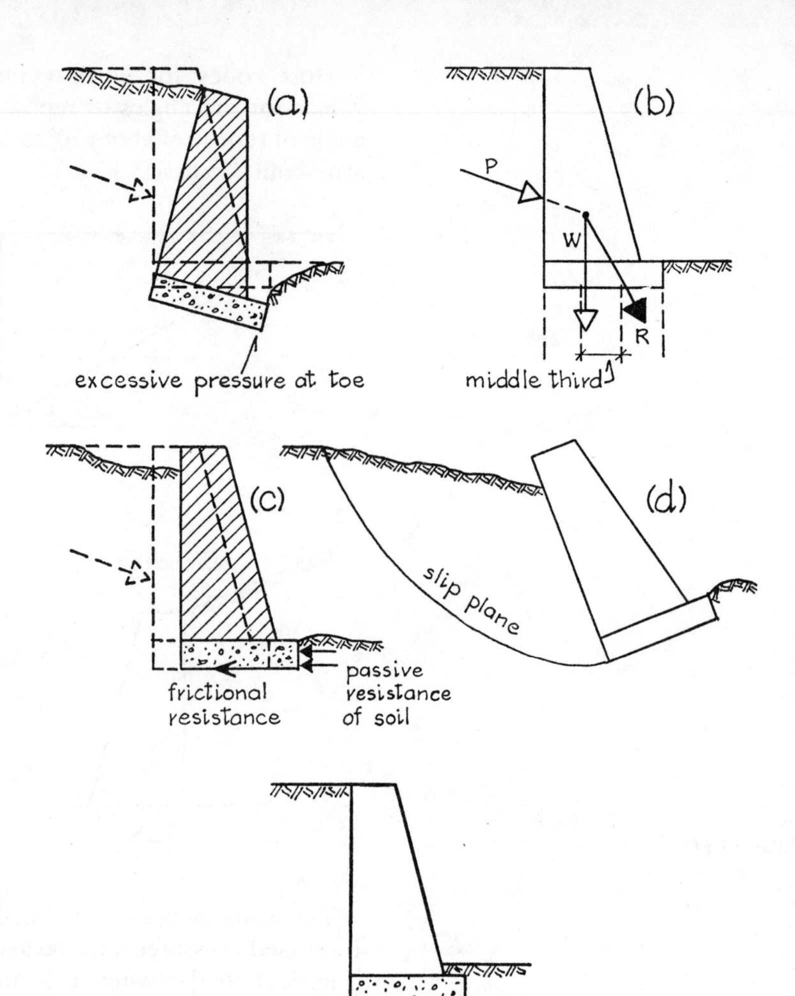

Fig. 18.12

Fig. 18.13

(3) That there is no tendency for the wall to slide [Fig. 18.12 (*c*)], or in the case of cohesive soils such as clay, to slip because of shear failure [Fig. 18.12 (*d*)].

A stepped foundation is sometimes used, as in Fig. 18.13, when an ordinary foundation would not give sufficient resistance to sliding.

Reinforced-concrete walls

These walls are chiefly used where space is restricted, and they are also more suitable than gravity walls when the height is considerable. There are two main types: *cantilever walls* and *walls with counterforts*.

Cantilever walls

These walls are suitable for heights up to about 8 m. The three main types are shown in Fig. 18.14. In all three examples the stem bends forwards as shown at (*d*) and is designed as a cantilever slab with main and distribution steel. (The main steel only is shown in the diagrams.) The thickness of the stem where it joins on to the base is about 75 mm or less for every metre of height of the wall. It is preferable that at least part of the base should project backwards under the earth filling so that the weight of earth pressing down on the base will help to give stability to the wall. The total length of the base will in these circumstances be equal to about one-half of the height of

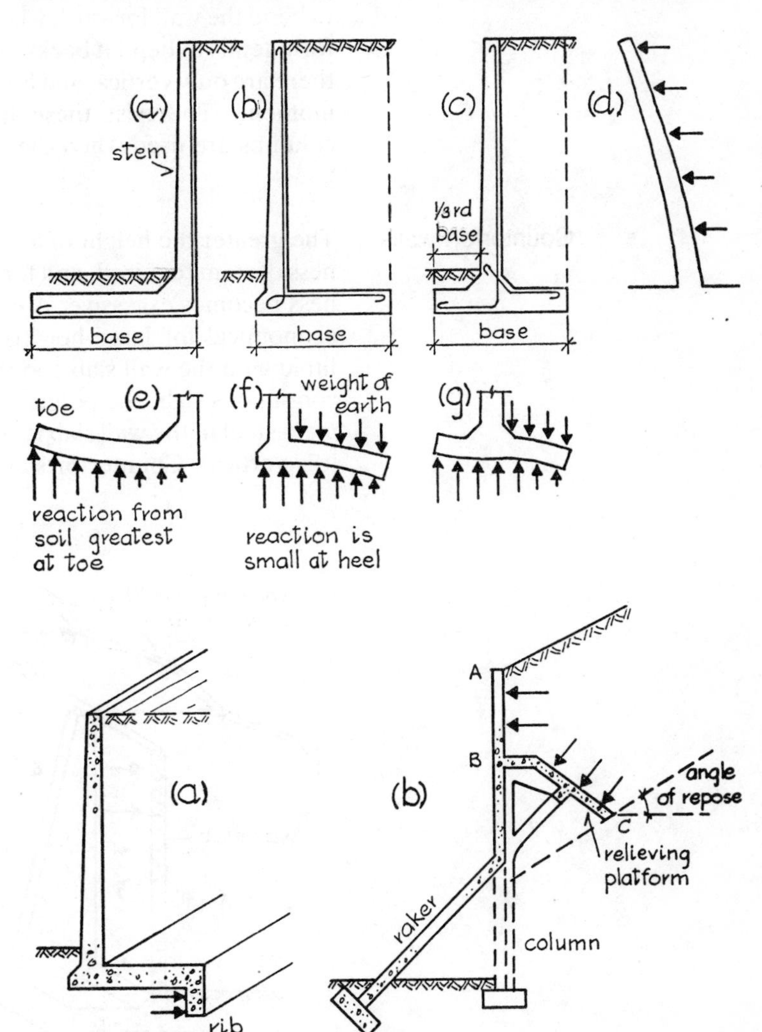

Fig. 18.14

Fig. 18.15

the wall. The bases will bend as shown at (*e*), (*f*) and (*g*). When it is impossible to project the base backwards a wall of the type shown at (*a*) can be used. Here the base must be longer than in the two other types (about two-thirds of the height of the wall) because there is much less vertical weight to resist overturning.

When the frictional resistance between the base and the soil (together with resistance from earth in front of the wall) is not sufficient to prevent sliding, ribs can be used as shown in Fig. 18.15 (*a*).

An interesting example of the intelligent use of material in cantilever walls comes from the U.S.S.R. A relieving platform is used to reduce the bending moments on the wall stem. This type of construction is recommended for walls exceeding about 6 m in height, and it is claimed that considerably less material is used than in the traditional designs. Figure 18.15 (*b*) is a simplified diagram which explains the principle. The pressure of the earth from *A* to *B* tends to bend the wall forward, whereas the pressure on the relieving platform tends to bend it backward. At some selected point, therefore, there are only vertical and horizontal forces to resist and no bending moment. To resist these forces, columns and rakers (sloping columns are used when the construction is reinforced concrete.

Counterfort walls

The greater the height of a cantilever wall, the greater is the thickness of stem required, and for walls higher than about 8 m the thickness becomes excessive. A counterforted wall (Fig. 18.16) is more economical for large heights. The *counterforts* are beams monolithic with the wall slab and the base, and the wall is designed as a continuous slab, the points of support being the counterforts. The main steel in the wall slab is horizontal, whereas in a cantilever wall it is vertical. Counterforts are cantilevers fixed at the base.

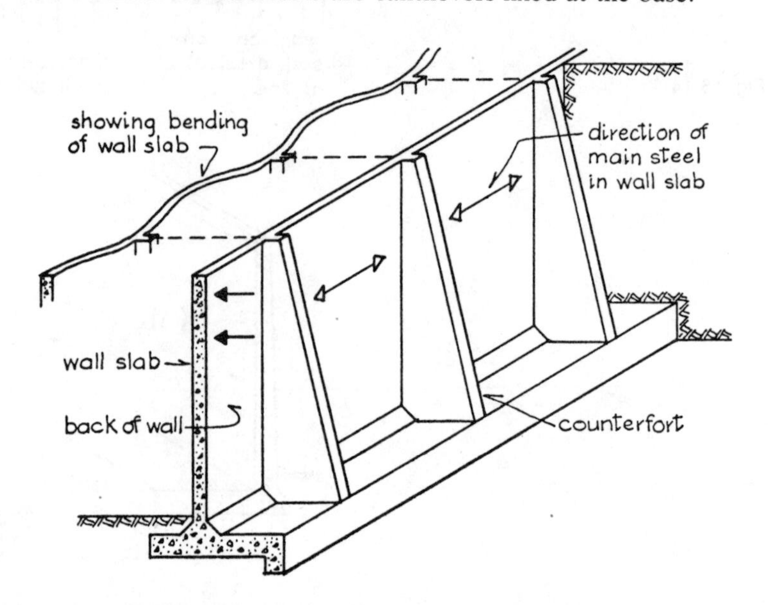

Basement walls

Usually, basement walls, in addition to resisting earth pressure, support beams or ground-floor slabs or even columns. Figure 18.17 (*a*) indicates a brick basement wall, and (*b*) shows a reinforced-concrete wall, which tends to bend as shown by dotted lines. [The reinforcement is not shown in (*b*).] Sometimes, the main columns of the building can be made to act as counterforts or buttresses and be designed to resist moments due to earth pressure in addition to resisting loads from floors above [Fig. 18.17 (*c*)]. The slab or wall behind the columns and against the earth can then be comparatively thin since it is spanning horizontally between columns.

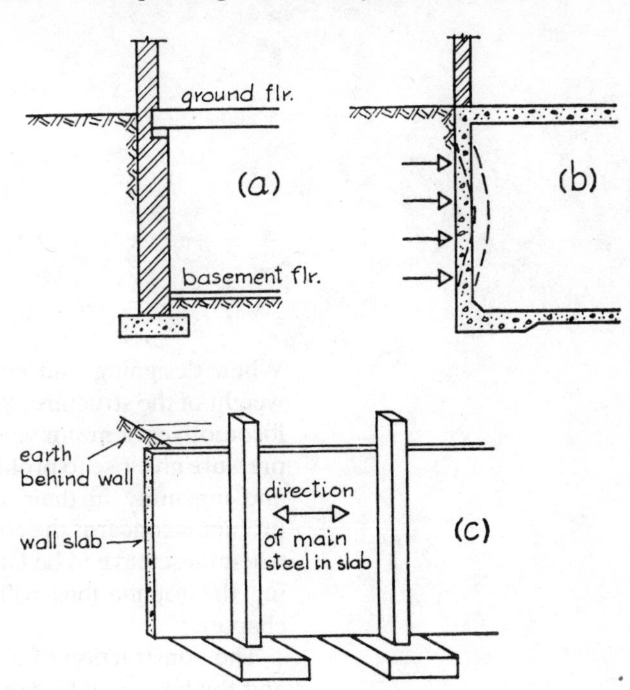

Fig. 18.17

19 Bridges

When designing and constructing a long-span bridge the great weight of the structure, the dynamic effects of moving loads such as locomotives or motor vehicles, and the aerodynamic effects of wind pressure give rise to problems which call for the greatest knowledge and ingenuity in their solution. Not the least important of the problems concerns the construction of the foundations, particularly when these have to be laid on the beds of rivers. Methods of building the foundations will be simply discussed at the end of the chapter.

The construction of a long-span bridge is a great achievement, and the history of bridge building includes many human, romantic and even tragic stories. It is impossible in one short chapter to deal adequately with this aspect of man's struggle with the forces of nature, and the reader is advised to consult books which are devoted exclusively to bridges. The titles of some of these books are given in the section on Further Reading on p. 247.

Summary of bridge types

All bridges may be broadly classified under three headings: *beam* (including the *truss* and the *cantilever*), *arch*, and *suspension*. Structural principles discussed in previous chapters apply equally to bridges.

Beam (simple and continuous) bridges

Figure 19.1 (*a*) illustrates a simple single-span bridge which may be of steel (probably a plate girder), reinforced concrete or prestressed concrete. In steel the maximum span for a simple beam bridge is usually about 30 m (although bridges with longer spans have been built). When, however, the spans are large, a continuous girder, as at (*b*), is usually adopted. A steel box girder bridge in California is also similar to that at (*b*) but has a central span of 230 m.

Beam (truss) bridges

Trussed bridges were once favoured for any span exceeding 45 m, but today the truss is reserved almost exclusively for long span structures, in the range 200 to 350 m. Here continuous trusses are

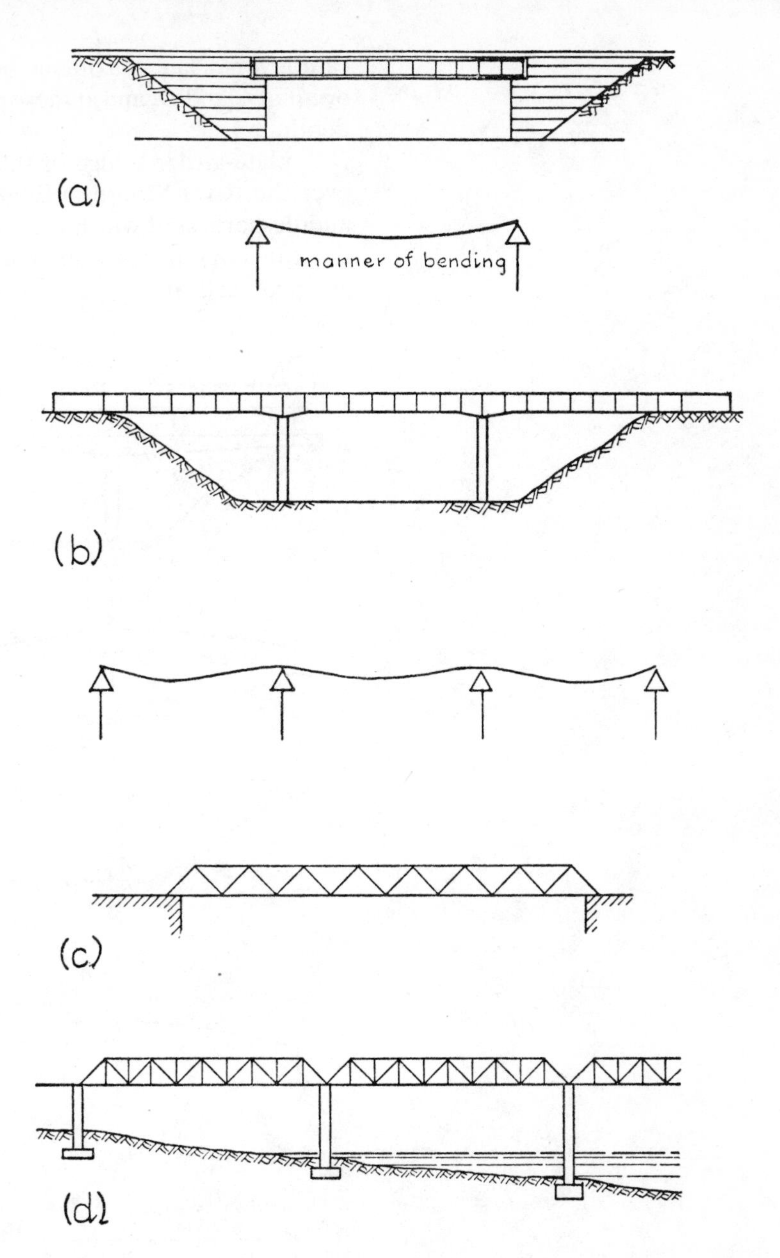

Fig. 19.1

almost always used but simply supported spans of the type shown in Fig. 19.1 (*c*) and (*d*) have been built to span up to 240 m.

Beam (cantilever) bridges

The principle of the cantilever bridge, with and without a suspended span, is illustrated in Figs 19.2 (*a*) and (*b*), although the latter is not common in solid beam or girder construction. The piers having been built, the bridge (anchored at *A* and *B*) is built out from each pier, and the middle portion of the bridge, called the *suspended span*, which is usually in one prefabricated unit, is then placed in position. The bridge therefore consists of two anchored cantilevers supporting [in type (*a*)] a beam 'suspended' from the ends of the cantilevers. The maximum bending moments and shear forces occur at C and D, and at these points the bridge is usually of greater depth.

A plate-girder bridge of the cantilever type is the Haw Bridge over the River Severn in England. The central span is 40 m (the middle portion of which is a suspended span of 21·4 m), and each side (anchor) span is 27 m. Roller-rocker bearings (*see* Chapter 10) are provided at each end of the bridge and at one end of the

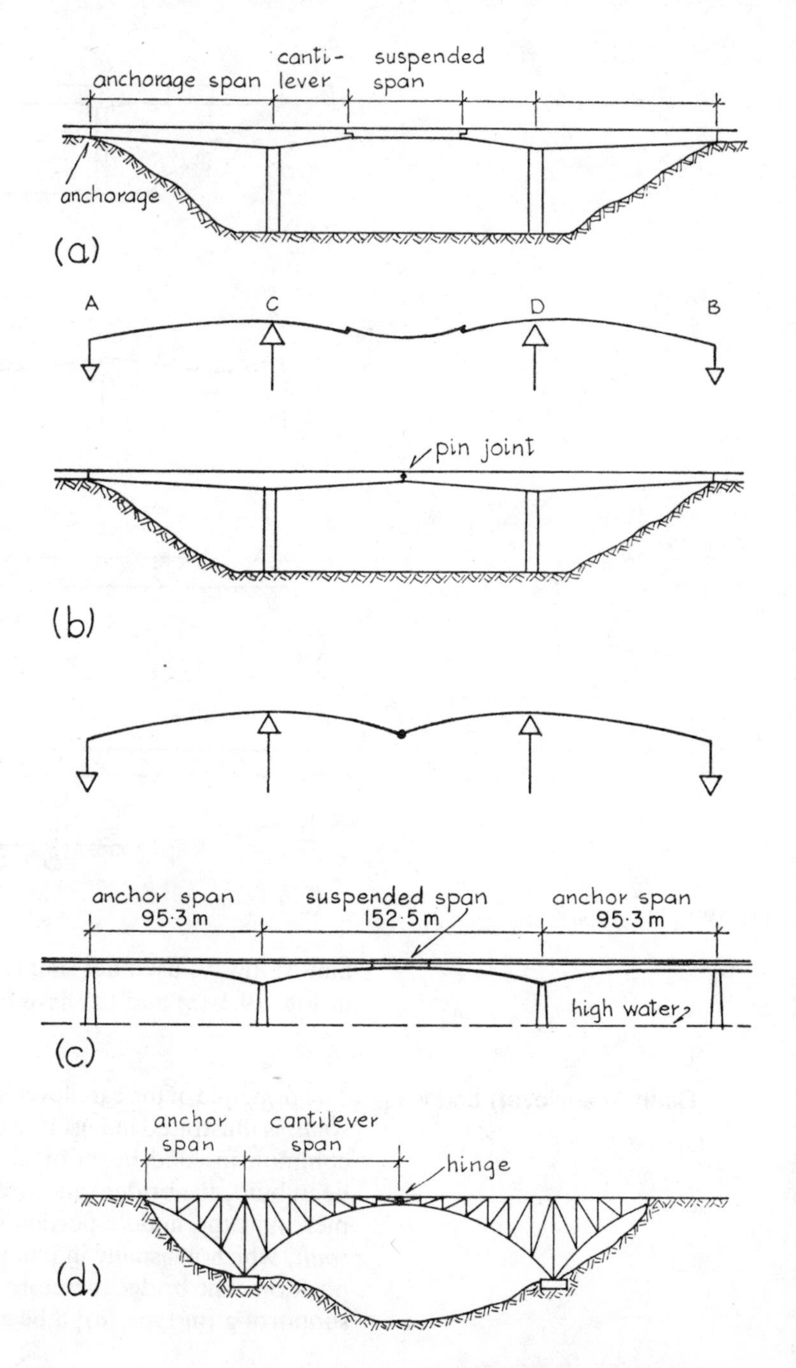

Fig. 19.2

suspended span to provide for expansion. Rocker bearings are used at the other end of the suspended span and at the two intermediate piers.

Another notable bridge in this category is the Medway Bridge, also in England. The overall length of the bridge with its approach viaducts is almost 1000 m. This is a prestressed-concrete bridge, and the river itself is bridged by three spans, as indicated in Fig. 19.2 (*c*),

General view of Haw Bridge over River Severn (*Reproduced from 'Building with Steel'*)

the longest span being 152 m. At the time of construction this was the largest span in the world for a prestressed concrete cantilevered bridge and although now exceeded in the Dominican Republic (198 m) and West Germany (208 m) it is still notable for its graceful appearance despite its impressive size.

Close-up of suspended span of Haw Bridge (*Reproduced from 'Building with Steel'*)

When spans are large, thus requiring a great depth of bridge, cantilever bridges are usually constructed of steel trusses (trussed girders). It is possible in this way to have spans of up to about 550 m between piers. Figure 19.2 (*d*) is an example where the cantilevers meet without a middle suspended span. Although this bridge may look like an arch, it is in fact a double-cantilever truss or trussed beam. It may be noted that in cantilever bridges the greatest depth of truss occurs at the main piers because it is at these points that the greatest stresses occur.

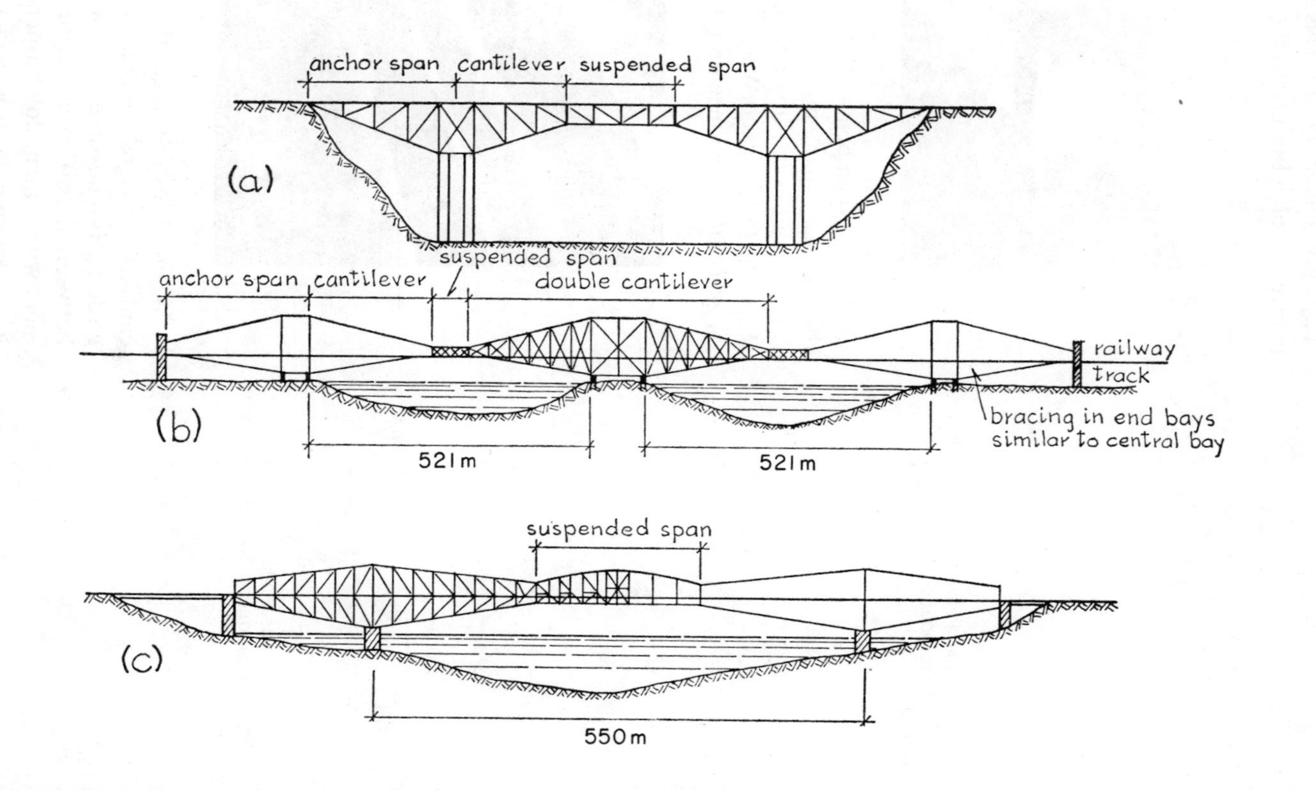
anchor span
cantilever
suspended span
(a)
suspended span
anchor span
cantilever
double cantilever
railway
track
bracing in end bays
similar to central bay
(b)
521 m
521 m
suspended span
(c)
550 m

Fig. 19.3

Figure 19.3 gives three examples of steel truss cantilever bridges with suspended spans. Two of the world's most famous bridges are illustrated in 19.3 (*b*) and (*c*). The Firth of Forth Railway Bridge in Lawrence River is shown in (*c*).

Firth of Forth Railway Bridge
(*Reproduced from 'Building with Steel'*)

The Auckland Harbour Bridge is a continuous truss with a cantilevered navigation span of 244 m. When the bridge was widened in 1969 continuous steel box girders were used rather than additional trusses. The box girders were supported from the original piers so

Quebec Bridge, St. Lawrence River
(*Capital Press Service, Ottawa*)

that the span arrangement was undisturbed and the clean functional appearance of the truss was preserved. At the same time it required one of the world's longest continuous spans for a steel box girder and the bridge is therefore a unique combination of the traditional and modern approaches to long span bridge building.

Auckland Harbour Bridge showing additional girders used to widen bridge in 1969 (*The High Commissioner for New Zealand*)

Arch bridges

Types of arch bridge are shown in Fig. 19.4; the arch is the main structural member and transmits the loads imposed on it to the abutments at the springing points. The part of the construction above the arch ring when the roadway or railway is at a higher level than the crown of the arch is called the *spandrel.* Figure 19.4 (*a*) is the traditional masonry arch with solid spandrel. Figure 19.4 (*b*) depicts a steel or reinforced-concrete arch with an open spandrel; the loads from the roadway are brought down to the arch ring as a number of point loads. A reinforced-concrete bridge of this type is the Sandö Bridge in Sweden, the horizontal span of the arch from

Reinforced-concrete arch at Berne, Switzerland

abutment to abutment being 264 m. The Aare Railway Bridge at Berne, Switzerland, has a reinforced-concrete arch of 152 m.

Since steel and reinforced concrete are capable of taking tension, the arch rings can be very much thinner than in masonry construction. The braced spandrel bridge [Fig. 19.4 (*c*)] is usually constructed in steel, as is also the bridge were the roadway is supported by hangers from the structural arch [Fig. 19.4 (*d*)].

Another type of arched bridge is the stiffened tied-arch, which is often called a *bow-string girder* [Fig. 19.4 (*e*)]. In an archery bow the

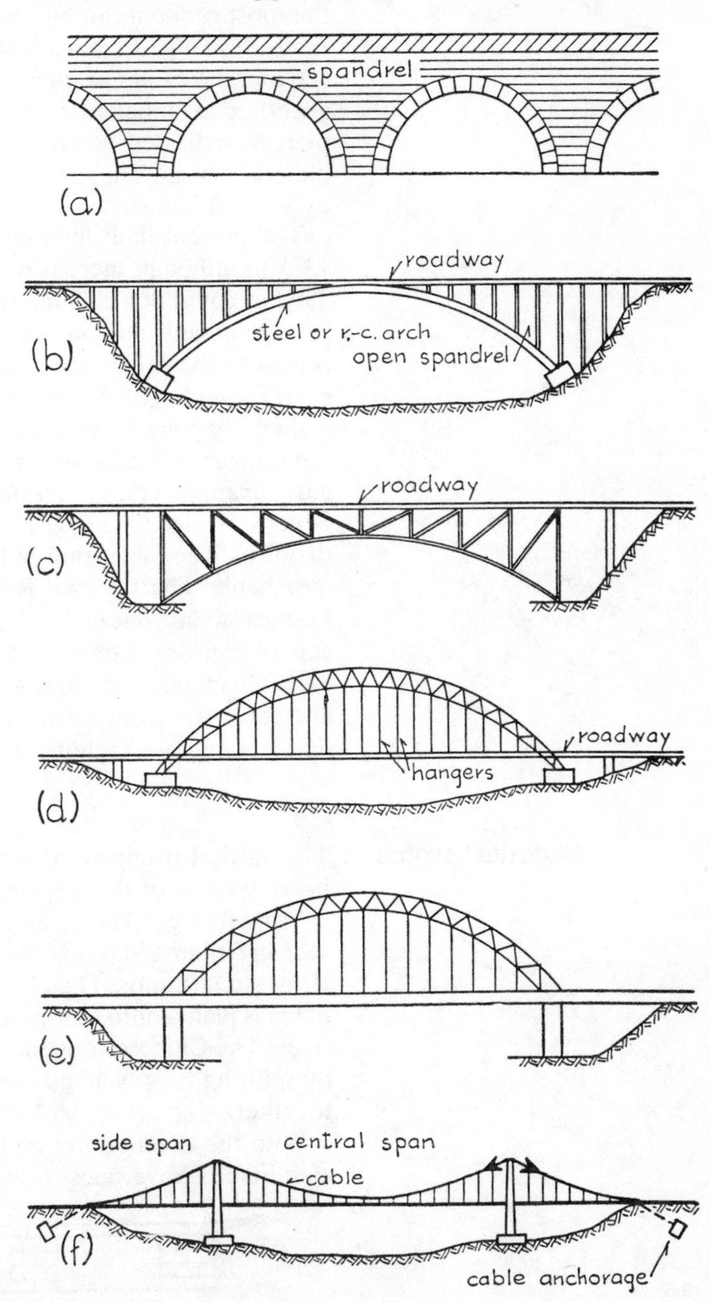

Fig. 19.4

string prevents the bow from flattening out. In a similar manner, the road-supporting horizontal girders are made strong enough to absorb the arch thrusts, and therefore the reactions on the piers and abutments are vertical. Although this type may appear similar to Fig. 19.4 (*d*), it is in fact different because the latter does not rely on the roadway girders to take the arch thrust.

Suspension bridges

When spans are large, about 600 m or more, suspension bridges are the most economical, but they can, of course, be used for smaller spans. Usually, there is a central span with two side spans [Fig. 19.4 (*f*)] and the cables passing over the top of the supporting piers are anchored in tunnels or by other means. Since the cables pull on each pier, as indicated by arrows, the load on the pier is almost completely vertical. The roadway is suspended from the inclined cables by vertical hangers.

The present limit for cable-suspended spans seems to be about 1375 m although there is no technical reason why single spans of 3000 m could not be built at some time in the future.

A relatively new development in recent years has been the return to the cable-braced bridge, first suggested by Poyet in the early example (1959) of a steel girder braced in this manner is the rather, they are continuous beams which are stiffened by cables radiating from a mast or tower at one or both ends of the span. An early example (1959) of a steel girder braced in this manner is the Severin Bridge at Cologne which has six cables supporting a span of 302 m. The cables radiate from the top of a single tower near the east bank. The General Rafael Urdaneta Bridge crossing Lake Maracaibo in Venezuela is an impressive cable braced prestressed concrete girder. Completed in 1961 the five navigation spans are each 236 m long and are supported by concrete trestles from below and cable ties from above. The stays pass over towers on each of six piers and are anchored to the girders near mid-span.

Historical sketch

The earliest primitive bridges were either of the stone or timber beam type or of the suspension type.

Stone bridges known as *clapper bridges* (Fig. 19.5) can still be seen in Britain today, as for instance at Eastleach Martin and at Post Bridge, Dartmoor. The earliest suspension bridges were probably of vines plaited into ropes and tied to trees at each side of the gulf or river. The Chinese constructed bridges of bamboo ropes obtained by splitting long bamboo canes into thin pieces and twisting them together.

With the exception of timber bridges and the primitive bridges mentioned above, most bridges until the advent of iron and steel

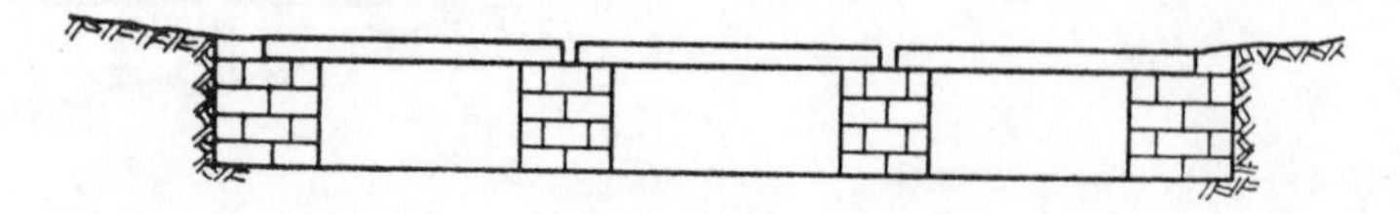

Fig. 19.5

had masonry arches. The Romans, who built many fine bridges, generally used the semicircular arch and the maximum span was about 24 m. The Pont du Gard at Nîmes, France [*see* Plate 17 (*c*)] has six arches in the bottom tier, the maximum clear span being 24·4 m, and is said to have been built about 19 B.C. The Ponte di Augusto at Rimini, Italy, has five semicircular arches, the maximum span being 8·5 m. This bridge is about the same age as the Pont du Gard.

The piers of Roman bridges were invariably massive, and their widths were usually about one-third of the span of the arch. No advantage appears to have been taken of the counterbalancing of thrusts from adjacent arches. During succeeding centuries, the Roman pattern was frequently followed, and in medieval times the

Pont du Gard, Nimes, France (*French Government Tourist Office*)

spans were usually between 12 and 24 m, although in the 14th century a bridge over the Adda at Trezzo is reputed to have had an arch span of over 70 m.

In some bridges a departure was made from the semicircular arch. The bridge at Avignon, France, had 21 elliptical arches, and the Rialto Bridge, Venice, constructed in the 16th century, has a segmental arch curve about one-third of a circle, the clear span being about 27 m.

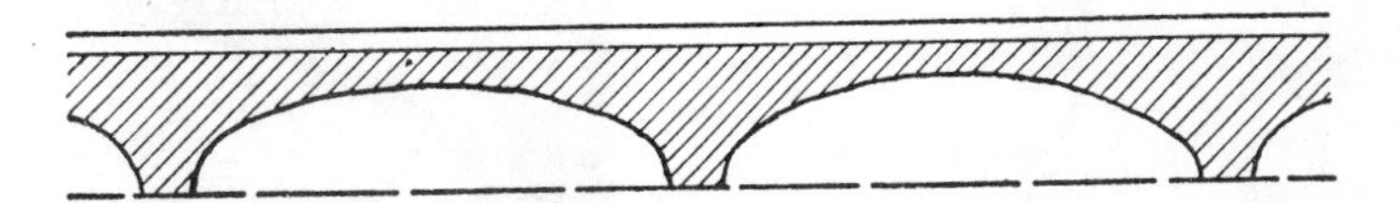

Fig. 19.6

Jean Perronet (1708–94), who was the first teacher and director of the École des Ponts et Chaussées (founded at Paris in 1747), was responsible for the design of several bridges, including the Pont de la Concorde in Paris. Perronet's arches are noted for the flatness of their curves and the narrowness of the supporting piers. He realized that piers of intermediate arches could be narrower than the abutment piers at either end of the bridge because of the counterbalancing effect of thrusts from adjacent arches. The pier width in Perronet's time was usually about one-fifth of the span of the arch, whereas he adopted a width of about one-tenth or one-twelfth of the span. An impression of part of one of Perronet's bridges is given in Fig. 19.6.

A notable bridge of the 18th century was built at Pontypridd, near Cardiff, by William Edwards, a local stonemason. This single-arch bridge has a span of 43 m and was completed about 1756. The bridge is remarkable not only for its span, which was not exceeded by an arched bridge until the new London Bridge was built (1831), but also for the thinness of the arch ring (750 mm) and for the perseverance of the builder. His first bridge, of three arches, was

William Edward's Bridge, Pontypridd, Glamorgan (*National Buildings Record*)

destroyed by floods about two years after completion; the second bridge, constructed soon after, was a single-span arch which failed shortly before completion owing to the weight of the haunches forcing up the crown of the arch. In his third and successful attempt, Edwards made the haunches lighter by leaving holes through them.

General View of old Waterloo Bridge, London (*Cement & Concrete Association*)

The first stone bridge across the Thames in London was the old London Bridge, which was begun in 1176 by a monk called Peter Colechurch. The bridge had 20 spans bridged by pointed arches, and the large number of piers greatly impeded the flow of the river. It was replaced by the new London Bridge, designed by one of the great engineers of the 19th century, John Rennie (1761–1821), and completed after his death by his two sons. This bridge, which was opened in 1831, has five semi-elliptical arches, the largest span being 45·8 m. Rennie also designed the old Waterloo Bridge, which was opened in June, 1817, on the second anniversary of the Battle of Waterloo. The span of each of the nine semi-elliptical arches was 36·6 m. This bridge was replaced in 1939–45 by the new Waterloo Bridge, which has five spans of 73·2 m each and consists of reinforced-concrete beams.

New Waterloo Bridge, London (*Cement & Concrete Association*)

John Rennie was also responsible for a cast-iron bridge at Southwark, London, completed in 1819, the centre of the three spans having a length of 73·2 m. This bridge lasted about 100 years, being replaced in 1921. A Swiss, Charles Labelye, built the fifteen-arch Westminster Bridge, completed in 1750, which became unsafe

Coalbrookdale Bridge (*National Buildings Record*)

before the lapse of 100 years owing to weakening of the piers. The engineer of the new Westminster Bridge was Thomas Page.

As evidenced by the London bridges mentioned above, masonry arches were still being used in the early years of the 19th century. The use of iron and steel was, however, increasing, and these materials, together with the new methods of construction (the trussed beam and the trussed arch) began to supersede the arch of masonry. As stated in Chapter 2, the first ferrous-metal bridge in the world was of cast iron—the Coalbrookdale Bridge over the River Severn. This bridge, of about 30 m span, followed the traditional arch form.

George Stephenson (1781–1848) and his son Robert (1803–59), who between them were responsible for the design of the famous steam locomotive, the 'Rocket,' were two of the great railway bridge builders of the 19th century. Robert Stephenson built a bridge of the bow-string type over the Tyne at Newcastle, and was also responsible for the famous Britannia Bridge over the Menai Straits in Wales, opened in 1849. This is a beam-type bridge consisting of two huge hollow rectangular tubes formed by riveting together wrought-iron plates (the trains run through the beams as through a tunnel). Robert Stephenson called in William Fairbairn and Professor Hodginson to collaborate in tests and calculations relating to the comparatively new material (wrought iron) and the proposed novel method of construction. The maximum span

Brooklyn Bridge (*United States Embassy*)

between piers of this bridge, which has a total span of 460 m, is 140 m.

The earliest suspension bridges employed chains built up of wrought-iron bars linked together, and a patent for flat bars was taken out by Sir Samuel Brown. This type of link was used by Thomas Telford (1757–1834) for the four chains of the Menai Straits suspension bridge, opened in 1826. These chains were replaced in 1939 by two chains of high-tensile steel. The Clifton suspension bridge near Bristol, England, was designed by a famous engineer, I. K. Brunel (1806–59), and completed in 1862. One of the first suspension bridges to use wire cables instead of solid links was built in Switzerland about 1835, each cable being formed of 1000 wires bound with wire wrapping. This bridge had a span of 265 m, which was exceeded in 1848 by a suspension bridge over the Ohio designed by Charles Ellett. The cables were formed of separate parallel wires connected together by iron bars. Several of Ellett's bridges were destroyed as a result of excessive swaying.

John Roebling (1806–69), who was born in Germany and emigrated to America in 1831, patented in 1841 a cable formed by squeezing strands together to form a cylindrical shape and wrapping the cable with light wire. This principle was adopted in his famous Brooklyn Bridge, completed in 1883. Each cable consisted of 19 strands with 286 wires in each strand. Roebling died in 1869 as the result of an injury sustained when making surveys connected with the bridge piers, but the Brooklyn Bridge project was taken over by his son, Colonel Washington A. Roebling. Col. Roebling contracted caisson disease as a result of working and directing operations in compressed air during the construction of the piers, and from 1872 until his death in 1927 was a permanent invalid. From his sickroom, however, he directed operations and was able to watch through a telescope the progress of the work. The central span of the bridge is nearly 490 m, and for the first time cables of steel instead of wrought iron were used. Remodelling and

Crumlin Viaduct, Monmouthshire (*E. G. Holt*)

reconstruction of the Brooklyn Bridge was carried out between 1948 and 1953, with David B. Steinman as consulting engineer.

As mentioned in Chapter 8, the 19th century saw the development of the truss, and this method of construction was used for many bridges, particularly in America. Quite a notable truss bridge, though seldom included in books on bridges, is the Crumlin Viaduct in Monmouthshire, opened in 1857. The main part of this railway bridge has seven spans of 45·8 m each bridged by Warren girders of wrought iron; the supports to the girders are cast-iron columns. The ill-fated bridge over the River Tay in Scotland, completed in 1878, was 3·2 km long, consisting of trusses, supported at intervals by piers. On the night of December 28th, 1879, eighteen months after completion of the bridge, the central portion of 900 m consisting of thirteen spans and known as the 'high girders,' collapsed in a great storm. A train with 75 passengers was thrown into the river below, and there were no survivors. The failure was attributed mainly to the ignoring of wind pressures in the design of the bridge. At the inquiry into the disaster, Sir Thomas Bouch, the engineer (knighted by Queen Victoria on the completion of the bridge), was asked:

> 'Sir Thomas, did you in designing this bridge make any allowance at all for wind pressure?'—'Not specially'.
> 'You made no allowance?'—'Not specially'.

In view of this admission, blame for the failure of the bridge was largely atttributed to Sir Thomas, and he died at the age of 58 only four months after the report of the inquiry was published.

At the time of the Tay Bridge disaster, Sir Thomas was engaged with preliminary work on a proposed suspension bridge over the Firth of Forth, but the design was now entrusted to Sir Benjamin Baker and Sir John Fowler, who adopted the cantilever principle. Extensive experiments on wind pressures were carried out, and the bridge was designed for a pressure of 2·4 kN/m^2. The Forth Bridge, which has two main spans of 521 m each, was completed in 1890 and for many years afterwards the longest-span bridge in the world. It was the first large-scale bridge to use open-hearth (Siemens–Martin) steel.

Space does not permit of detailed descriptions of the many notable bridges which have been built during the 19th and the present centuries; a few examples are given below.

Eades Bridge

This bridge over the Mississippi at St Louis was built by Captain J. B. Eades (1820–87) and was completed in 1874. It is a steel-arch bridge of three spans, the greatest of which is 158 m, and is notable for the first extensive use of steel in bridge building as well as for the first use (in America) of pneumatic caissons in the construction of the piers.

Hell Gate Bridge, New York

This two-hinged steel arch [of the type shown in Fig. 19.4 (*d*)] has a span between abutments of 310 m and was completed in 1915. The engineer, who was born in Austria, was Gustave Lindenthal (1850–1935).

Bayonne Bridge, New York

With a span of 503·9 m, this steel-arch bridge was opened in 1931, the chief engineer being O. H. Ammann.

Sydney Harbour Bridge

The main span of this steel arch is 503·6 m, and R. Freeman was chiefly responsible for the design. The bridge was started before the Bayonne Bridge, and would have been the largest arch bridge in the world had it not been beaten by the small distance of 600 mm. Sydney Harbour Bridge was opened in 1932, four months after the Bayonne Bridge.

Sydney Harbour Bridge

Quebec Bridge

This bridge of the cantilever type was the first to exceed (with a span of 550 m) the span of the Forth Bridge. It was completed in 1917 after several major accidents. On August 20th, 1907, during erection of the cantilever arms, 200 MN of steel collapsed into the river carrying 82 men with it. The designer, Theodore Cooper, was broken in health and spirits by the disaster and died within a few years. Building was resumed, but another major accident occurred in 1916 when collapsing steelwork killed 11 men.

George Washington Bridge, New York

The span between piers of this suspension bridge is 1067 m. It was designed and built by the Port of New York Authority under the direction of O. H. Ammann, with Cass Gilbert as consulting architect. This bridge has four cables of 915 mm diameter.

Golden Gate Bridge, San Francisco

When Joseph Strauss completed this bridge in 1937 he set a new world record for span length (1280 m) which was not exceeded for another 27 years. Even today the towers at 227 m are still the highest for any suspended structure anywhere in the world. Some idea of the immense size of this bridge can be gauged from the fact that due to the curvature of the earth the tops of the towers are 45 mm further apart than their foundations. Under the worst combination of traffic loading, wind and temperature effects, the total deflection at the top of a tower could be as much as 560 mm in the spanwise direction. Even though the deck is 27 m wide and the stiffening trusses are 8 m deep, it is still expected to sway 6 m sideways at midspan during a hurricane with wind speeds of 160 km/hr.

Tacoma Narrows Bridge

This bridge over Puget Sound in the State of Washington, with a main span of 854 m, was opened in July, 1940. The depth of the stiffening girders of the roadway of this suspension bridge was 2·44 m, only 1/350 of the span, whereas recommended minimum depths of from 1/90 to 1/50 of the span were usually adopted. From the time of its completion, the slender bridge had exhibited a tendency to oscillate, and on November 7th, 1940, a wind of comparatively small velocity (68 km/hr) set up a rocking motion of the bridge deck leading to collapse. One of the chief engineers, watching the death throes of the bridge, had to be restrained from committing suicide. The bridge was redesigned to give greater resistance to aerodynamic effects. The new stiffening girder is 10·07 m deep and consists of open-work trusses instead of the original solid-steel girder 2·44 m deep. The Tacoma Bridge disaster taught valuable lessons regarding the limits of slenderness of stiffening girders for suspension bridges. Present practice is to conduct wind-tunnel experiments to determine the optimum shape for the stiffening girders so as to minimize aerodynamic oscillations. Many of these new ideas were incorporated in the Mackinac Bridge across Lake Michigan. Designed by D. B. Steinman and completed in 1957, this bridge is noted for its relatively long side spans which combine to give a total suspended length of 2255 m even though the main span is only 1158 m.

Forth Road Bridge, Scotland

Running almost parallel to the Forth Railway Bridge, this suspension bridge spans 1006 m with a stiffening girder that is 9 m deep. Designed by Freeman, Fox and Partners and completed in 1964 almost 75 years after the railway bridge, it is a worthy companion to the cantilevered steel truss of Fowler and Baker.

Verrazano Narrows Bridge, New York

Also completed in 1964, the main span of this suspension bridge exceeds that of Golden Gate by 18 m. O. H. Ammann was the chief structural engineer for this monumental structure which has stiffening girders 10 m deep and a deck 35 m wide.

Severn Bridge, England–Wales

Plans for a new bridge across the Severn River were well advanced when the Forth Road Bridge was given priority. Freeman, Fox and Partners adapted their Severn design for the Forth site and took the opportunity to modify the stiffening girders for the Severn Bridge. The 9 m deep trussed girders were replaced by a single steel box section 32 m wide and 3 m deep. The box was streamlined so as to be aerodynamically stable and the suspender cables supporting the box were inclined rather than hung vertically so as to further dampen wind oscillations. Significant savings in material were achieved since not only was the deck shallower and therefore lighter, but the loads in the cables and towers were also reduced with consequent savings in section sizes and material content. When compared with the Forth Bridge, which is of similar span and completed only 2 years earlier, the Severn Bridge used almost a third less steel and saved approximately £1.33 million (about 15% of the total cost). Completed in 1966 the Severn was a turning point in the history of suspension bridge design. The ideas developed here by its designer Sir Gilbert Roberts for Freeman, Fox and Partners have since been applied to the Bosphorus Bridge in Istanbul (span 1074 m, completed 1973) and the Humber Bridge near Hull in England (span 1410 m, expected completion date 1978).

Gladesville Bridge, Sydney

Reinforced-concrete and prestressed-concrete bridges

The extensive use of reinforced concrete for bridge construction did not begin until well into the present century. Among the finest examples of early reinforced concrete arches and box girders are those of Robert Maillart (1872–1940) who was responsible for promoting this new material in Europe in the 1920s. He demonstrated exceptional skill for solving difficult technical problems with outstanding architectural success. Even today many of his bridges in Switzerland are the wonder and admiration of engineers and laymen around the world.

Since Maillart's time many notable medium and long span bridges have been built in both reinforced and prestressed concrete but maximum spans are still well below those possible with steel. Nevertheless prestressed concrete arches have been built to span 305 m (for example, the Gladesville bridge across the Parramatta River in Sydney, completed in 1964) and prestressed cantilevered box girders are commonly used for continuous spans extending up to about 210 m (for example the Medway Bridge discussed on page 223).

Foundations

As in all other structures, the whole weight of a bridge must be transmitted to soil or rock capable of supporting it. If a bridge has only two supports, one at each end, the total weight is taken by these supports. When there are intermediate piers as well as end supports, each pier and its foundation must be designed to resist a certain proportion of the load. When the bridge is over a river or lake, the piers may extend for a considerable distance below water level. For the construction of the piers either *cofferdams* or *caissons* are used.

Cofferdams

A dam is a structure for holding back water and a coffer is a box, so a *cofferdam* may be thought of as a box-like structure, usually rectangular or circular, but without a top or bottom, constructed at a pier location to enble the pier to be built within it. The principle is simple (Fig. 19.7), but there are various types of cofferdam according to the depth of water to be retained and the nature of the bed of the river.

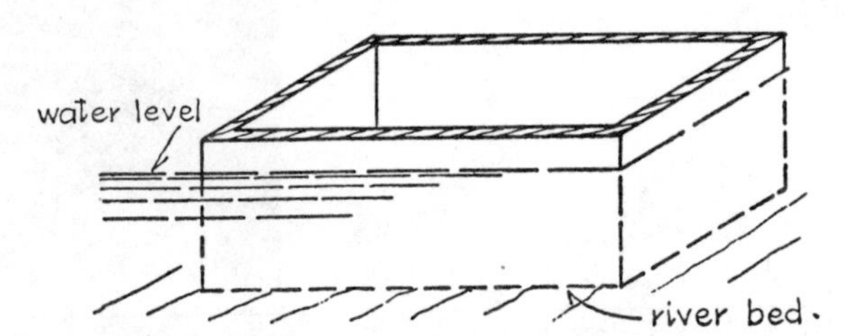

Fig. 19.7

A cofferdam is a temporary structure and is usually built in place; it is made practically watertight and by pumping is kept dry inside to enable workmen to place the concrete or masonry. The walls usually consist of interlocking sheet piling of timber, reinforced concrete or steel, as shown in Fig. 19.8, and each wall of the box may consist of a single layer of piles or a double layer with puddle filling.

When the walls consist of a single layer of piles, internal bracing [Fig. 19.9 (*a*)] is often required to resist the pressure of the water outside the cofferdam. Where internal bracing is not economical, cofferdams with double walls can be used [Fig. 19.9 (*b*)]. Each wall consists of an inner and outer skin of interlocking sheet piles, the space between being filled with a *puddle,* which is a mixture of sand and clay or gravel and clay.

In order that water may be pumped from inside a cofferdam, the piles are driven down if possible to an impervious stratum so that there is little or no leakage of water into the bottom of the excavation. If this condition does not apply, concrete can be deposited through the water to form a layer at the bottom of the excavation. (Concrete is capable of hardening under water.) Water can then be pumped out and the remainder of the construction completed in the dry.

Fig. 19.8

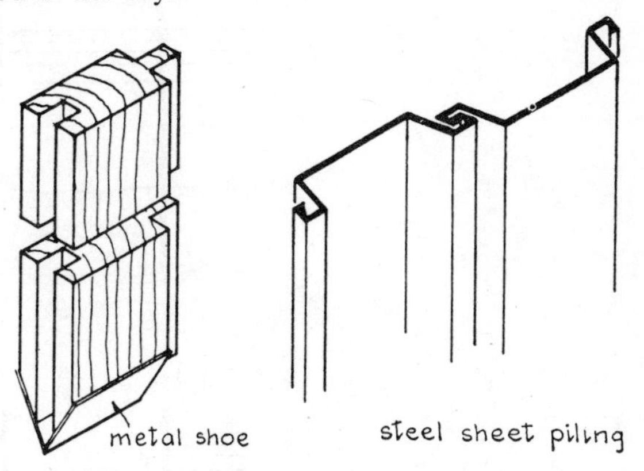

Briefly, the procedure is as follows.

(1) Construction of the cofferdam by driving down sheet piles through the water until they reach rock or other suitable load-bearing stratum [Fig. 19.10 (*a*)].

(2) Either—

(*a*) Pumping out the water from inside the cofferdam and excavating material down to the bearing stratum; or

(*b*) Excavating material by bucket, placing a layer of concrete to act as a water seal and then pumping out the water [Fig. 19.10 (*b*) and (*c*)].

(3) Construction of the pier [Fig. 19.10 (*d*)].

(4) Removal of the sheet piling to dismantle the cofferdam.

When a firm bearing stratum is not found at a suitable depth it may be necessary to use piles [Fig. 19.10 (*c*)], which may have to be driven through water when the cofferdam cannot be pumped dry in the initial stages. Before starting to concrete 'in the wet,' divers are often sent down to inspect the bottom of the excavation.

Fig. 19.9

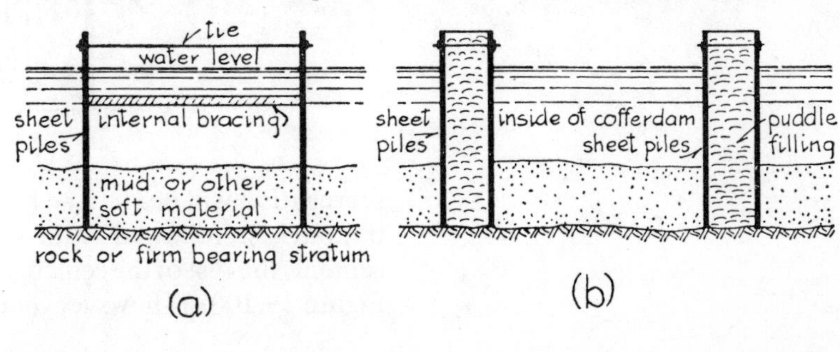

Various methods are adopted for depositing concrete through water, depending on the magnitude of the work. If concrete is allowed merely to drop through the water from surface,

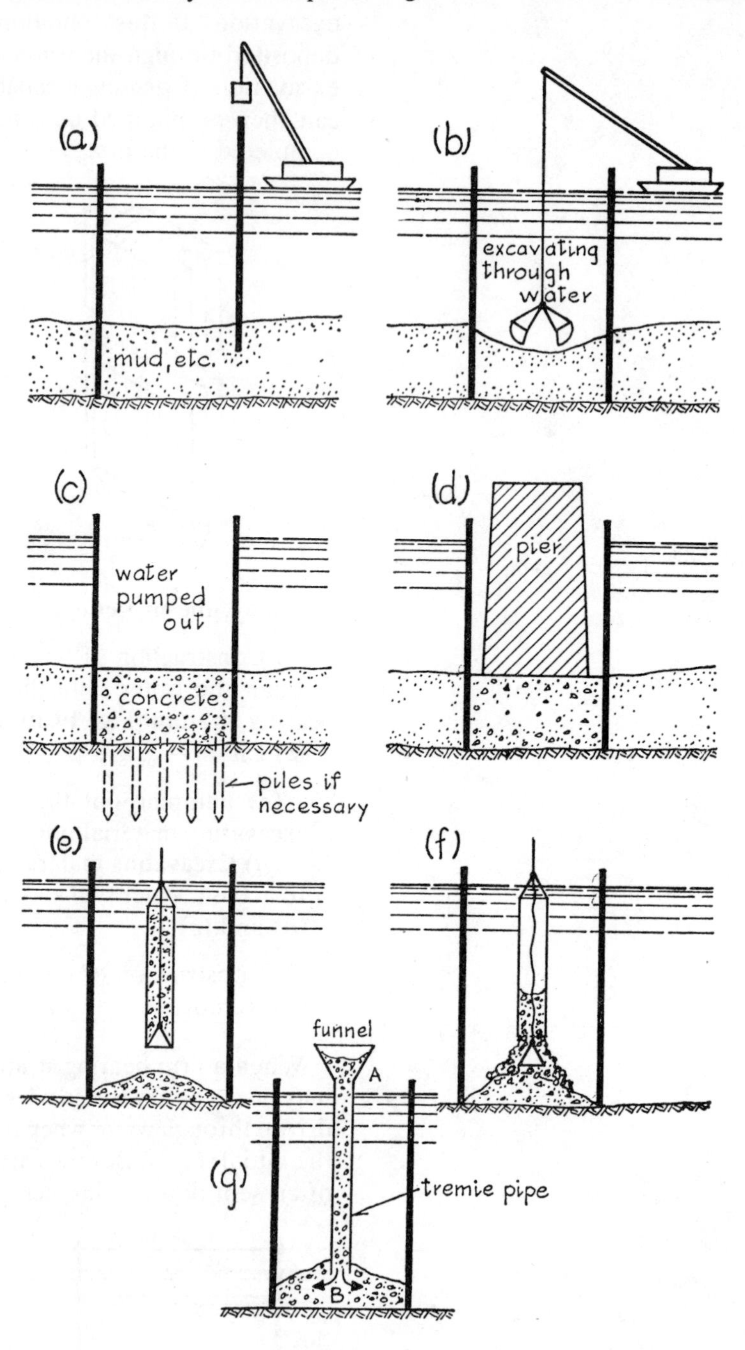

Fig. 19.10

segregation will occur. The heavier gravel particles will sink to the bottom first, followed by the sand particles and then by some of the cement, the rest of the cement being held in suspension in the water. Figure 19.10 (*e*) shows a cylinder, filled with wet concrete, with a stopper at the lower end. When the cylinder reaches the bottom of the excavation (or the top of previously placed concrete) the rope holding the stopper in place is allowed to go slack and the cylinder is pulled upwards [Fig. 19.10 (*f*)]. The concrete slides out as the cylinder rises, and when the cylinder is empty the cable attached to the stopper is tightened and pulls the stopper back into place, thus closing the cylinder in readiness to receive a fresh batch of wet concrete.

The *tremie* method of concreting is illustrated in Fig. 19.10 (*g*). A pipe with a hopper, *A*, at the top end is kept filled with wet concrete the weight of which forces it out at the lower end, *B*. As the height of concrete in the excavation increases, the tremie pipe is gradually withdrawn, but the end *B* is always kept below the surface of the concrete. The tremie method is more suitable for the placing of large quantities of concrete than the method illustrated in Fig. 19.10 (*f*).

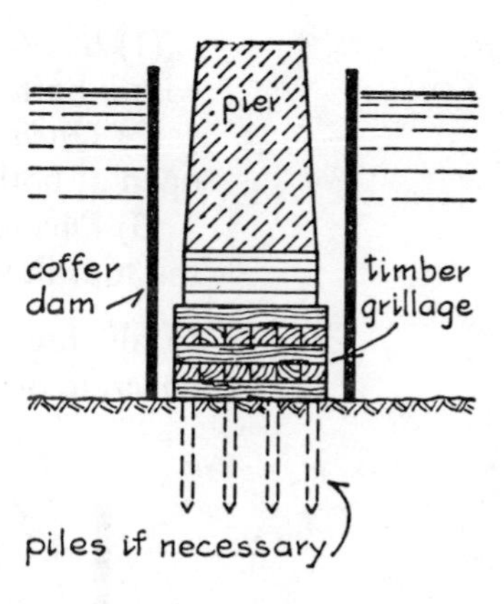

Fig. 19.11

In the construction of the river piers for the Medway Bridge, 6·4 m thick layers of concrete were placed under water by lowering special bottom-opening skips each containing 3 m³.

Cofferdams have been used for many hundreds of years; the Chinese used double-wall cofferdams of bamboo piles with clay filling between the walls. Until the advent of steel, timber piles were used for the construction of cofferdams and the foundations of the piers were usually piles, or timber grillages, or large blocks of masonry laid directly on the bearing stratum. Figure 19.11 shows a wooden foundation on piles.

In the construction of a bridge across the Loire, France, in the 18th century, piles were driven to refusal—i.e. until they could be driven no further—in the bed of the river. Oak beams to form a grillage were built up on the piles to about 2 m below water level to form a base for the masonry pier. For the piers of the Cincinnati Bridge (by Roebling), timber foundations rested directly on a bed of gravel and extended up to low-water mark where the masonry piers commenced.

As an example of the dimensions of cofferdams, one of the towers of the George Washington Bridge had two cofferdams, each 30 m ×

33 m in plan and 24 m deep. For one of the piers of the San Francisco Oakland Bridge a braced single-wall steel cofferdam, 16 m × 37 m in plan, was nearly 31 m deep.

Caissons

The word *caisson* comes from the French word *caisse*, meaning a chest or case. When the plan area of the required construction is large and the water is not too deep, a cofferdam is likely to be the more economical solution. Caissons are likely to be used when the required foundation has a small plan area in relation to the depth of water, or when it would be difficult to drive piles to form a cofferdam because of boulders in the soil.

A cofferdam is constructed at the actual site of a pier and is afterwards dismantled. A caisson is usually constructed wholly or in part on shore, floated or conveyed into position and sunk at the site, where it remains as an integral part of the foundation. Caissons can be broadly divided into three types—

(1) *Box Caisson.* A watertight timber or reinforced-concrete box with a bottom but no top.
(2) *Open Caisson.* A timber, steel or reinforced-concrete box open at both top and bottom.
(3) *Pneumatic Caisson.* A box, open at the bottom and closed at the top, in which compressed air is used.

In all three types, the caisson forms a shell which is filled with concrete or masonry to form the bridge pier.

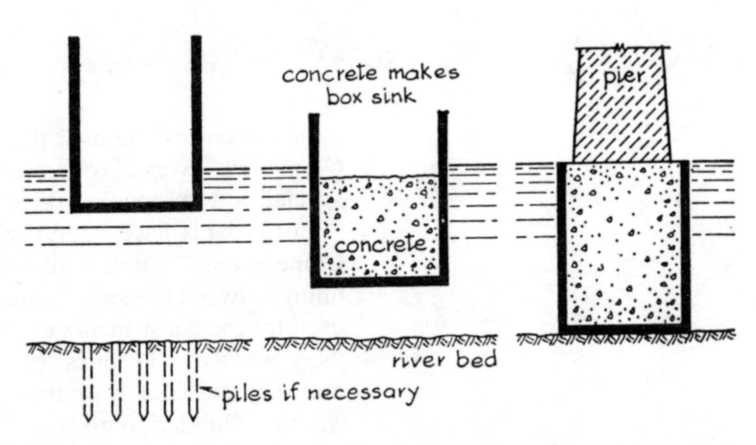

Fig. 19.12

Box caissons

These caissons are suitable where a firm bearing stratum is available without excavation being required. Alternatively, it is possible to rest the box on previously driven piles. The box may be sunk by weighting it or by placing in position the permanent concrete or masonry (Fig. 19.12).

Open caissons

These caissons may be rectangular or circular in plan and may have single or double walls. The bottom of the wall is provided with a cutting edge, usually of steel, to enable the caisson to sink through the soil. When a single-wall caisson is used it is necessary to load it with *kentledge* (weights such as steel rails) to force the caisson through the soil as the material is being excavated. When the caisson reaches a firm bearing stratum, either several metres thickness of concrete is placed to form a seal so that the water can be

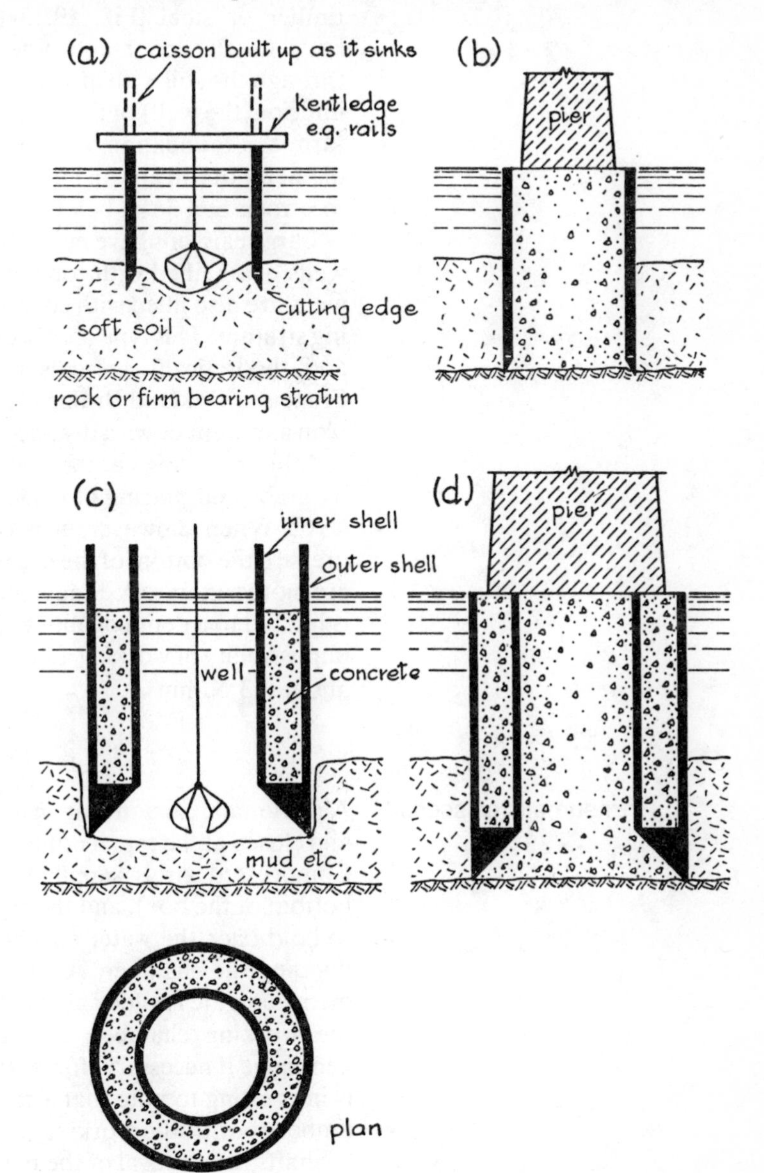

Fig. 19.13

pumped out and the remainder of the work completed in the dry, or all the concrete is deposited through water. The caisson is usually surmounted with a cofferdam, sometimes of the same construction as the caisson, so that the cofferdam can be eventually removed. (This is because the caisson, which is left in place, should not extend above low-water level.) Figure 19.13 (*a*) illustrates a single-wall caisson.

Small cylindrical caissons of the single-wall type are frequently used when the loads to be supported by piers are not large and it is necessary to go down a considerable distance to avoid the scouring action of the water. Larger caissons, whether rectangular or circular, are provided with double walls of reinforced concrete, timber or steel [Fig. 19.13 (*c*)]. Concrete placed in the space between the inner and outer shells enables the caisson to sink through the soil with little or no added kentledge. The whole of the interior (the well) of the caisson is eventually filled with concrete to form the foundation for the pier. One such caisson had an inner diameter of 6 m and an outer diameter of 12 m, and was taken down to a rock bed 36 m below low-water level.

Large caissons have more than one well, the pockets between the wells being filled with concrete to make the caissons sink, and the wells are also filled with concrete when the caissons reach the bearing stratum. This type of caisson is used for the deepest foundations, and theoretically there is no limit to the depth. For the San Francisco Oakland Bridge (1937), one caisson was 28 m × 60 m in plan and went down a distance from the water surface of 67 m.

A big advantage of the open caisson is that all work (excavating by grabs and placing of concrete) can be done from above water level. When, however, it is considered necessary to inspect and prepare the bottom of the excavation in the dry, pneumatic caissons are more suitable. A disadvantage of the pneumatic caisson is that it is limited to a depth of about 35 m below water level because of the impossibility of working in compressed air at a pressure greater than about 0·3 N/mm^2.

Pneumatic caissons

A pneumatic caisson is an inverted box, i.e. it is open at the bottom and closed at the top; it is illustrated diagrammatically in Fig. 19.14. Compressed air is supplied to the working chamber (the open bottom of the box), and the pressure exerted by the air is sufficient to hold back the water outside the caisson and to prevent it from flooding the chamber. As the material is excavated by the workmen, concrete is added above the roof (the closed top of the box) of the working chamber, and the weight of this (together with kentledge if necessary) forces the cutting edges through the soil. (It is interesting to note that a caisson for the Brooklyn Bridge had a timber roof to the working chamber 4·6 m thick.)

Shafts for removal of the muck and for enabling workmen to get into and out of the working chamber must be provided with *air-locks*. If a man, after working in high pressure, comes back too suddenly to normal atmospheric conditions he gets a disease called the *bends*. This is due to an excess of nitrogen in the blood and can cause paralysis and death. On leaving the working chamber it is necessary to rest in a room (the air-lock) in which the pressure is gradually reduced to normal. When circumstances (such as deep foundations) require a high air pressure it may be necessary for workmen to spend 50 minutes in the air-lock at the end of a shift of only one hour.

When the caisson reaches the rock or other firm bearing stratum the working chamber is filled with concrete.

The pneumatic method was used by John Wright for a bridge at Rochester, England, in 1851, and was also used by Brunel for his Royal Albert Bridge at Saltash. James B. Eades introduced the pneumatic caisson to America and used it for his St Louis Bridge, where, at the East Pier, rock was 29 m down (4 m of water and 25 m of sand).

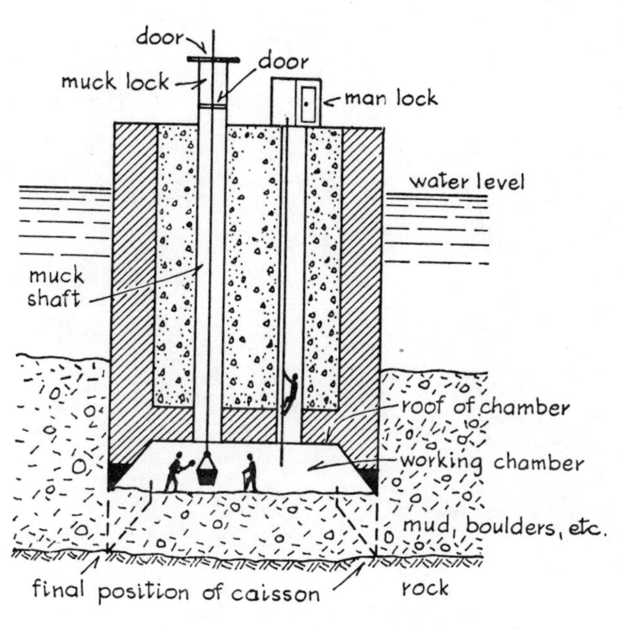

Fig. 19.14

Further Reading

Listed below are several books that might be consulted if more information and greater detail are required in particular topics. The list is certainly not exclusive since there are literally hundreds of books that might have been included. Therefore when selecting titles for this list, those with a professional basis have been chosen in preference to detailed academic texts.

The list is broadly subdivided in four areas for easier reference, and texts of a more elementary nature are marked with an asterisk.

A. Analysis

These books range from the very elementary that are suitable for architects wishing to learn simplified structural analysis to the more advanced that describe analytical techniques suitable for computer-aided analysis. A gradual progression through these would give an excellent grounding in Structural Analysis.

*1. *Architectural Structures,* H. J. Cowan, Elsevier 2nd Edition, 1976, 448 pp.
*2. *Structural Design in Architecture,* M. Salvadori and M. Levy, Prentice Hall, 1967, 457 pp.
*3. *Statics,* J. L. Meriam, Wiley 2nd Edition, 1975.
4. *Structural Mechanics,* W. Morgan and D. T. Williams, Pitman SI Edition, 1971.
5. *Elementary Structural Analysis,* C. H. Norris and J. B. Wilbur, McGraw Hill 2nd Edition, 1960.
6. *The Analysis of Engineering Structures,* A. J. S. Pippard and J. F. Baker, Arnold 4th Edition, 1968.
7. *Theory and Problems of Structural Analysis,* J. J. Tuma, McGraw Hill, 1969, 292 pp.
8. *Matrix and Digital Computer Methods in Structural Analysis,* W. M. Jenkins, McGraw Hill, 1969.

B. Design

Three basic texts on the design of steel, reinforced concrete and prestressed concrete structures are included here and would be suitable for those well grounded in analysis and now ready for the design details.

9. *Design of Steel Structures,* B. Bresler, T. Y. Lin and J. B. Scalzi, Wiley 2nd Edition, 1968, 830 pp.
10. *Design of Concrete Structures,* G. Winter, L. C. Urquhart, C. E. O'Rourke, and A. H. Nilson, McGraw Hill 8th Edition, 1973, 659 pp.
11. *Design of Prestressed Concrete Structures,* T. Y. Lin, Wiley 2nd Edition, 1963.

C. Manuals

Books that summarize formulae, code requirements, analytical methods and give tables of section properties are always useful and deserve a place on an engineer's or architect's bookshelf. Beware of the danger of selecting formulae at random from such books and always check the assumptions made in their derivation to be sure they are intended for the situation in hand.

12. *Formulas for Stress and Strain,* R. J. Roark, McGraw Hill 4th Edition, 1965.
13. *Steel Designers Manual,* Constructional Steel Research and Development Organisation, Wiley 4th Edition (SI), 1972, 1089 pp.
14. *Handbook of Concrete Engineering,* M. Fintel (Editor), Van Nostrand Reinhold, 1974, 801 pp.
*15. *Simplified Engineering for Architects and Builders,* H. Parker, Wiley Interscience 5th Edition, 1975, 411 pp.

D. General

Books on shells, membranes, cable roofs, space frames, geodesic domes and bridges have been included here for general information. Most contain a bibliography if further reading in a particular area is required.

16. *Thin Shell Concrete Structures,* D. P. Billington, McGraw Hill, 1965, 322 pp.
17. *Tensile Structures,* F. Otto (Editor), Vols. 1 and 2, MIT Press, 1973, 320 pp. and 170 pp. combined edition.
18. *Steel Space Structures,* Z. Makowski, Michael Joseph (London), 1965, 214 pp.
19. *The Dymaxion World of Buckminster Fuller,* R. W. Marks and Buckminster R. Fuller, Doubleday, 1973.
*20. *A Span of Bridges,* H. J. Hopkins, Praeger (New York), 1970, 288 pp.
*21. *Bridges, The Spans of North America,* D. Plowden, Viking Press (New York), 1974, 328 pp.
*22. *Structure in Architecture,* M. Salvadori and R. Heller, Prentice Hall 2nd Edition, 1975, 414 pp.

Index